Perspectives in Mathematical Sciences II

Pure Mathematics

Statistical Science and Interdisciplinary Research

Series Editor: Sankar K. Pal *(Indian Statistical Institute)*

Description:
In conjunction with the Platinum Jubilee celebrations of the Indian Statistical Institute, a series of books will be produced to cover various topics, such as Statistics and Mathematics, Computer Science, Machine Intelligence, Econometrics, other Physical Sciences, and Social and Natural Sciences. This series of edited volumes in the mentioned disciplines culminate mostly out of significant events — conferences, workshops and lectures — held at the ten branches and centers of ISI to commemorate the long history of the institute.

Vol. 1 Mathematical Programming and Game Theory for Decision Making
edited by S. K. Neogy, R. B. Bapat, A. K. Das & T. Parthasarathy (Indian Statistical Institute, India)

Vol. 2 Advances in Intelligent Information Processing: Tools and Applications
edited by B. Chandra & C. A. Murthy (Indian Statistical Institute, India)

Vol. 3 Algorithms, Architectures and Information Systems Security
edited by Bhargab B. Bhattacharya, Susmita Sur-Kolay, Subhas C. Nandy & Aditya Bagchi (Indian Statistical Institute, India)

Vol. 4 Advances in Multivariate Statistical Methods
edited by A. SenGupta (Indian Statistical Institute, India)

Vol. 5 New and Enduring Themes in Development Economics
edited by B. Dutta, T. Ray & E. Somanathan
(Indian Statistical Institute, India)

Vol. 6 Modeling, Computation and Optimization
edited by S. K. Neogy, A. K. Das and R. B. Bapat (Indian Statistical Institute, India)

Vol. 7 Perspectives in Mathematical Sciences I: Probability and Statistics
edited by N. S. N. Sastry, T. S. S. R. K. Rao, M. Delampady and B. Rajeev (Indian Statistical Institute, India)

Vol. 8 Perspectives in Mathematical Sciences II: Pure Mathematics
edited by N. S. N. Sastry, T. S. S. R. K. Rao, M. Delampady and B. Rajeev (Indian Statistical Institute, India)

Platinum Jubilee Series

Statistical Science and Interdisciplinary Research — Vol. 8

Perspectives in Mathematical Sciences II

Pure Mathematics

Editors

N. S. Narasimha Sastry

T. S. S. R. K. Rao

Mohan Delampady

B. Rajeev

Indian Statistical Institute, India

Series Editor: ***Sankar K. Pal***

NEW JERSEY • LONDON • SINGAPORE • BEIJING • SHANGHAI • HONG KONG • TAIPEI • CHENNAI

Published by

World Scientific Publishing Co. Pte. Ltd.
5 Toh Tuck Link, Singapore 596224
USA office: 27 Warren Street, Suite 401-402, Hackensack, NJ 07601
UK office: 57 Shelton Street, Covent Garden, London WC2H 9HE

British Library Cataloguing-in-Publication Data
A catalogue record for this book is available from the British Library.

Statistical Science and Interdisciplinary Research — Vol. 8
PERSPECTIVES IN MATHEMATICAL SCIENCES II
Pure Mathematics

ISBN-13 978-981-4273-64-0
ISBN-10 981-4273-64-3

Printed in Singapore.

Foreword

The Indian Statistical Institute (ISI) was established on 17th December, 1931 by a great visionary Professor Prasanta Chandra Mahalanobis to promote research in the theory and applications of statistics as a new scientific discipline in India. In 1959, Pandit Jawaharlal Nehru, the then Prime Minister of India introduced the ISI Act in the parliament and designated it as an *Institution of National Importance* because of its remarkable achievements in statistical work as well as its contribution to economic planning.

Today, the Indian Statistical Institute occupies a prestigious position in the academic firmament. It has been a haven for bright and talented academics working in a number of disciplines. Its research faculty has done India proud in the arenas of Statistics, Mathematics, Economics, Computer Science, among others. Over seventy five years, it has grown into a massive banyan tree, like the institute emblem. The Institute now serves the nation as a unified and monolithic organization from different places, namely Kolkata, the Headquarters, Delhi, Bangalore, and Chennai, three centers, a network of five SQC-OR Units located at Mumbai, Pune, Baroda, Hyderabad and Coimbatore, and a branch (field station) at Giridih.

The platinum jubilee celebrations of ISI have been launched by Honorable Prime Minister Prof. Manmohan Singh on December 24, 2006, and the Government of India has declared 29th June as the "Statistics Day" to commemorate the birthday of Professor Mahalanobis nationally.

Professor Mahalanobis, was a great believer in interdisciplinary research, because he thought that this will promote the development of not only Statistics, but also the other natural and social sciences. To promote interdisciplinary research, major strides were made in the areas of computer science, statistical quality control, economics, biological and social sciences, physical and earth sciences.

The Institute's motto of "unity in diversity" has been the guiding principle of all its activities since its inception. It highlights the unifying role of statistics in relation to various scientific activities.

In tune with this hallowed tradition, a comprehensive academic programme, involving Nobel Laureates, Fellows of the Royal Society, Abel prize winner and other dignitaries, has been implemented throughout the Platinum Jubilee year, highlighting the emerging areas of ongoing frontline research in its various scientific divisions, centers, and outlying units. It includes international and national-level seminars, symposia, conferences and workshops, as well as series of special lectures. As an outcome of these events, the Institute is bringing out a series of comprehensive volumes in different subjects under the title *Statistical Science and Interdisciplinary Research*, published by the World Scientific Press, Singapore.

The present volume titled *Perspectives in Mathematical Sciences II: Pure Mathematics* is the eighth one in the series. The volume consists of fifteen chapters, written by eminent mathematicians from different parts of the world. These chapters cover a wide range of topics and provide a current perspective of different areas of research, emphasizing the major challenging issues. Some of the articles are written keeping the students and the young researchers of mathematics in mind. I believe the state-of-the art studies presented in this book will be very useful to both researchers as well as practitioners.

Thanks to the contributors for their excellent research contributions, and to the volume editors Profs. N. S. Narasimha Sastry, T. S. S. R. K. Rao, M. Delampady and B. Rajeev for their sincere effort in bringing out the volume nicely in time. Initial design of the cover by Mr. Indranil Dutta is acknowledged. Sincere efforts by Prof. Dilip Saha and Dr. Barun Mukhopadhyay for editorial assistance are appreciated. Thanks are also due to World Scientific for their initiative in publishing the series and being a part of the Platinum Jubilee endeavor of the Institute.

December 2008
Kolkata

Sankar K. Pal
Series Editor and
Director

Preface

Indian Statistical Institute, a premier research institute founded by Professor Prasanta Chandra Mahalanobis in Calcutta in 1931, celebrated its platinum jubilee during the year 2006-07. On this occasion, the institute organized several conferences and symposia in various scientific disciplines in which the institute has been active.

From the beginning, research and training in probability, statistics and related mathematical areas including mathematical computing have been some of the main activities of the institute. Over the years, the contributions from the scientists of the institute have had a major impact on these areas.

As a part of these celebrations, the Division of Theoretical Statistics and Mathematics of the institute decided to invite distinguished mathematical scientists to contribute articles, giving "a perspective of their discipline, emphasizing the current major issues". A conference entitled "Perspectives in Mathematical Sciences" was also organized at the Bangalore Centre of the institute during February 4-8, 2008.

The articles submitted by the speakers at the conference, along with the invited articles, are brought together here in two volumes (Part I and Part II).

Part I consists of articles in Probability and Statistics. Articles in Statistics are mainly on statistical inference, both frequentist and Bayesian, for problems of current interest. These articles also contain applications illustrating the methodologies discussed. The articles on probability are based on different "probability models" arising in various contexts (machine learning, quantum probability, probability measures on Lie groups, economic phenomena modelled on iterated random systems, "measure free martingales", and interacting particle systems) and represent active areas of research in probability and related fields.

Part II consists of articles in Algebraic Geometry, Algebraic Number Theory, Functional Analysis and Operator Theory, Scattering Theory, von Neumann Algebras, Discrete Mathematics, Permutation Groups, Lie Theory and Super Symmetry.

All the authors have taken care to make their exposition fairly self-contained. It is our hope that these articles will be valuable to researchers at various levels.

The editorial committee thanks all the authors for writing the articles and sending them in time, the speakers at the conference for their talks and various scientists who have kindly refereed these articles. Thanks are also due to the National Board for Higher Mathematics, India, for providing partial support to the conference. Finally, we thank Ms. Asha Lata for her help in compiling these volumes.

October 16, 2008

N. S. Narasimha Sastry
T. S. S. R. K. Rao
Mohan Delampady
B. Rajeev

Contents

Chapter 1

Use of Resultants and Approximate Roots for Doing the Jacobian Problem

Shreeram S. Abhyankar

Mathematics Department,
Purdue University, West Lafayette, IN 47907, USA
ram@cs.purdue.edu

This is an expository article giving a modified version of my talks at ISI-Kolkata and ISI-Bangalore. After sketching the history of the jacobian problem, I shall discuss the two basic tools which are employed in attacking this problem. The first is the theory of resultants which are usually coupled with discriminants. The second is the theory of approximate roots of polynomials which are inspired by the construction of square roots of positive real numbers.

1.1. Introduction

Two given bivariate polynomials are said to form a jacobian pair if their jacobian equals a nonzero constant, and they are said to form an automorphic pair if the variables can be expressed as polynomials in the given polynomials. By the chain rule we see that every automorphic pair is a jacobian pair. The jacobian problem asks if conversely every jacobian pair is an automorphic pair. It turns out that a useful method for attacking this problem is to study the similarity of polynomials. Two bivariate polynomials are similar means their degree forms, i.e., highest degree terms, are powers of each other when they are multiplied by suitable nonzero constants. Geometrically this amounts to saying that the corresponding plane curves have the same points at infinity counting multiplicities. At any rate, the points at infinity correspond to the distinct irreducible factors of the degree form.

Before getting into technicalities, I shall first give a short history of the problem or rather the history of my acquaintance with the problem. For that we have to go back to 1965 when a German mathematician, Karl Stein who created Stein Manifolds, wrote me a letter asking a question. He said that there was an interesting 1955 paper in the Mathematische Annalen by Engel [9]. In this paper Engel claims to prove the jacobian theorem or what is now known as the jacobian problem or the jacobian conjecture or whatever. Karl Stein said to me that it is an interesting theorem but he cannot understand the proof. Can I help him? He also reduced it, or generalized it, to a conjecture about complex spaces. I wrote back to Stein giving a counterexample to his complex space conjecture. But I did not look at the Engel paper. Then in 1968, Max Rosenlicht of Berkeley asked me the same question and still I did not look at the Engel paper. Finally in 1970, my own guru (= venerable teacher) Oscar Zariski asked me the same question. Then, following the precept that one must obey one's guru, I looked up the Engel paper and found it full of mistakes and gaps.

The primary mistake in the Engel paper, which was repeated in a large number of published and unpublished wrong proofs of the jacobian problem in the last thirty-five years, is the presumed "obvious fact" that the order of the derivative of a univariate function is exactly one less than the order of the function. Being a prime characteristic person I never made this mistake. Indeed, the "fact" is correct only if the order of the function in nondivisible by the characteristic. Of course you could say that the jacobian problem is a characteristic zero problem, and zero does not divide anybody. But zero does divide zero. So the "fact" is incorrect if the order of the function is zero, i.e., if the value of the function is nonzero. Usually this mistake is well hidden inside a long argument, because you may start with a function which has a zero or pole at a given point and your calculation may lead to a function having a nonzero value at a resulting point.

A gap is a spot where you are not sure of the argument because of imprecise definitions or what have you. The gap in the Engel paper seems to be the uncritical use of the Zeuthen-Segre invariant. For this invariant of algebraic surfaces see the precious 1935 book of Zariski [12]. Over the years I have made several attempts to understand the somewhat mysterious theory of this invariant, and I still continue to do so.

In 1970-1977 I discussed the matter in my courses at Purdue and also in India and Japan. Mostly I was suggesting to the students to fix the proof and, to get them started, I proved a few small results. Notes of my lectures were taken down by Heinzer, van der Put, Sathaye, and Singh. These appeared in [1] and [4]. Then I put the matter aside for thirty years. Seeing that the problem has remained unsolved in spite of a continuous stream of wrong proofs announced practically every six months, I decided to write up my old results, together with some enhancements obtained recently, in the form of a series of three long papers [6], [7], [8], in the Journal of Algebra, dedicated to the fond memory of my good friend Walter Feit. The ISI Jubilee Volume has given me a welcome opportunity of introducing these papers to the young students with an invitation to further investigate the problem.

Now one of my old results says that the jacobian conjecture is equivalent to the implication that each member of a jacobian pair can have only one point at infinity. Another says that each member of any jacobian pair has at most two points at infinity. Note that the first result is a funny statement; it only says that to prove the jacobian conjecture, it suffices to show that each member of any jacobian pair has only one point at infinity. The second result is of a more definitive nature, and it remains true even when we give weights to the variables which are different from the normal weights. Very recently I noticed that, and this is one of the enhancements, the weighted two point theorem yields a very short new proof of Jung's 1942 automorphism theorem [10]. This automorphism theorem says that every automorphism of a bivariate polynomial ring is composed of a finite number of linear automorphisms and elementary automorphisms. In a linear automorphsim both variables are sent to linear expressions in them. In an elementary automorphism, one variable is unchanged and a polynomial in it is added to the second variable. In his 1972 lecture notes [11], Nagata declared the automorphism theorem to be very profound and so it did come as a pleasant surprise to me that the weighted two point theorem yields a five line proof of the automorphism theorem. For other recent enhancements let me refer to my Feit memorial papers cited above.

The present paper is only meant to whet the student's appetite. At any rate the material of this paper is based on my recent talks in various places such as ISI-Kolkatta and ISI-Bangalore.

1.2. Basic Technique

Our basic technique in attacking the jacobian problem is the use of resultants and approximate roots. Resultants are usually coupled with discriminants; the theory of these two objects will be discussed in Section 3. Approximate roots are polynomial concepts coming out of the construction of square roots in the theory of real numbers; these two topics will be discussed in Section 4. Here, assuming resultants and approximate roots, let us very briefly see how they are used in trying to do the jacobian problem.

Given any jacobian pair $F(X,Y)$ and $G(X,Y)$, by making a homogeneous linear transformation we can arrange that they are monic polynomials of positive degrees N and M in Y respectively. Adjoin W to the ground field k and consider the algebraic closure K of $k(W)$. Now eliminate Z by using the Z-resultant to get

$$\phi(X,Y) = (-1)^{M+N}\mathrm{Res}_Z(F(W,Z) - X, G(W,Z) - Y).$$

Then $\phi(X,T)$ is a monic polynomial of degree N in Y with coefficients in $K[X]$. Let $\rho_0 = N$ and $\rho_1 = M$. Let $d_1 = N$ and $d_2 = \mathrm{GCD}(\rho_0, \rho_1)$. Let $\mathrm{App}_D\Phi$ denote the D-th approximate root of a monic polynomial Φ whose Y-degree is a multiple E of D, i.e., $\mathrm{App}_D\Phi$ is the unique monic polynomial Ψ of Y-degree E/D such that $\deg_Y(\Phi - \Psi^D) < E - (E/D)$. Let $\psi_1(X,Y) = Y$. For $2 \le i \le h$ let us inductively define

$$\psi_i(X,Y) = \mathrm{App}_{d_i}\phi(X,Y)$$

with

$$\rho_i = \deg_X \mathrm{Res}_Y(\phi(X,Y), \psi_i(X,Y))$$

and

$$d_{i+1} = \mathrm{GCD}(\rho_0, \dots, \rho_i)$$

so that $d_2 > d_3 > \cdots > d_{h+1} = 1$. Here h is called the number of characteristic pairs.

By manipulating with this data, we can show that the polynomials F and G are similar. We can also show that if either $h \le 2$ or $h = 3$ with d_h even then the given jacobian pair (F, G) is an automorphic pair. As a consequence we can settle the jacobian conjecture when $\min(M,N) \le 52$. For details see [1] to [8].

1.3. Resultants and Discriminants

The material of this Section is taken from pages 100-104 of my new Algebra Book [5]. Details of proof can be found on pages 172-188 of that Book.

Beginning algebra students encounter discriminants of quadratic polynomials, but resultants are less well known. They were introduced by Sylvester around 1840. Actually in some sense we need to start with Descartes who introduced coordinates around 1637. Bézout built on these ideas to introduce his version of resultants for the original proof of Bézout's Theorem.

Bézout's Theorem, proved by him around 1770, is one of the oldest theorems of algebraic geometry. It says that a curve of degree m and a curve of degree n meet in mn points provided they have no common component and provided the intersections are counted properly.

One way of proving Bézout's Theorem is by using resultants. Vertical tangents can be located by using discriminants which are special cases of resultants.

Assuming n, m to be nonnegative integers, the Y-Resultant of two polynomials

$$f(Y) = a_0 Y^n + a_1 Y^{n-1} + \cdots + a_n$$
$$g(Y) = b_0 Y^m + b_1 Y^{m-1} + \cdots + b_m$$

is the determinant

$$\mathrm{Res}_Y(f, g) = \det(\mathrm{Resmat}_Y(f, g))$$

of the $n + m$ by $n + m$ matrix

$$\mathrm{Resmat}_Y(f,g) = \begin{pmatrix} a_0 & a_1 & \cdots & & a_n & 0 & \cdots & 0 \\ 0 & a_0 & a_1 & \cdots & & a_n & 0 \cdots & 0 \\ \cdot & \cdot & \cdot & \cdots & \cdot & \cdot & \cdots & \cdot \\ \cdot & \cdot & \cdot & \cdots & \cdot & \cdot & \cdots & \cdot \\ \cdot & \cdot & \cdot & \cdots & \cdot & \cdot & \cdots & \cdot \\ 0 & 0 & \cdots & a_0 & a_1 & \cdots & & a_n \\ b_0 & b_1 & \cdots & & b_m & 0 & \cdots & 0 \\ 0 & b_0 & b_1 & \cdots & & b_m & 0 \cdots & 0 \\ \cdot & \cdot & \cdot & \cdots & \cdot & \cdot & \cdots & \cdot \\ \cdot & \cdot & \cdot & \cdots & \cdot & \cdot & \cdots & \cdot \\ \cdot & \cdot & \cdot & \cdots & \cdot & \cdot & \cdots & \cdot \\ 0 & 0 & \cdots & b_0 & b_1 & \cdots & & b_m \end{pmatrix}$$

where the first m rows consist of the coefficients of f and the last n rows consist of the coefficients of g. More precisely, the first row starts with the coefficients of f, these are shifted one step to the right to get the second row, shifted two steps to the right to get the third row, and so on for the first m rows, then the $(m+1)$-st row starts with the coefficients of g, these are shifted one step to the right to get the $(m+2)$-nd row, and so on for the next n rows. The matrix is completed by stuffing zeroes elsewhere. The determinant $\mathrm{Res}_Y(f, g)$ is sometimes called the Sylvester resultant of f and g because it was introduced by Sylvester in his 1840 paper where he enunciated the following:

BASIC FACT (T1). If the coefficients a_i, b_j belong to a domain R then we have: $\mathrm{Res}_Y(f,g) = 0 \Leftrightarrow n+m \neq 0$ and either $a_0 = 0 = b_0$ or f and g have a common root in some overfield of R.

In case $n > 0$, the Y-Discriminant of f is defined to be the Y-Resultant of f and f_Y, i.e.,

$$\mathrm{Disc}_Y(f) = \mathrm{Res}_Y(f, f_Y).$$

where we view f_Y to be the polynomial

$$f_Y(Y) = na_0Y^{n-1} + (n-1)a_1Y^{n-2} + \cdots + a_{n-1}$$

i.e., we let the discriminant to be the determinant of the appropriate $2n-1$ by $2n-1$ matrix without considering whether na_0 equals zero or not.

From the Basic Fact (T1) we deduce the following:

COROLLARY (T2). If $n > 1$ and the coefficients a_i belong to a domain R then: $\mathrm{Disc}_Y(f) = 0 \Leftrightarrow$ either $a_0 = 0$ or f has a multiple root in some overfield of R.

OBSERVATION (O1). [**Resultant and Projection**]. If $X_1, \ldots, X_N$ are indeterminates over a field k with $N \in \mathbb{N}_+$ and R is either the polynomial ring $k[X_1, \ldots, X_N]$ or the power series ring $k[[X_1, \ldots, X_N]]$, then $\mathrm{Res}_Y(f,g)$ equals a polynomial or power series $\Phi = \Phi(X_1, \ldots, X_N)$. If a_0 and b_0 are in $k^\times$ with $nm \neq 0$ and k is algebraically closed then, in the polynomial case, by the Basic Fact it follows that the hypersurface $\Phi = 0$ in the N-space of $(X_1, \ldots, X_N)$ is the projection of the intersection of the hypersurfaces $f = 0$ and $g = 0$ in the $(N+1)$-space of $(X_1, \ldots, X_N, Y)$. Moreover, without assuming k to be algebraically closed but assuming that a_0 and b_0 are nonzero constants, in the polynomial case as well as the power series case, by the Basic Fact it follows that: Φ is identically zero (i.e., Φ is the zero element of R) $\Leftrightarrow$ f and g have a nonconstant common factor in $R[Y]$.

OBSERVATION (O2). [**Discriminant and Projection**]. Again if $X_1, \ldots, X_N$ are indeterminates over a field k with $N \in \mathbb{N}_+$ and R is either the polynomial ring $k[X_1, \ldots, X_N]$ or the power series ring $k[[X_1, \ldots, X_N]]$, then $\mathrm{Disc}_Y(f)$ equals a polynomial or power series $\Delta = \Delta(X_1, \ldots, X_N)$. Now if a_0 is in $k^\times$ with $n > 1$ and k is algebraically closed then, in the polynomial case, for all values $(U_1, \ldots, U_N)$ of $(X_1, \ldots, X_N)$ in k, the equation $f = 0$ has n roots which may or may not be distinct, and by the Corollary it follows that these roots are not distinct iff $\Delta(U_1, \ldots, U_N) = 0$. In other words, when we project the hypersurface $f = 0$ in $(N+1)$-space onto the N-space, above most points there lie n points, and $\Delta = 0$ is the locus of those points above which there lie less than n points. Moreover, without assuming k to be algebraically closed but assuming that a_0 is a nonzero constant, in the polynomial case as well as the power series case, by the Corollary it follows that: Δ is identically zero $\Leftrightarrow f$ has a nonconstant multiple factor in $R[Y]$.

OBSERVATION (O3). [**Isobaric Property**]. View the coefficients a_i, b_j as indeterminates over $\mathbb{Z}$. Give weight i to a_i, and j to b_j. Then $0 \neq \text{Res}_Y(f,g) \in \mathbb{Z}[a_0,\ldots,a_n,b_0,\ldots,b_m]$ is isobaric of weight mn, i.e., for the weight of any monomial $a_0^{i_0}\ldots a_n^{i_n} b_0^{j_0}\ldots b_m^{j_m}$ occurring in $\text{Res}_Y(f,g)$ we have $(\sum_{0\le r\le n} ri_r) + (\sum_{0\le s\le m} sj_s) = mn$. In particular, the principal diagonal $a_0^m b_m^n$ has weight mn, and it does not cancel out because there is no other term of b_m-degree n in the resultant; the principal diagonal of an $N\times N$ matrix (A_{ij}) is the term $A_{11}A_{22}\ldots A_{NN}$. The resultant being isobaric of weight mn is the fundamental fact behind various cases of Bézout's Theorem. The following two Observations, where we use the set-up of Observation (O1), illustrate this for plane curves and general hypersurfaces respectively.

OBSERVATION (O4). [**Plane Bézout**]. Let $N = 1$ with $X = X_1$, and assume that a_0, b_0 are nonzero elements in k, and f, g are polynomials of total (X,Y)-degrees n and m respectively. By the isobaric property we see that then always $\deg_X \Phi \le mn$ and "in general" $\deg_X \Phi = mn$. Hence the n-degree plane curve $f = 0$ meets the m-degree plane curve $g = 0$ in mn points "counted properly." The possibility of $\deg_X \Phi < mn$ is explained by saying that some intersections have "gone to infinity."

OBSERVATION (O5). [**Hyperspatial Bézout**]. Let N be general and assume that a_0, b_0 are nonzero elements in k, and f, g are polynomials of total $(X_1,\ldots,X_N,Y)$-degrees n and m respectively. By the isobaric property we see that then always $\deg_{(X_1,\ldots,X_N)} \Phi \le mn$ and "in general" $\deg_{(X_1,\ldots,X_N)} \Phi = mn$. Hence, in the $(N+1)$-dimensional space, the n-degree hypersurface $f = 0$ and the m-degree hypersurface $g = 0$ meet along a "secundum" (= a subvariety of dimension two less than dimension of the ambient space) which projects onto the (mn)-degree hypersurface $\Phi = 0$ in N-dimensional space. Again the possibility of $\deg_{(X_1,\ldots,X_N)} \Phi < mn$ says that some intersections have "gone to infinity."

EXAMPLE (X1). [**Resultant and Discriminant in Terms of Roots**]. If the coefficients a_i, b_j belong to a domain R and $a_0 \neq 0 \neq b_0$ then, upon writing

$$f(Y) = a_0 \prod_{1\le i\le n} (Y - \alpha_i) \quad \text{and} \quad g(Y) = b_0 \prod_{1\le j\le m} (Y - \beta_j)$$

with $\alpha_1, \ldots, \alpha_n$ and $\beta_1, \ldots, \beta_m$ in an overfield of R, we have

$$\begin{aligned}\mathrm{Res}_Y(f,g) &= a_0^m b_0^n \prod_{1\le i\le n}\prod_{1\le j\le m} (\alpha_i - \beta_j)\\ &= a_0^m \prod_{1\le i\le n} g(\alpha_i) = (-1)^{mn} b_0^n \prod_{1\le j\le m} f(\beta_j)\end{aligned}$$

and

$$\mathrm{Disc}_Y(f) = (-1)^{n(n-1)/2} a_0^n \prod_{1\le i<j\le n} (\alpha_i - \alpha_j)^2.$$

EXAMPLE (X2). [**Quadratic Resultant**]. Considering the quadratic polynomials

$$f(Y) = aY^2 + bY + c \quad \text{and} \quad g(Y) = a'Y^2 + b'Y + c'$$

and calculating the 4×4 determinant

$$\begin{pmatrix} a & b & c & 0 \\ 0 & a & b & c \\ a' & b' & c' & 0 \\ 0 & a' & b' & c' \end{pmatrix}$$

we get $\mathrm{Res}_Y(f,g) = (a^2c'^2 + a'^2c^2) + (b^2a'c' + b'^2ac) - (abb'c' + a'b'bc) - 2aca'c'$.

EXAMPLE (X3). [**Quadratic Discriminant**]. Considering the quadratic

$$f(Y) = aY^2 + bY + c$$

and calculating the 3×3 determinant

$$\begin{pmatrix} a & b & c \\ 2a & b & 0 \\ 0 & 2a & b \end{pmatrix}$$

we get $\mathrm{Disc}_Y(f) = -a(b^2 - 4ac)$.

EXAMPLE (X4). [**Cubic Discriminant**]. Considering the cubic

$$f(Y) = a_0Y^3 + a_1Y^2 + a_2Y + a_3$$

and calculating an appropriate 5×5 determinant we get

$$\mathrm{Disc}_Y(f) = -a_0(a_1^2a_2^2 - 4a_0a_2^3 - 4a_1^3a_3 - 27a_0^2a_3^2 + 18a_0a_1a_2a_3).$$

EXAMPLE (X5). [**Special Quartic Discriminant**]. Considering the quartic

$$f(Y) = Y^4 + pY^2 + qY + r$$

and calculating an appropriate 7×7 determinant we get

$$\mathrm{Disc}_Y(f) = 16p^4r - 4p^3q^2 - 128p^2r^2 + 144pq^2r - 27q^4 + 256r^3.$$

EXAMPLE (X6). [**General Quartic Discriminant**]. Considering the quartic

$$f(Y) = a_0Y^4 + a_1Y^3 + a_2Y^2 + a_3Y + a_4$$

and calculating an appropriate 7×7 determinant we get

$$\begin{cases} \mathrm{Disc}_Y(f) = a_0(256a_0^3a_4^3 - 192a_0^2a_1a_3a_4^2 - 128a_0^2a_2^2a_4^2 + 144a_0^2a_2a_3^2a_4 \\ \qquad - 27a_0^2a_3^4 + 144a_0a_1^2a_2a_4^2 - 6a_0a_1^2a_3^2a_4 - 80a_0a_1a_2^2a_3a_4 \\ \qquad + 18a_0a_1a_2a_3^3 + 16a_0a_2^4a_4 - 4a_0a_2^3a_3^2 - 27a_1^4a_4^2 \\ \qquad + 18a_1^3a_2a_3a_4 - 4a_1^3a_3^3 - 4a_1^2a_2^3a_4 + a_1^2a_2^2a_3^2). \end{cases}$$

1.4. Real Numbers and Approximate Roots

The material of this Section is taken from pages 52-59 of my new Algebra Book [5].

Historically, the process of counting gave rise to the set $\mathbb{N}_+$ of all positive integers. Augmenting it by zero, this set was enlarged to get the set $\mathbb{N}$ of all nonnegative integers. Inserting the negative of everybody gave the full set $\mathbb{Z}$ of all integers. Finally the process of division gave rise to the set $\mathbb{Q}$ of all rational numbers.

Taking this much for granted, we shall describe the limiting processes which gave rise first to the set $\mathbb{R}$ of real numbers and then to the set $\mathbb{C}$ of all complex numbers. The passage from $\mathbb{Q}$ to $\mathbb{R}$ was made precise by Cauchy by means of Cauchy Sequences around 1830, and by Dedekind by means of Dedekind Cuts around 1880. After sketching this development, we show how the construction of square roots of positive integers led us, around 1975, to the construction of approximate roots of polynomials. Briefly, this solves the problem as to how close we can come to finding the D-the root of a one-variable polynomial whose degree is a multiple of the positive integer D.

As in the quoted Algebra Book, we shall divide the material into a series of Definitions and Exercises. For the basic definitions of terms and standard set theory symbols which we shall use, such as $\in$ for "element of," the reader may consult the first few pages of the said Algebra Book, or any other current text-book of college mathematics.

DEFINITION (D1). [**Real and Complex Numbers**]. Real numbers may be defined as equivalence classes of Cauchy sequences of rational numbers, and the complex number field $\mathbb{C}$ may then be defined as the splitting field of the quadratic polynomial Y^2+1 over the real number field $\mathbb{R}$. By analogy with $\mathbb{N}_+$, by $\mathbb{Q}_+$ we denote the set of all positive rationals; moreover, by $\mathbb{Q}_{0+}$, $\mathbb{Q}_-$, and $\mathbb{Q}_{0-}$ we denote the set of all nonnegative rationals, negative rationals, and nonpositive rationals respectively; as usual, the absolute value of any $r \in \mathbb{Q}$ is denoted by $|r|$, i.e., $|r| = r$ or $-r$ according as $r \in \mathbb{Q}_{0+}$ or $r \in \mathbb{Q}_-$; by context, this will not be confused with the size $|S|$ of a set S. A sequence $x = (x_i)_{1\le i<\infty}$ in $\mathbb{Q}$ is Cauchy means for every $\epsilon \in \mathbb{Q}_+$ there exists $N_\epsilon \in \mathbb{N}_+$ such that for all $i > N_\epsilon$ and $j > N_\epsilon$ we have $|x_i - x_j| < \epsilon$. This is equivalent to the Cauchy sequence $x' = (x'_i)_{1\le i<\infty}$, in symbols $x \sim x'$, if for every $\epsilon \in \mathbb{Q}_+$ there exists $M_\epsilon \in \mathbb{N}_+$ such that for all $i > M_\epsilon$ we have $|x_i - x'_i| < \epsilon$. Now $\mathbb{R}$ may be defined to be the quotient $C_{\mathbb{Q}}/\sim$ of the set $C_{\mathbb{Q}}$ of all Cauchy sequences in $\mathbb{Q}$ by the equivalence relation $\sim$. We "identify" $\mathbb{Q}$ with a subset of $\mathbb{R}$ by sending any $q \in \mathbb{Q}$ to the equivalence class of the Cauchy sequence $(q_i)_{1\le i<\infty}$ with $q_i = q$ for all i. If $y = (y_i)_{1\le i<\infty}$ is another Cauchy sequence in $\mathbb{Q}$ then the sequences $(x_i + y_i)_{1\le i<\infty}$ and $(x_iy_i)_{1\le i<\infty}$ are Cauchy sequences whose equivalence classes are unchanged if x and y are replaced by equivalent Cauchy sequences; this makes $\mathbb{R}$ into a ring and in fact an overfield of $\mathbb{Q}$.

DEFINITION (D2). [**Ordered Fields**]. The order relation $\le$ can be extended from $\mathbb{Q}$ to $\mathbb{R}$ by declaring that the equivalence class of x is $\le$ the equivalence class of y if the Cauchy sequences x and y in their equivalence classes can be chosen so that $x_i \le y_i$ for all i; see (E3) below. Like $\mathbb{Z}$ and $\mathbb{Q}$, this makes $\mathbb{R}$ an ordered domain, i.e., a domain whose underlying additive abelian group is an ordered abelian group and in which the product of any positive elements (i.e., elements which are greater than zero) is again positive. Out of these $\mathbb{Q}$ and $\mathbb{R}$ are ordered fields, i.e., fields whose underlying domains are ordered domains. We extend the notation $\mathbb{Q}_+$, $\mathbb{Q}_{0+}$, $\mathbb{Q}_-$, and $\mathbb{Q}_{0-}$ to any ordered abelian group G (and hence to ordered domains and ordered fields) by putting $G_+ = \{g \in G : g > 0\}$, $G_{0+} = G_+ \cup \{0\}$, $G_- = \{g \in G : g < 0\}$, and $G_{0-} = G_- \cup \{0\}$; also we define the absolute value of any $g \in G$ by putting $|g| = g$ or $-g$ according as $g \in G_{0+}$ or $g \in G_-$. In particular this defines the sets R_+, R_{0+}, R_-, R_{0-}, and defines the absolute value $|r|$ of any $r \in \mathbb{R}$. Note that now $\mathbb{Z}_+ = \mathbb{N}_+$ and $\mathbb{Z}_{0+} = \mathbb{N}$. An ordered abelian group G is archimedean means for all x, y in G_+ we have $nx > y$ for some $n \in \mathbb{N}_+$. Clearly $\mathbb{R}$ is an archimedean ordered field, i.e., an ordered field whose underlying additive ordered abelian group is archimedean. A sequence $x = (x_i)_{1 \le i < \infty}$ in an ordered abelian group G is Cauchy means for every $\epsilon \in G_+$ there exists $N_\epsilon \in \mathbb{N}_+$ such that for all $i > N_\epsilon$ and $j > N_\epsilon$ we have $|x_i - x_j| < \epsilon$; the sequence x is convergent means it converges to a limit $\xi \in G$, i.e., for every $\epsilon \in G_+$ there exists $M_\epsilon \in N_+$ such that for all $i > M_\epsilon$ we have $|\xi - x_i| < \epsilon$; we indicate this by some standard notation such as $x_i \to \xi$ as $i \to \infty$ or $\lim_{i \to \infty} x_i = \xi$. An ordered abelian group is complete means in it every Cauchy sequence is convergent. Clearly $\mathbb{R}$ is a complete field, i.e., an ordered field whose underlying additive group is complete as an ordered abelian group.

DEFINITION (D3). [**Torsion Subgroups and Divisible Groups**]. By the subgroup of a group G generated by elements $x_1, x_2, \dots$ in G we mean the smallest subgroup of G which contains these elements. The order of an element in a group is the order of the subgroup generated by it. The subgroup of an additive abelian group G generated by all of its elements of finite order is called the torsion subgroup of G; if this is zero then G is said to be torsion free. An additive abelian group G is divisible means for every $g \in G$ and $n \in \mathbb{Z}^\times$ there is $h \in G$ with $nh = g$.

EXERCISE (E1). Show that for any elements a, b in a torsion free additive abelian group and any nonzero integer n we have: $na = nb \Rightarrow a = b$. Show that any ordered abelian group is torsion free, and hence any ordered field is of characteristic zero. Show that for all x, y in an ordered field we have $|xy| = |x||y|$. Show that the usual order on $\mathbb{Q}$ is the only order on it which makes it an ordered field.

EXERCISE (E2). Show that for all x, y in an ordered abelian group G we have $|x + y| \le |x| + |y|$, and from this deduce that any convergent sequence in G is Cauchy and has a unique limit.

DEFINITION (D4). [**Rational Completions**]. Given any torsion free additive abelian group G, let $G^* = G \times \mathbb{Z}^\times / \sim$ where the equivalence relation $\sim$ is given by: $(g, n) \sim (g', n') \Leftrightarrow n'g = ng'$, and embed G in G^* by identifying every $g \in G$ with the equivalence class containing $(g, 1)$. Define addition in G^* by taking equivalence classes in the proposed equation $(g, n) + (h, m) = (mg + nh, nm)$. Note that then G^* is a divisible additive abelian group such that for every $\overline{g} \in G^*$ we have $n\overline{g} \in G$ for some $n \in \mathbb{Z}^\times$. We call G^* the rational completion of G. Note that if G is divisible then $G^* = G$.

EXERCISE (E3). In (D2) show that the induced relation $\leq$ on the equivalence classes of Cauchy sequences in $\mathbb{Q}$ is a linear order.

EXERCISE (E4). In (D4) show that the induced relation $\leq$ on the equivalence classes of Cauchy sequences in an ordered abelian group is a linear order.

EXERCISE (E5). Let G be any nonzero archimedean ordered abelian group. Show that given any $g > 0$ in G and $x > 0$ in $\mathbb{R}$, there exists a unique order monomorphism (i.e., a group monomorphism which is order preserving) $\phi : G \to \mathbb{R}$ such that $\phi(g) = x$, and there exists a unique order isomorphism (i.e., a group isomorphism which is order preserving) $\psi : \overline{G^*} \to \mathbb{R}$ such that $\psi(g) = x$. Moreover, for these maps we always have $\phi(h) = \psi(h)$ for all $h \in G$.

EXERCISE (E6). Show that $x \mapsto x^2$ gives a surjection $\mathbb{R} \to \mathbb{R}_{0+}$.

HINT. The usual method of finding the decimal expansion $1.14142\ldots$ of $\sqrt{2}$ can be explained in terms of decimal expansions of integers by saying that $1^2 < 2 < 2^2$, $14^2 < 2 \times 10^2 < 15^2$, $141^2 < 2 \times 10^4 < 142^2$, $1414^2 < 2 \times 10^6 < 1415^2$, $14142^2 < 2 \times 10^8 < 14143^2$, and so on. More generally let $n > 1$ and $d > 1$ be any integers, and let $y > 0$ and $i > 0$ be any integers. Then clearly there is a unique integer x_i such that $x_i^d \leq yn^{di} < (x_i + 1)^d$; x_i is nothing but the n-adic expansion of the largest integer $\leq y^{1/d}n^i$. Obviously the sequence (x_i/n^i) is Cauchy and for its limit x in $\mathbb{R}_+$ we have $x^d = y$. Any positive rational can be written in the form y/z^d where y and z are positive integers, and then we get $x/z \in \mathbb{R}_+$ with $(x/z)^d = y/z^d$. Finally, any $\eta \in \mathbb{R}_+$ can be written as the limit of a sequence (η_j) in $\mathbb{Q}_+$ and then taking $\xi_j \in \mathbb{R}_+$ with $\xi_j^d = \eta_j$ we get a Cauchy sequence (ξ_j) for whose limit $\xi \in \mathbb{R}_+$ we have $\xi^d = \eta$.

DEFINITION (D5). [**Rational Ranks**]. In view of the first sentence of (E1), any torsion free divisible additive abelian group may clearly be regarded as a $\mathbb{Q}$-vector-space. The $\mathbb{Q}$-vector-space dimension of the rational completion of a torsion free additive abelian group G is called the rational rank of G and is denoted by $r(G)$. Alternatively, $r(G)$ may be characterized as the cardinal of a maximal (= nonenlargeable) $\mathbb{Z}$-linearly independent subset H of G, where independent means that for any finite number of distinct elements $x_1, \ldots, x_d$ in H and any integers $n_1, \ldots, n_d$ we have: $n_1x_1 + \cdots + n_dx_d = 0 \Rightarrow n_1 = \cdots = n_d = 0$.

DEFINITION (D6). [**Dedekind Cuts**]. Instead of using Cauchy sequences to prove (E5), we can use Dedekind Cuts. So let G be a nonzero divisible archimedean ordered abelian group. A Dedekind cut of G is a pair (L, U) of nonempty subsets of G with $U = G \setminus L$ such that for all $l \in L$ and $u \in U$ we have $l < u$ and there is no $u' \in U$ with $U = \{u \in G : u' \leq u\}$. For any $t \in G$ we get a Dedekind cut (L_t, U_t) with $L_t = \{l \in G : l \leq t\}$ and $U_t = \{u \in G : t < u\}$. Let D_G be the set of all Dedekind cuts of G. It can be shown that G is complete $\Leftrightarrow t \mapsto (L_t, R_t)$ gives a surjection $G \to D_G$. Indeed, D_G may be defined to be the completion of G. At any rate, for proving (E5), given any $h \in G$ let $\theta(h)$ be the real number which corresponds to the Dedekind cut (L, U) of $\mathbb{Q}$ where $L = \{m/n \in \mathbb{Q}$ with $m \in \mathbb{Z}$ and $n \in \mathbb{N}_+ : nh \leq mg\}$ and $U = \mathbb{Q} \setminus L$, and take $\phi(h) = x\theta(h)$.

DEFINITION (D7). [**Approximate Roots**]. In the Hint to (E6) we showed how to use n-adic expansions of positive integers to find successive approximations to the d-th root of a positive integer. Mixing a generalization of this with a generalization of the completing the square method of solving quadratic equations leads us to the concept of approximate roots of polynomials. So consider a monic polynomial

$$F = F(Y) = Y^N + \sum_{1 \leq i \leq N} A_i Y^{N-i}$$

of degree $N > 0$ in Y with coefficients A_i in a ring R. If N is a unit in R then we can generalize the completing the square idea to completing the N-th power by writing

$$F(Y) = (Y + A_1/N)^N + \sum_{2 \leq i \leq N} A'_i (Y + A_1/N)^{N-i}$$

with $A'_i \in R$, i.e., by killing the coefficient of Y^{N-1}. [On page 58 of my Lectures on Algebra Volume I, inadvertently the two plus signs in the above display have been printed as minus signs and, three lines above that, $N > 0$ has been printed as $N \geq 0$]. To generalize this further let $D > 0$ be an integer which divides N. Instead of assuming N to be a unit in R, assume D to be a unit in R; note that in case of a field R this is equivalent to assuming that the characteristic of R does not divide D and so characteristic zero is always ok. Now we look for a monic polynomial

$$G = G(Y) = Y^{N/D} + \sum_{1 \le i \le N/D} B_i Y^{(N/D)-i}$$

of degree N/D in Y with coefficients B_i in R such that G^D is as close to being equal to F as possible. As (E7) below shows, if we interpret this as requiring $\deg_Y(F-G^D) < N-(N/D)$ then a unique G exists, and we call it the approximate D-th root of F (relative to Y) and denote it by $\mathrm{App}_D(F)$ or $\mathrm{App}_{D,Y}(F)$. Recall that for any m, n in $\mathbb{N}$ with $n > 1$, the n-adic expansion of m consists of writing $m = \sum_{i \ge 0} m_i n^i$ where integers $0 \le m_i < n$ are the digits of the expansion. Likewise for any f, g in $R[Y]$ with g monic of positive Y-degree, the g-adic expansion of f consists of writing $f = \sum_{i \ge 0} f_i g^i$ where $f_i \in R[Y]$ with $\deg_Y f_i < \deg_Y g$ are the digits of the expansion. By (E8) below these expansions exist and are unique. Moreover, if f is monic of Y-degree $N > 0$ and the Y-degree of g is $N/D \in \mathbb{N}_+$ where $D \in \mathbb{N}_+$ is a unit in R, then $f_D = 1$ and $f_i = 0$ for all $i > D$; we try completing the D-th power by putting $\tau_f(g) = \tau_{f,Y}(g) = g + (f_{D-1}/D)$ and calling it the f-Tschirnhausen of g (relative to Y). This references to the 1683 work of Tschirnhausen who was a friend of Leibnitz. By (E9) below, starting with any monic g of degree N/D and applying τ_f to it N/D times will produce the approximate D-th root of f.

EXERCISE (E7). Let F be a monic polynomial of degree $N > 0$ in Y over a ring R. Let $D > 0$ be an integer such that D divides N and D is a unit in R. Show that there exists a unique monic polynomial G of degree N/D in Y over R such that $\deg_Y(F-G^D) < N-(N/D)$. Hint: With display as in (D7), the last condition gives the equations $A_i = DB_i + P_i(B_1, \ldots, B_{i-1})$ for $1 \le i \le N/D$ where the coefficient of Y^{N-i} in G^D equals $DB_i+P_i(B_1, \ldots, B_{i-1})$ with P_i a polynomial over $\mathbb{Z}$; since D is a unit in R, these can be solved successively (in a unique manner).

EXERCISE (E8). Given integers $m \ge 0$ and $n > 1$, show the unique existence of the n-adic expansion of m. Given univariate polynomials f, g over a ring R with g monic of positive degree, show the unique existence of the g-adic expansion $f = \sum_{i \ge 0} f_i g^i$ of f. Show that if f is monic of degree $N > 0$ and the degree of g is N/D where D is a positive integer factor of N, then $f_D = 1$ and $f_i = 0$ for all $i > D$, and moreover: $\mathrm{App}_D(f) = g \Leftrightarrow f_{D-1} = 0$.

EXERCISE (E9). Let f, g be univariate monic polynomials of positive degrees N and N/D over a ring R where D is a positive integer which divides N, and is a unit in R. Let $\tau_f(g) = \overline{g}$, and let $f = \sum_{0 \le i \le D} f_i g^i$ and $f = \sum_{0 \le i \le D} \overline{f}_i \overline{g}^i$ be the g-adic and $\overline{g}$-adic expansions of f respectively. Show that if $f_{D-1} \neq 0 \neq \overline{f}_{D-1}$ then $\deg(\overline{f}_{D-1}) < \deg(f_{D-1})$. From this deduce that $\tau_f^{N/D}(g) = \mathrm{App}_D(f)$.

Epilogue

MANGALACHARAN
ATA VISHVATMAKE DEVE | YENE VAGYADNYE TOSHAVE
TOSHONI MAJA DYAVE | PASAYDANA HE
GANITAVIDYECHEE JAGRUTEE | KARONIYA SARVA JAGATEE
PRADNYASURYE UJALATEE | SUKHAVAYA SAKALA JANA

Here is a free Paraphrase of the above MANGALACHARAN = INVOCATION in my mother tongue MARATHI whose founding father DHYANESHVAR composed the first two lines around 1250 A.D. to which I added the last two lines.

PARAPHRASE. May the Lord God of the Universe be pleased with my recounting of the story of algebra and geometry which are the essence of our beloved subject of mathematics. Being pleased may he shower his blessings upon us and make our endeavor pleasurable.

References

[1] Abhyankar, S. S. (1977). *Expansion Techniques in Algebraic Geometry.* Tata Institute of Fundamental Research, Bombay.
[2] Abhyankar, S. S. (1977). On the semigroup of a Meromorphic Curve (Part I). *Proceedings of the International Symposium on Algebraic Geometry* (Kyoto), Kinokuniya, Tokyo, pp. 249-414.
[3] Abhyankar, S. S. (1990). *Algebraic Geometry for Scientists and Engineers.* American Mathematical Society, 1990.
[4] Abhyankar, S. S. (1994). Some Remarks on the Jacobian Question. Purdue Lecture Notes, pp. 1-20 (1971); Published in the *Proceedings of the Indian Academy of Sciences.* **104** 515-542.
[5] Abhyankar, S. S. (2006). *Lectures on Algebra I.* World Scientific.
[6] Abhyankar, S. S. (2008). Some Thoughts on the Jacobian Conjecture, Part I. *Journal of Algebra* **319** 493-548.
[7] Abhyankar, S. S. (2008). Some Thoughts on the Jacobian Conjecture, Part II. *Journal of Algebra.* **319** 1154-1248.
[8] Abhyankar, S. S. (2008). Some Thoughts on the Jacobian Conjecture, Part III. *Journal of Algebra.* To Appear.

[9] Engel, W. (1955). Ein Satz über ganze Cremona Transformationes der Ebene. *Mathematische Annalen.* **130** 11-19.
[10] Jung, H. W. E. (1942). Über ganze birationale Transformatione der Ebene. *Crelle Journal.* **184** 161-174.
[11] Nagata, M. (1972). *On the automorphism group of* $k[X,Y]$, Lecture Notes in Mathematics, Tokyo.
[12] Zariski, O. (1935). *Algebraic Surfaces.* Springer-Verlag.

Chapter 2

Monodromy of Principal Bundles

Indranil Biswas and A. J. Parameswaran
School of Mathematics,
Tata Institute of Fundamental Research,
Homi Bhabha Road, Mumbai 400005, India
indranil@math.tifr.res.in and param@math.tifr.res.in

2.1. Introduction

Let M be a smooth manifold and E_G a C^∞ principal G–bundle over M, where G is a Lie group. The Lie algebra of G will be denoted by $\mathfrak{g}$. Let ∇ be a connection on E_G. Therefore, ∇ is a $\mathfrak{g}$–valued smooth one–form on the total space of E_G satisfying the following two conditions:

- the map $TE_G \longrightarrow \mathfrak{g}$ defined by ∇ intertwines the action of G on E_G and the adjoint action of G on $\mathfrak{g}$, and
- the restriction of ∇ to any fiber of E_G is the Maurer–Cartan form.

The connection ∇ defines parallel transport on E_G. In other words, given a smooth curve

$$\gamma : [0,1] \longrightarrow M$$

and a point z in the fiber $(E_G)_{\gamma(0)}$ of E_G over the point $\gamma(0)$, there is a unique smooth map

$$\widetilde{\gamma}_z : [0,1] \longrightarrow E_G \tag{2.1}$$

that satisfies the following three conditions:

- $p \circ \widetilde{\gamma}_z = \gamma$, where $p : E_G \longrightarrow M$ is the natural projection,
- $\widetilde{\gamma}_z$ is horizontal with respect to ∇, which means that the image of the differential $d\widetilde{\gamma}_z$ lies in the kernel of the form ∇, and
- $\widetilde{\gamma}_z(0) = z$.

Any $\widetilde{\gamma}_z$ as in Eq. (2.1) satisfying the first two of the above three conditions is called a *horizontal lift* of γ.

Fix a point $x_0 \in M$, and also fix a point $z_0 \in (E_G)_{x_0}$. Consider all curves γ as above with $\gamma(0) = x_0$, and consider their horizontal lifts $\widetilde{\gamma}$ with $\widetilde{\gamma}(0) = z_0$. Let Z denote the closure of the subset of E_G consisting of all points of the form $\widetilde{\gamma}(1)$, with $\widetilde{\gamma}$ of this type. There is a closed subgroup H of G such that a point $z \in (E_G)_{x_0}$ lies in Z if and only if $z = z_0 g$ for some $g \in H$. More precisely,

$$E_H := Z \subset E_G \tag{2.2}$$

is a smooth reduction of structure group of E_G to H, and H is a closed subgroup of G.

Note that if we replace the base point z_0 by any other point of Z, then both H and Z remain unchanged. If we replace z_0 by $z_0 g_0$, where $g_0 \in G$ is a fixed point, then Z gets replaced by $Z g_0$, and H gets replaced by the group $g_0^{-1} H g_0$. Hence the conjugacy class of the subgroup $H \subset G$ is independent of x_0 and z_0 (it depends only on ∇).

Let $\mathrm{Ad}(E_G) = E_G \times^G G$ be the adjoint bundle. So $\mathrm{Ad}(E_G)$ is the fiber bundle over M associated to E_G for the adjoint action of G on itself. Hence the fibers of $\mathrm{Ad}(E_G)$ are groups isomorphic to G. More precisely, each fiber of $\mathrm{Ad}(E_G)$ is identified with G uniquely up to an inner automorphism of G. Consider

$$\mathrm{Ad}(E_H) \subset \mathrm{Ad}(E_G), \tag{2.3}$$

where E_H is defined in Eq. (2.2). Since H gets replaced by $g_0^{-1} H g_0$ if z_0 is replaced by $z_0 g_0$, the subgroup

$$\mathrm{Ad}(E_H)_{x_0} \subset \mathrm{Ad}(E_G)_{x_0}$$

is independent of the choice of z_0; it depends only on ∇ and x_0. The subbundle $\mathrm{Ad}(E_H)$ of $\mathrm{Ad}(E_G)$ in Eq. (2.3) depends only on ∇.

The subbundle $\mathrm{Ad}(E_H)$ in Eq. (2.3) is called the *holonomy bundle*, and the subgroup $\mathrm{Ad}(E_H)_{x_0}$ of $\mathrm{Ad}(E_G)_{x_0}$ is called the *holonomy group* for ∇. Note that the holonomy group depends on x_0.

The conjugacy class of subgroups of G defined by $\mathrm{Ad}(E_H)_{x_0}$ (recall that $\mathrm{Ad}(E_G)_{x_0}$ is identified with G up to an inner automorphism) is independent of the choice of x_0.

Now let X be a smooth projective variety defined over a field k, and let G be a linear algebraic group defined over k. Let E_G be a principal G–bundle over X. Under some semistability assumption on E_G, and also an assumption that X admits a k–rational point, a canonical holonomy group of E_G exists, and also a holonomy bundle of E_G exists. Note that no reference to any connection on E_G is made.

The constructions of the above mentioned holonomy group and the holonomy bundle are based on a work of Nori. Our aim here is to describe these constructions.

In [5] and [6], the term monodromy is used instead of holonomy. We will stick to the terminology of [5] and [6].

In [1] an interesting construction of holonomy group (= monodromy) and holonomy bundle of polystable vector bundles over complex projective varieties is given (see also [2], [3]). An application of holonomy is given is [4].

We will mainly describe results from [6]. In the last section, we will very briefly describe some work in progress.

2.2. Tannakian Category

In this section we will briefly recall some results on Tannakian category that we need. This section is reproduced from [5].

Let k be any field and H an affine algebraic group scheme defined over k. We will denote by H–*mod* the category of finite dimensional rational left representations of H. The tensor product operation of two H–modules will be denoted by $\widehat{\otimes}$. We denote by k–*mod* the category of finite dimensional vector spaces over k, and we denote by $\otimes$ the usual tensor product operation on k–*mod*. Let

$$T : H\text{–}mod \longrightarrow k\text{–}mod$$

be the forgetful functor. Let $\mathcal{O}$ be a trivial one dimensional representation of H. The quadruple $(H\text{–}mod, \widehat{\otimes}, T, \mathcal{O})$ determines H in the following sense (see [14, Theorem 1], and [11, Section 2.1, Theorem 1.1]).

Theorem 2.1. ([11], [14]) *Let $\mathcal{C}$ be a category with a distinguished object $\mathcal{O}$ equipped with an operation*

$$\widehat{\otimes} : \mathcal{C} \times \mathcal{C} \longrightarrow \mathcal{C}$$

and $T : \mathcal{C} \longrightarrow k$–mod a functor satisfying the following eight conditions:

(C1) *$\mathcal{C}$ is an abelian category with direct sums;*
(C2) *isomorphism classes of objects of $\mathcal{C}$ form a set;*
(C3) *T is an additive, faithful and exact functor;*

(C4) *$\widehat{\otimes}$ is k-linear in each variable, and $T \circ \widehat{\otimes} = \otimes \circ (T \times T)$;*
(C5) *$\widehat{\otimes}$ is associative preserving T;*
(C6) *$\widehat{\otimes}$ is commutative preserving T;*
(C7) *the object $\mathcal{O}$ of $\mathcal{C}$ is equipped with an isomorphism $\phi : k \longrightarrow T(\mathcal{O})$ such that $\mathcal{O}$ is an identity object of $\widehat{\otimes}$ preserving T;*
(C8) *for every object L of $\mathcal{C}$ such that $T(L)$ is one dimensional, there is an object L^{-1} of $\mathcal{C}$ such that $L\widehat{\otimes}L^{-1}$ is isomorphic to $\mathcal{O}$.*

Then there exists a unique affine algebraic group scheme H defined over k such that the quadruple $(\mathcal{C}, \widehat{\otimes}, T, \mathcal{O})$ is identified with the quadruple (H–mod, $\widehat{\otimes}, T, \mathcal{O}$).

Any quadruple $(\mathcal{C}, \widehat{\otimes}, T, \mathcal{O})$ as in Theorem 2.1 satisfying all the conditions in Theorem 2.1 will be called a *neutral Tannakian category* (see [11, p. 76]).

Let H_1 and H_2 be two affine algebraic group schemes. An algebraic group homomorphism $H_1 \longrightarrow H_2$ is defined by a morphism of neutral Tannakian categories

$$C_I : H_2\text{–}mod \longrightarrow H_1\text{–}mod \tag{2.4}$$

compatible with $\widehat{\otimes}$, T and $\mathcal{O}$. Such a morphism C_I defines a closed embedding of H_1 in H_2 if and only if every object of H_1–*mod* is isomorphic to a subquotient (in H_1–*mod*) of the image of an object of H_2–*mod* under C_I. This is proved in [7, p. 139, Proposition 2.21(b)].

Next we recall a description of principal bundles over a scheme S defined over k using the language of categories and functors in a similar spirit. Let H be an affine algebraic group scheme defined over k and $E_H \longrightarrow S$ a principal H-bundle over S. Then for every object

$$\rho : H \longrightarrow \mathrm{GL}(V)$$

in H–*mod* we can construct the associated vector bundle $E_\rho := E_H \times^H V$ over S. Here $E_H \times^H V = (E_H \times V)/H$ with the action of any $h \in H$ sending any $(z, v) \in E_H \times V$ to $(zh, \rho(h^{-1})v)$ (a point of a group scheme H over k means a R-valued point for some k-algebra R). This defines a functor $\mathcal{F}_E$ from the category H–*mod* to the category of vector bundles $\mathrm{Vect}(S)$ on S. Nori proved, in [11, Section 2.2], that this functor determines the principal H-bundle in the following sense:

Theorem 2.2. ([11]) *Let $\mathcal{F} : H$–mod $\longrightarrow \mathrm{Vect}(S)$ be a functor satisfying the following:*

(F1) *$\mathcal{F}$ is a k–additive exact functor;*
(F2) $\mathcal{F} \circ \widehat{\otimes} = \otimes \circ (\mathcal{F} \times \mathcal{F})$;
(F3)$_1$ *the functor $\mathcal{F}$ preserves commutativity, or in other words, if c is the canonical isomorphism of $V\widehat{\otimes}W$ with $W\widehat{\otimes}V$ in H–mod, then $\mathcal{F}(c)$ is the canonical isomorphism of the corresponding vector bundles;*
(F3)$_2$ *the functor $\mathcal{F}$ preserves associativity, which means that if a is the canonical isomorphism of $U\widehat{\otimes}(V\widehat{\otimes}W)$ with $(U\widehat{\otimes}V)\widehat{\otimes}W$ in H–mod, then $\mathcal{F}(a)$ is the canonical isomorphism of the corresponding vector bundles;*
(F3)$_3$ *the vector bundle $\mathcal{F}(\mathcal{O})$ is the trivial line bundle $\mathcal{O}_S$ on S;*
(F4) *for any $V \in H$–mod of dimension n, the vector bundle $\mathcal{F}(V)$ is of rank n.*

Then there exists a unique principal H-bundle $E \longrightarrow S$ such that $\mathcal{F}$ is identified with $\mathcal{F}_E$.

Let

$$\rho : H_1 \longrightarrow H_2 \tag{2.5}$$

be a group homomorphism defined by the functor

$$C_I : H_2\text{–}mod \longrightarrow H_1\text{–}mod$$

as in Eq. (2.4). Suppose $E_1 \longrightarrow S$ is a principal H_1-bundle defined by a functor $\mathcal{F}_1 : H_1\text{–}mod \longrightarrow \text{Vect}(S)$. We define

$$\mathcal{F}_2 : H_2\text{–}mod \longrightarrow \text{Vect}(S)$$

by $\mathcal{F}_2 := \mathcal{F}_1 \circ C_I$. Since both $\mathcal{F}_1$ and C_I are additive, exact, preserving commutativity and associativity it follows that $\mathcal{F}_2$ satisfies all the conditions $(F1)$, $(F2)$, $(F3)_1$ and $(F3)_2$ in Theorem 2.2. For the remaining conditions in Theorem 2.2, recall that C_I takes $\mathcal{O}$ to $\mathcal{O}$, and it preserves the rank of the modules. In view of this, the functor $\mathcal{F}_2$ satisfies both $(F3)_3$ and $(F4)$ as the functor $\mathcal{F}_1$ does so. Consequently, $\mathcal{F}_2$ defines a principal H_2-bundle over S; the principal H_2-bundle constructed this way will be denoted by E_2. This principal H_2-bundle E_2 is identified with the principal H_2-bundle $E_1 \times^{H_1} H_2$ obtained by extending the structure group of E_1 using the homomorphism ρ in Eq. (2.5). We recall that $E_1 \times^{H_1} H_2 = (E_1 \times H_2)/H_1$, where the action of any $h \in H_1$ sends any $(z, h_2) \in E_1 \times H_2$ to $(zh, \rho(h^{-1})h_2)$.

Definition 2.1. Let E_2 be a principal H_2-bundle over S given by a functor $\mathcal{F}_2$ as in Theorem 2.2. A *reduction of structure group* to a closed subgroup scheme $H_1 \subset H_2$ is a principal H_1-bundle E_1 given by a functor $\mathcal{F}_1$ that satisfies the identity

$$\mathcal{F}_2 = \mathcal{F}_1 \circ C_I \,,$$

where C_I is the functor as in Eq. (2.4) defining the subgroup $H_1 \hookrightarrow H_2$.

Giving a reduction of structure group of a principal H_2-bundle E_2 over S to a closed subgroup scheme $H_1 \subset H_2$ is equivalent to giving a subscheme

$$E_1 \subset E_2$$

such that the action of H_1 on E_2 preserves E_1 making it a principal H_1-bundle over S.

2.3. A Tannakian Category for a Pointed Curve

Fix a field k. Let X be a geometrically irreducible smooth projective curve defined over this field k.

The notion of a semistable vector bundle was introduced by Mumford; we will recall the definition.

A vector bundle E over X is called *semistable* if for every subbundle $V \subset E$ of positive rank, the inequality

$$\frac{\text{degree}(V)}{\text{rank}(V)} \leq \frac{\text{degree}(E)}{\text{rank}(E)} \tag{2.6}$$

holds. If

$$\frac{\text{degree}(V)}{\text{rank}(V)} < \frac{\text{degree}(E)}{\text{rank}(E)} \tag{2.7}$$

for all subbundles V with $0 < \text{rank}(V) < \text{rank}(E)$, then E is called *stable.* The rational number $\frac{\text{degree}(E)}{\text{rank}(E)}$ is called the *slope* of E, and it is customary to denote it by $\mu(E)$.

Let ℓ be a field extension of k. If a vector bundle E over X admits a subbundle V with $\mu(V) > \mu(E)$, then for the subbundle $V \bigotimes_k \ell$ of the vector bundle $E \bigotimes_k \ell$ over $X \times_k \ell$, the inequality $\mu(V \bigotimes_k \ell) > \mu(E \bigotimes_k \ell)$ holds. In other words, if E is not semistable, then $E \bigotimes_k \ell$ is not semistable. The following proposition shows that the converse is also valid.

Proposition 2.1. *Let ℓ be a field extension of k. A vector bundle E over X is semistable if and only if the vector bundle $E \bigotimes_k \ell$ over $X \times_k \ell$ is semistable.*

The above proposition is proved in [9] under the assumption that k is infinite (see [9, p. 97, Proposition 3]), and it is proved in [8] under the assumption that k is perfect (see [8, p. 222]). Hence the proposition holds for all fields.

If the characteristic of k is p, with $p > 0$, then we have the Frobenius map

$$F_k : k \longrightarrow k$$

defined by $\lambda \longmapsto \lambda^p$. Consider the Cartesian diagram

$$\begin{array}{ccccc} X & \xrightarrow{\pi} & X_1 & \xrightarrow{\varphi} & X \\ & & \downarrow & & \downarrow \\ & & \mathrm{Spec}(k) & \xrightarrow{F_k} & \mathrm{Spec}(k) \end{array} \tag{2.8}$$

where and π is the relative Frobenius map (see [13, p. 118]). The composition $\varphi \circ \pi$ will be denoted by F_X. If the characteristic of k is zero, then F_X will denote the identity map of X.

A semistable vector bundle W over X is called *strongly semistable* if the iterated pull back

$$(\overbrace{F_X \circ \cdots \circ F_X}^{n-\text{times}})^* W$$

is semistable for all $n \geq 1$.

So if the characteristic of k is zero, then any semistable vector bundle is strongly semistable. We note that strongly semistable vector bundles are usually defined under the assumption that the base field is perfect. In view of Proposition 2.1, the above definition is compatible with it.

Let $\mathcal{S}_X$ denote the category of all strongly semistable vector bundles on X. Let $\mathcal{C}_X$ denote the space of all maps

$$f : \mathbb{Q} \longrightarrow \mathrm{Obj}(\mathcal{S}_X) \tag{2.9}$$

satisfying the following two conditions:

- $f(\lambda) = 0$ for all but finitely many rational numbers, and
- For any $\lambda \in \mathbb{Q}$ with $f(\lambda) \neq 0$,

$$\mu(f(\lambda)) := \lambda .$$

We will define direct sum and tensor product, as well as dual, of objects of $\mathcal{C}_X$.

If $E, F \in \mathcal{S}_X$ (see Eq. (2.9)) with $\mu(E) = \mu(F)$, then the direct sum $E \bigoplus F$ is also strongly semistable with $\mu(E \bigoplus F) = \mu(E)$. Also, the dual vector bundle E^* is strongly semistable with $\mu(E^*) = -\mu(E)$.

For any $f, g \in \mathcal{C}_X$, define $f \bigoplus g \in \mathcal{C}_X$ by

$$\lambda \longmapsto f(\lambda) \oplus g(\lambda).$$

Define the *dual* f^* of f by

$$\lambda \longmapsto f(-\lambda)^*.$$

For $V, W \in \mathcal{S}_X$ of positive ranks, the vector bundle $V \bigotimes W$ is also strongly semistable ([12, p. 288, Theorem 3.23]). We also note that

$$\mu(V \bigotimes W) = \mu(V) + \mu(W). \tag{2.10}$$

For any $f, g \in \mathcal{C}_X$, define

$$(f \otimes g)(\lambda) := \bigoplus_{z \in \mathbb{Q}} f(z) \otimes g(\lambda - z). \tag{2.11}$$

For any two objects $f, g \in \mathcal{C}_X$, an element of

$$\mathrm{Hom}(f, g) := \bigoplus_{\gamma \in \mathbb{Q}} H^0(X, \mathcal{H}om(f(\gamma), g(\gamma))) \tag{2.12}$$

will be called a *homomorphism* from f to g. Any $h \in \mathrm{Hom}(f, g)$ will also be considered as a map

$$h : \mathbb{Q} \longrightarrow \bigoplus_{\gamma \in \mathbb{Q}} H^0(X, \mathcal{H}om(f(\gamma), g(\gamma)))$$

such that $h(\gamma) \in H^0(X, \mathcal{H}om(f(\gamma), g(\gamma)))$ for all γ.

A homomorphism h from f to g will be called an *isomorphism* if

$$h(\gamma) : f(\gamma) \longrightarrow g(\gamma)$$

is an isomorphism for all $\gamma \in \mathbb{Q}$.

Take any object f in the category $\mathcal{C}_X$. A *sub-object* of f is an object f' in $\mathcal{C}_X$ such that for each $\lambda \in \mathbb{Q}$, the vector bundle $f'(\lambda)$ is a subbundle of the vector bundle $f(\lambda)$. If f' is a sub-object of f, then the object of $\mathcal{C}_X$ that sends any $\lambda \in \mathbb{Q}$ to the quotient vector bundle $f(\lambda)/f'(\lambda)$ will be called a *quotient-object* of f. For any object f of the category $\mathcal{C}_X$, a *sub-quotient* of f is a sub-object of some quotient-object of f.

Let $\varphi : V \longrightarrow W$ be a homomorphism between strongly semistable vector bundles V and W over X with

$$\mu(V) = \mu(W). \tag{2.13}$$

Then either φ is injective, or kernel(φ) is a strongly semistable vector bundle over X with $\mu(\text{kernel}(\varphi)) = \mu(V)$. Similarly, either φ is surjective or cokernel(φ) $:= W/\varphi(V)$ is a strongly semistable vector bundle over X with $\mu(\text{cokernel}(\phi)) = \mu(W)$.

Take any $f, f' \in \mathcal{C}_X$ and any $h \in \text{Hom}(f, f')$ (see Eq. (2.12)). Consider the function $\mathbb{Q} \longrightarrow \mathcal{S}_X$ that sends any $\lambda \in \mathbb{Q}$ to the kernel of the homomorphism

$$h(\lambda) : f(\lambda) \longrightarrow f'(\lambda).$$

This function defines an object of $\mathcal{C}_X$, which we will call the *kernel* of h, and it will be denoted by kernel(h). So kernel(h) is a sub–object of f.

Similarly, the object of $\mathcal{C}_X$ defined by function $\mathbb{Q} \longrightarrow \mathcal{S}_X$ that sends any $\lambda \in \mathbb{Q}$ to the cokernel of the homomorphism

$$h(\lambda) : f(\lambda) \longrightarrow f'(\lambda)$$

will be called the *cokernel* of h, and it will be denoted by cokernel(h). So cokernel(h) is a quotient–object of f'.

We note that $\mathcal{C}_X$ is an abelian category.

We will henceforth assume that the curve X admits a k–rational point. Fix a k–rational point x of X.

For any vector bundle V over X, the fiber of V over x will be denoted by V_x.

The category of finite dimensional vector spaces over the field k will be denoted by *k–mod.*

We have a functor

$$\omega : \mathcal{C}_X \longrightarrow k\text{–}mod \tag{2.14}$$

defined by

$$f \longmapsto \bigoplus_{z \in \mathbb{Q}} f(z)_x .$$

Let $\mathcal{O}_X$ be the trivial line bundle over X defined by the structure sheaf of X. The object in $\mathcal{C}_X$ that sends any $\lambda \in \mathbb{Q} \setminus \{0\}$ to the vector bundle of rank zero and sends 0 to $\mathcal{O}_X$ will also be denoted by $\mathcal{O}_X$.

The triple $(\mathcal{C}_X, \mathcal{O}_X, \omega)$, where ω is defined in Eq. (2.14) and $\mathcal{O}_X \in \mathcal{C}_X$ is the object defined above, together form a neutral Tannakian category over k; see [7], [10] for neutral Tannakian category. Any neutral Tannakian category over k gives an affine group scheme defined over k (see Theorem 2.1). Hence the triple $(\mathcal{C}_X, \mathcal{O}_X, \omega)$ produce an affine group scheme defined over k.

Let $\mathcal{G}_X$ denote the group scheme defined over k given by the neutral Tannakian category $(\mathcal{C}_X, \mathcal{O}_X, \omega)$.

We will now show that there is a tautological principal $\mathcal{G}_X$–bundle over X.

Let Vect(X) denote the category of vector bundles over X. We have a functor

$$\mathcal{F}_X : \mathcal{C}_X \longrightarrow \mathrm{Vect}(X), \tag{2.15}$$

defined by

$$f \longmapsto \bigoplus_{\lambda=-\infty}^{\infty} f(z).$$

Using [11, Lemma 2.3, Proposition 2.3], the functor $\mathcal{F}_X$ in Eq. (2.15) defines a principal $\mathcal{G}_X$–bundle over X, where $\mathcal{G}_X$ is the affine group scheme defined above (see Theorem 2.2).

Let $E_{\mathcal{G}_X}$ denote the tautological $\mathcal{G}_X$–bundle over X constructed above.

2.4. Monodromy of a Strongly Semistable Principal Bundles

As before, k is any field, X is a geometrically irreducible smooth projective curve defined over k and x is a k–rational point of X.

Let G be a linear algebraic group defined over the field k. Let $Z'_0(G)$ denote the connected component, containing the identity element, of the reduced center of G.

We assume that G satisfies the following condition: there is no nontrivial character of G which is trivial on $Z'_0(G)$.

Let

$$Z_0(G) \subset Z'_0(G) \tag{2.16}$$

be the (unique) maximal split torus contained in $Z'_0(G)$. Let $Z_0(G)^*$ denote the group of all characters of $Z_0(G)$. The subgroup $Z_0(G)$ gives a decomposition of any G–module, which we will describe now.

Let V be a finite dimensional left G–module. Let be the decomposition

$$V = \bigoplus_{\chi \in Z_0(G)^*} V_\chi, \tag{2.17}$$

into isotypical components of the $Z_0(G)$–module V. So for any character χ of $Z_0(G)$,

$$V_\chi \subset V$$

is the subspace on which $Z_0(G)$ acts as scalar multiplications through the character χ. It is easy to see that each V_χ is left invariant by the action of G on V.

See [6] for the proof of the following lemma.

Lemma 2.1. *Take any character $\chi \in Z_0(G)^*$. Let V and W be two nonzero finite dimensional left G–modules such that $Z_0(G)$ acts on both V and W as scalar multiplications through the character χ. Let E_G be a principal G–bundle over X. Let E_V (respectively, E_W) be the vector bundle over X associated to the principal G–bundle E_G for the G–module V (respectively, W). Then*

$$\mu(E_V) = \mu(E_W).$$

The following is a corollary of [7, p. 139, Proposition 2.21].

Corollary 2.1. *All characters of $Z_0(G)$ arise from the indecomposable representations of G. In other words, for any character χ of $Z_0(G)$, there is some nonzero finite dimensional indecomposable left G–module V such that $Z_0(G)$ acts on V as scalar multiplications through the character χ.*

If E and F are two vector bundles over X, then $\mu(E \bigotimes F) = \mu(E) + \mu(F)$. Therefore, Lemma 2.1 and Corollary 2.1 combine together to give the following corollary:

Corollary 2.2. *Fix any principal G–bundle E_G over X. Then there is a homomorphism to the additive group*

$$\delta_{E_G} : Z_0(G)^* \longrightarrow \mathbb{Q}$$

that sends any character χ to $\mu(E_V)$, where V is a finite dimensional nonzero left G–module on which $Z_0(G)$ acts as scalar multiplications through the character χ, and E_V is the vector bundle over X associated to the principal G–bundle E_G for the G–module V.

Definition 2.2. Let G be any affine group scheme defined over k. A principal G–bundle E_G over a geometrically irreducible smooth projective curve X will be called *strongly semistable* if for any indecomposable finite dimensional left G–module $V \in G\text{–}mod$, the vector bundle over X associated to E_G for V is strongly semistable.

See [13], [12] for the definition of a (strongly) semistable principal bundle with a reductive group as the structure group. We will show that the above definition coincides with the usual definition a (strongly) semistable principal G–bundle when G is reductive. For that we will need the following theorem. We note that Definition 2.2 is justified by the following lemma which is proved in [6].

Lemma 2.2. *Let H be a reductive linear algebraic group and X a geometrically irreducible smooth projective curve defined over k. A principal H–bundle E_H over X is strongly semistable if and only if for every indecomposable H–module V, the vector bundle $E_V = E_H(V)$ over X associated to the principal H–bundle E_H for V is strongly semistable.*

Let E_G be a strongly semistable principal G–bundle over X. To each G–module we will associate an object of the neutral Tannakian category $\mathcal{C}_X$ that we constructed in Section 2.3.

Let V be a finite dimensional left G–module. We noted earlier that each V_χ in Eq. (2.17) is a G–module. Let E_{V_χ} be the vector bundle over X associated to the principal G–bundle E_G for the above G–module V_χ.

The following lemma is proved in [6].

Lemma 2.3. *The vector bundle E_{V_χ} is strongly semistable, and if $V_\chi \neq 0$, then*

$$\mu(E_{V_\chi}) = \delta_{E_G}(\chi),$$

where δ_{E_G} is the homomorphism constructed in Corollary 2.2.

Lemma 2.3 has the following corollary:

Corollary 2.3. *For any $\lambda \in \mathbb{Q}$ and any $V \in G$–mod, the direct sum*

$$E^\lambda_G(V) := \bigoplus_{\{\chi \in Z_0(G)^* | \delta_{E_G}(\chi) = \lambda\}} E_{V_\chi}$$

is either zero, or it is a strongly semistable vector bundle with

$$\mu(E^\lambda_G(V)) = \lambda .$$

As before, G–*mod* denote the category of finite dimensional G–modules. As before, by $\mathrm{Vect}(X)$ we will denote the category of vector bundles over X. Consider the function

$$f_{E_G,V} : \mathbb{Q} \longrightarrow \mathrm{Vect}(X)$$

defined by

$$f_{E_G,V}(\lambda) := E^\lambda_G(V), \tag{2.18}$$

where $V \in G$–*mod*, and $E_G^{\lambda}(V)$ is defined in Corollary 2.3. From Corollary 2.3 it follows immediately that this function $f_{E_G,V}$ is an object of the category $\mathcal{C}_X$ constructed in Section 2.3. Therefore, to each object of G–*mod* we have associated an object of $\mathcal{C}_X$.

Let $\mathcal{C}_{E_G}$ denote the subcategory of $\mathcal{C}_X$ defined by all objects f of $\mathcal{C}_X$ such that there exists some $V \in G$–*mod* with the property that f is isomorphic to a sub-quotient of $f_{E_G,V}$, where $f_{E_G,V}$ is the object of $\mathcal{C}_X$ constructed from V in Eq. (2.18). The morphisms remain unchanged. In other words, for any two objects f and f' in $\mathcal{C}_{E_G}$, the morphisms from f to f' are the morphisms from f to f' considered as objects of $\mathcal{C}_X$.

It is straight–forward to check that $\mathcal{C}_{E_G}$ is a neutral Tannakian subcategory of $\mathcal{C}_X$. Therefore, the neutral Tannakian category $\mathcal{C}_{E_G}$ gives an affine group scheme defined over k (see Theorem 2.1).

Definition 2.3. The affine group scheme defined over k given by the neutral Tannakian category $\mathcal{C}_{E_G}$ will be called the *monodromy group scheme* of E_G. The monodromy group scheme of E_G will be denoted by M.

Since $\mathcal{C}_{E_G}$ is a Tannakian subcategory of $\mathcal{C}_X$, the monodromy group scheme M is a quotient of the group scheme $\mathcal{G}_X$ constructed in Section 2.3 (see [7, Proposition 2.21]). Let

$$\phi_{E_G} : \mathcal{G}_X \longrightarrow M \tag{2.19}$$

be the quotient map.

Just as we have the tautological $\mathcal{G}_X$–bundle $E_{\mathcal{G}_X}$, there is a tautological principal M–bundle over X.

Definition 2.4. Let E_M denote the tautological principal M–bundle over X. This principal M–bundle E_M will be called the *monodromy bundle* for E_G.

The principal M–bundle E_M is evidently the one obtained by extending the structure group of the principal $\mathcal{G}_X$–bundle $E_{\mathcal{G}_X}$ using the homomorphism ϕ_{E_G} in Eq. (2.19).

We will next show that there is a tautological embedding of the monodromy group scheme M into the fiber, over the fixed k–rational point $x \in X$, of the adjoint bundle for E_G.

Let $\mathrm{Ad}(E_G)$ be the adjoint bundle for the principal G–bundle E_G over X. Let $\mathrm{Ad}(E_G)_x$ be the fiber of $\mathrm{Ad}(E_G)$ over the fixed k–rational point x of X.

If ω is the fiber functor for the principal G–bundle E_G over X, then the group $\mathrm{Ad}(E_G)_x$ defined over k represents the functor $\underline{\mathrm{Aut}}^{\otimes}(\omega)$. Using Theorem 2.11 in [7, p. 130], we get a natural homomorphism from the group scheme $\mathcal{G}_X$ (constructed in Section 2.3) to $\mathrm{Ad}(E_G)_x$. Let

$$\Phi(E_G) : \mathcal{G}_X \longrightarrow \mathrm{Ad}(E_G)_x \tag{2.20}$$

be this natural homomorphism.

It is easy to see that the image of the homomorphism $\Phi(E_G)$ in Eq. (2.20) coincides with the monodromy group scheme M in Definition 2.3.

Therefore, we have the following proposition:

Proposition 2.2. *The monodromy group scheme M for E_G (introduced in Definition 2.3) is identified with the image of the homomorphism $\Phi(E_G)$ in Eq. (2.20). In other words, the kernel of the homomorphism $\Phi(E_G)$ coincides with the kernel of the homomorphism ϕ_{E_G} in Eq. (2.19).*

There is a natural inclusion $M \hookrightarrow \mathrm{Ad}(E_G)_x$ obtained from the fact that the quotients of $\mathcal{G}_X$ for the two homomorphisms $\Phi(E_G)$ and ϕ_{E_G} coincide.

We will now investigate the behavior of the monodromy group and the monodromy bundle under the extensions of structure group.

Let

$$\rho : G \longrightarrow G_1 \tag{2.21}$$

be an algebraic homomorphism between linear algebraic groups defined over k. Let $Z_0(G_1)$ denote the (unique) maximal split torus contained in the reduced center of G_1. We assume the following:

- The group G_1 does not admit any nontrivial character which is trivial on $Z_0(G_1)$.
- The homomorphism ρ in Eq. (2.21) satisfies the condition

$$\rho(Z_0(G)) \subset Z_0(G_1). \tag{2.22}$$

The following lemma in proved in [6].

Lemma 2.4. *Let E_G be a strongly semistable principal G–bundle over X. Then the principal G_1–bundle $E_{G_1} := E_G(G_1)$, obtained by extending the structure group of E_G using ρ (defined in Eq. (2.21)) is also strongly semistable.*

Let E_G be a strongly semistable principal G–bundle over X. Hence by Lemma 2.4, the principal G_1–bundle E_{G_1}, obtained by extending the structure group of the principal G–bundle E_G using the homomorphism ρ, is also strongly semistable. The homomorphism ρ in Eq. (2.21) induces a homomorphism of group schemes

$$\widetilde{\rho} : \mathrm{Ad}(E_G) \longrightarrow \mathrm{Ad}(E_{G_1}) \tag{2.23}$$

over X.

The following lemma is proved in [6].

Lemma 2.5. *The monodromy group scheme $M_1 \subset \mathrm{Ad}(E_{G_1})_x$ for E_{G_1} is the image $\widetilde{\rho}(x)(M)$, where $\widetilde{\rho}(x)$ is the homomorphism in Eq. (2.23) restricted to the k–rational point x of X, and $M \subset \mathrm{Ad}(E_G)_x$ is the monodromy group scheme of E_G (see Proposition 2.2). Furthermore, the monodromy bundle E_{M_1} for E_{G_1} is the extension of structure group of the monodromy bundle E_M for E_G by the homomorphism $M \longrightarrow M_1$ obtained by restricting $\widetilde{\rho}(x)$.*

2.5. More on Monodromy

For a subgroup scheme $H \subset G$, let $(H \bigcap Z_0(G))_{\mathrm{red}}$ denote the reduced intersection; let $(H \bigcap Z_0(G))_0$ denote the (unique) maximal split torus contained in the abelian group $(H \bigcap Z_0(G))_{\mathrm{red}}$.

As before, let E_G be a principal G–bundle E_G over the curve X.

Definition 2.5. A reduction of structure group

$$E_H \subset E_G \tag{2.24}$$

of E_G to a subgroup scheme $H \subset G$ will be called *balanced* if for every character χ of H trivial on $(H \bigcap Z_0(G))_0 \subset H$ (see the above definition), we have

$$\mathrm{degree}(E_H(\chi)) = 0\,,$$

where $E_H(\chi)$ is the line bundle over X associated to the principal H–bundle E_H for the character χ.

Since any character of $(H \bigcap Z_0(G))_{\mathrm{red}}/(H \bigcap Z_0(G))_0$ is of finite order, if a character χ of H is trivial on $(H \bigcap Z_0(G))_0$, then there is a positive integer n such that the character χ^n of H is trivial on $(H \bigcap Z_0(G))_{\mathrm{red}}$. Therefore, a reduction of structure group $E_H \subset E_G$ as in Definition 2.5 is balanced if and only if for every character χ of H trivial on $(H \bigcap Z_0(G))_{\mathrm{red}}$ we have

$$\mathrm{degree}(E_H(\chi)) = 0\,.$$

Since the quotient $(H \bigcap Z_0(G))/(H \bigcap Z_0(G))_{\text{red}}$ is a finite group scheme, if a character χ of H is trivial on $(H \bigcap Z_0(G))_{\text{red}}$, then there is a positive integer n such that the character χ^n of H is trivial on $H \bigcap Z_0(G)$. Therefore, a reduction $E_H \subset E_G$ as in Definition 2.5 is balanced if and only if for every character χ of H trivial on $H \bigcap Z_0(G)$ we have

$$\text{degree}(E_H(\chi)) = 0 .$$

The following proposition is proved in [6].

Proposition 2.3. *Let E_G be a strongly semistable principal G–bundle over X and $E_H \subset E_G$ a balanced reduction of structure group of E_G to a subgroup scheme $H \subset G$. Then the principal H–bundle E_H over X is strongly semistable.*

Let E_G be a strongly semistable principal G–bundle over X. In Proposition 2.2 we saw that the monodromy group scheme M (constructed in Definition 2.3) is canonically embedded in $\text{Ad}(E_G)_x$. For national convenience, we will denote by $\widetilde{G}$ the group $\text{Ad}(E_G)_x$ defined over k. Let $E_{\widetilde{G}}$ be the principal $\widetilde{G}$–bundle over X obtained by extending the structure group of the monodromy bundle E_M (see Definition 2.4) using the inclusion of M in $\widetilde{G}$. Therefore,

$$E_M \subset E_{\widetilde{G}} \tag{2.25}$$

is a reduction of structure group of $E_{\widetilde{G}}$ to M.

Let $Z_0(\widetilde{G})$ denote the unique maximal split torus contained in the reduced center of $\widetilde{G}$.

The following theorem is proved in [6].

Theorem 2.3. *Assume that the group $\widetilde{G} := \text{Ad}(E_G)_x$ does not admit any nontrivial character which is trivial on $Z_0(\widetilde{G})$. Then the reduction of structure group in Eq. (2.25) is a balanced reduction of structure group of $E_{\widetilde{G}}$ to M. In particular, the principal M–bundle E_M is strongly semistable.*

Assume that the fiber of the principal G–bundle E_G, over the k–rational point x of X, admits a rational point. If we fix a rational point in the fiber of E_G over x, then $\widetilde{G}$ gets identified with G, and the principal bundle $E_{\widetilde{G}}$ gets identified with E_G.

If $E_H \subset E_G$ is a reduction of structure group, to a subgroup scheme $H \subset G$, of a principal G–bundle E_G over X, then the adjoint bundle $\text{Ad}(E_H)$ is a subgroup scheme of the group scheme $\text{Ad}(E_G)$ over X.

The following theorem is proved in [6].

Theorem 2.4. *Let E_G be a strongly semistable principal G–bundle over a geometrically irreducible smooth projective curve X defined over k, where G is a linear algebraic group defined over k with the property that G does not admit any nontrivial character which is trivial on $Z_0(G)$. Fix a k–rational point x of X. Let $H \subset G$ be a subgroup scheme and $E_H \subset E_G$ a balanced reduction of structure group of E_G to H. Then the image in $\mathrm{Ad}(E_G)_x$ of the monodromy group scheme M (image by the homomorphism in Proposition 2.2) is contained in the subgroup scheme $\mathrm{Ad}(E_H)_x \subset \mathrm{Ad}(E_G)_x$.*

More properties of the monodromy group and the monodromy reduction can be found in [5].

2.6. Bundles on Higher Dimensional Varieties

In this section we will assume k to be an algebraically closed field, but we do not put any assumptions on the characteristic of k. Let X be an irreducible smooth projective variety equipped with a very ample line bundle ξ.

The *degree* of a torsionfree coherent sheaf F on X is defined to be the degree of F restricted to the general complete intersection curve obtained by intersecting hyperplanes from the complete linear system $|\xi|$.

A torsionfree coherent sheaf E over X is called *semistable* (respectively, *stable*) if the inequality Eq. (2.6) (respectively, Eq. (2.7)) holds for all coherent subsheaves $V \subset E$ with $0 < \mathrm{rank}(V) < \mathrm{rank}(E)$. A semistable sheaf is called *polystable* if it is a direct sum of stable sheaves.

As before, a semistable vector bundle E is called *strongly semistable* if its iterated pullbacks by the self–map F_X of X are all semistable. We recall F_X is the Frobenius map of X when the characteristic of k is positive, and it is the identity map of X when the characteristic of k is zero.

Let E be a strongly semistable vector bundle over X. There is a filtration of coherent subsheaves

$$0 = V_0 \subset V_1 \subset \cdots \subset V_{n-1} \subset V_n = E$$

such that each successive quotient V_i/V_{i-1}, $i \in [1, n]$, is polystable, and $\mu(V_i/V_{i-1}) = \mu(V)$. More precisely, V_i/V_{i-1} is the maximal polystable subsheaf of E/V_{i-1}, which is also called the *socle* of E/V_{i-1}.

Let

$$\mathcal{U}_E \subset X \tag{2.26}$$

be the maximal Zariski open dense subset such that each V_i/V_{i-1}, $i \in [1, n]$, is locally free. It can be shown that E possesses a monodromy group as well as a monodromy bundle over the open subset $\mathcal{U}_E$.

Compared to the case of curves, the construction of the monodromy group and the monodromy bundle for higher dimensional X is technically more complicated.

Let G be a connected reductive linear algebraic group defined over k. Let E_G be a strongly semistable principal G–bundle over X. Consequently, the adjoint vector bundle $\mathrm{ad}(E_G)$ is strongly semistable ([12, p. 288, Theorem 3.23]).

Fix a point $x \in \mathcal{U}_{\mathrm{ad}(E_G)}$, where $\mathcal{U}_{\mathrm{ad}(E_G)}$ is constructed as in Eq. (2.26). There is a canonical monodromy group $H \subset \mathrm{Ad}(E_G)_x$ as well as a monodromy reduction over the open subset $\mathcal{U}_{\mathrm{ad}(E_G)}$.

References

[1] Balaji, V. and Kollár, J. (2008). Holonomy groups of stable vector bundles. *Publ. Res. Inst. Math. Sci.* (to appear), math.AG/0601120.

[2] Biswas, I. (2005). Stable bundles and extension of structure group. *Diff. Geom. Appl.* **23** 67–78.

[3] Biswas, I. (2007). On the algebraic holonomy of stable principal bundles, preprint.

[4] Biswas, I., Parameswaran, A. J. and Subramanian, S. (2004). Numerically effective line bundles associated to a stable bundle over a curve. *Bull. Sci. Math.* **128** 23–29.

[5] Biswas, I., Parameswaran, A. J. and Subramanian, S. (2006). Monodromy group for a strongly semistable principal bundle over a curve. *Duke Math. Jour.* **132** 1–48.

[6] Biswas, I. and Parameswaran, A. J. (2008). Monodromy group for a strongly semistable principal bundle over a curve, II. *Jour. K-Theory* **1** 583–607.

[7] Deligne, P. and Milne, J. S. (1982). Tannakian Categories. In *Hodge cycles, motives, and Shimura varieties* (eds. P. Deligne, J. S. Milne, A. Ogus and K.-Y. Shih), pp. 101–228, Lecture Notes in Mathematics, **900**, Springer-Verlag, Berlin-Heidelberg-New York.

[8] Harder, G. and Narasimhan, M. S. (1975). On the cohomology groups of moduli spaces of vector bundles on curves. *Math. Ann.* **212** 215–248.

[9] Langton, S. G. (1975). Valuative criteria for families of vector bundles on algebraic varieties. *Ann. of Math.* **101** 88–110.

[10] Nori, M. V. (1976). On the representations of the fundamental group scheme. *Compos. Math.* **33** 29–41.

[11] Nori, M. V. (1982). The fundamental group scheme. *Proc. Ind. Acad. Sci. (Math. Sci.)* **91** 73–122.

[12] Ramanan, S. and Ramanathan, A. (1984). Some remarks on the instability flag. *Tôhoku Math. Jour.* **36** 269–291.
[13] Ramanathan, A. (1975). Stable principal bundles on a compact Riemann surface. *Math. Ann.* **213** 129–152.
[14] Saavedra Rivano, N. (1972). *Catégories Tannakiennes,* Lecture Notes in Mathematics, **265** Springer-Verlag, Berlin-Heidelberg-New York.

Chapter 3

Oligomorphic Permutation Groups

Peter J. Cameron

School of Mathematical Sciences,
Queen Mary, University of London,
London E1 4NS, U.K.
P.J.Cameron@qmul.ac.uk

A permutation group G (acting on a set Ω, usually infinite) is said to be *oligomorphic* if G has only finitely many orbits on Ω^n (the set of n-tuples of elements of Ω). Such groups have traditionally been linked with model theory and combinatorial enumeration; more recently their group-theoretic properties have been studied, and links with graded algebras, Ramsey theory, topological dynamics, and other areas have emerged.

This paper is a short summary of the subject, concentrating on the enumerative and algebraic aspects but with an account of group-theoretic properties. The first section gives an introduction to permutation groups and to some of the more specific topics we require, and the second describes the links to model theory and enumeration. We give a spread of examples, describe results on the growth rate of the counting functions, discuss a graded algebra associated with an oligomorphic group, and finally discuss group-theoretic properties such as simplicity, the small index property, and "almost-freeness".

3.1. Introduction

Despite the history and importance of group theory, we have very little idea what an arbitrary group looks like. We have made important strides in understanding finite groups, by determining the finite simple groups; but we can only study general groups under some very strong condition, usually a finiteness condition. We have theories of finitely generated groups, locally finite groups, residually finite groups, groups of finite cohomological dimension, linear groups, profinite groups, and so forth, but no theory of general groups.

Oligomorphic groups satisfy a rather different kind of finiteness condition; paradoxically, one which makes them "large" rather than "small". A permutation group G (a subgroup of the symmetric group on a set Ω) is said to be *oligomorphic* if G has only finitely many orbits on Ω^n for every natural number n. (An element of G acts componentwise on the set Ω^n of all n-tuples of points of Ω.)

Thus, by definition, an oligomorphic group G gives us a sequence of natural numbers, the numbers of orbits on n-tuples for $n = 0, 1, 2, \ldots$. Not surprisingly, the theory is connected with counting problems in various parts of mathematics (combinatorics, model theory, graded algebras). Curiously, there is rather less we can say about the groups themselves. If G is oligomorphic and H is a proper subgroup having the same orbits on n-tuples as G for all n, then counting cannot distinguish between G and H, even though they may be very different as groups (for example, G may be simple while H is a free group).

In the remainder of this section, we introduce some basics of permutation group theory and of the counting functions associated with oligomorphic groups. For further information about permutation groups, see [7, 14]. Note also that there are many connections between parts of the theory of oligomorphic groups and that of (combinatorial) species, as developed by [21].

3.1.1. *Permutation groups*

This section is a brief introduction to permutation groups. For more details, see [7].

The *symmetric group* $\mathrm{Sym}(\Omega)$ on a set Ω is the group of all permutations of Ω. If Ω is finite, say $\Omega = \{1, 2, \ldots, n\}$, we write the symmetric group as S_n. We write permutations on the right, so that the image of α under g is written αg.

An *orbit* of G is an equivalence class of the relation $\alpha \sim \beta$ if $\alpha g = \beta$ for some $g \in G$; in other words, a set of the form $\{\alpha g : g \in G\}$. We say that G is *transitive* if it has only one orbit. In a sense, any permutation group can be "resolved" into transitive groups.

There is a partial converse. If G_i is a transitive permutation group on Ω_i for $i \in I$, where I is some index set, the *cartesian product* $\prod_{i\in I} G_i$ is the set of functions $f : I \to \bigcup_{i\in I} G_i$ satisfying $f(i) \in G_i$ for all $i \in I$. It has two natural actions:

- The *intransitive action* on the disjoint union of the sets Ω_i: if $\alpha \in \Omega_i$,

then $\alpha f = \alpha f(i)$. If each group G_i is transitive, then the sets Ω_i are the orbits of the cartesian product.
- There is also a *product action*, componentwise on the cartesian product of the sets Ω_i.

If I is finite, we speak of the *direct product*, and write it as (for example) $G_1 \times \cdots \times G_k$, if $I = \{1, \ldots, k\}$.

Thus, if $G_1 = S_n$ and $G_2 = S_m$, then $G_1 \times G_2$ has an intransitive action on $m+n$ points, and a product action on mn points (which can be regarded as a rectangular grid with G_1 permuting the rows and G_2 the columns).

Now let G be transitive on Ω. A *congruence* is an equivalence relation on Ω which is G-invariant. There are two "trivial" congruences on Ω: the relation of equality, and the "universal" relation with a single equivalence class. We say that G is *primitive* if there are no other congruences. For example, the symmetric group is primitive.

Important examples of imprimitive groups are the *wreath products*, defined as follows. Let H be a permutation group on Γ, and K a permutation group on Δ. Let $\Omega = \Gamma \times \Delta$, regarded as a union of copies of Γ indexed by Δ: thus $\Gamma_\delta = \{(\gamma, \delta) : \gamma \in \Gamma\}$, for each $\delta \in \Delta$. Take a copy H_δ of H for each $\delta \in \Delta$, where H_δ acts on Γ_δ. Then the cartesian product $B = \prod_{\delta \in \Delta} H_\delta$ acts on Ω (in its intransitive action). Moreover, the group K acts on Ω by permuting the second components (i.e. by permuting the "fibres" Γ_δ. The *wreath product* $G = H \mathrm{Wr} K$ is the group generated by B and K; we call B the *base group* (it is a normal subgroup) and K the *top group* of the wreath product.

We note that there are different notations and conventions in other areas of mathematics. For example, in experimental design in statistics (cf. [2]), direct product (product action) is called *crossing*, and wreath product is called *nesting*, but nesting is written with the top structure before the bottom one, e.g. Δ/Γ in our case.

If $|\Gamma| > 1$ and $|\Delta| > 1$, the wreath product is imprimitive: the relation $(\gamma, \delta) \equiv (\gamma', \delta')$ if $\delta = \delta'$ is a congruence. For this reason it is called the *imprimitive action.* Any imprimitive permutation group can be embedded in a wreath product in a natural way.

There is another action of the wreath product, the *power action*, on the set Γ^Δ of functions from Δ to Γ. It bears a similar relation to the imprimitive action as the product action does to the intransitive action for the cartesian product. If we regard the domain of the imprimitive action as a fibred space, with fibres Γ_δ isomorphic to Γ indexed by Δ, then the domain for the power action is the set of global sections (subsets containing one point from each fibre).

3.1.2. *Oligomorphic permutation groups*

For a natural number n, a permutation group is *n-transitive* if it acts transitively on the set of n-tuples of distinct elements of Ω, and *n-set-transitive* if it acts transitively on the set of n-element subsets of Ω. We say a permutation group is *highly transitive* if it is n-transitive for all n, and *highly set-transitive* if it is n-set-transitive for all n.

Oligomorphic permutation groups generalise these classes. Thus, we let $F_n(G)$ denote the number of orbits of G on the set of n-tuples of distinct elements, and $f_n(G)$ the number of orbits on n-element subsets. So G is n-transitive (resp. n-set-transitive) if $F_n(G) = 1$ (resp. $f_n(G) = 1$). If the group G is clear, we drop it and write simply F_n, f_n.

The definition speaks of orbits on Ω^n, the set of all n-tuples. Let $F_n^*(G)$ denote the number of these orbits. It is clear that, for given n, one of these numbers is finite if and only if the others are; indeed, we have

- $F_n^* = \sum_{k=1}^{n} S(n,k)F_k$, where $S(n,k)$ is the Stirling number of the second kind (the number of partitions of an n-set into k parts);
- $f_n \leq F_n \leq n!\, f_n$.

As an example for the first point, if G is highly transitive, then

$$F_n^*(G) = \sum_{k=0}^{n} S(n,k) = B(n),$$

the nth *Bell number* (the number of partitions of an n-set).

In the second point, both bounds are attained:

- Let $G = \mathrm{Sym}(\Omega)$ (we will always denote this group by S). Then $f_n(S) = F_n(S) = 1$ for all n.
- Let G be the group of order-preserving permutations of the rational numbers (we will always denote this group by A). We can map any n rational numbers in increasing order to any other n such by an order-preserving permutation (by filling in the gaps to produce a piecewise-linear map); so $f_n(A) = 1$ and $F_n(A) = n!$.

Highly set-transitive groups must resemble the above types. All we can do is to modify the total order slightly. The following is proved in [3].

Theorem 3.1. *Let G be a highly set-transitive but not highly transitive permutation group. Then there is a linear or circular order preserved by G. In particular, G is not 4-transitive.*

3.1.3. *Topology*

We will only consider permutation groups of countable degree in this article. If we are interested in the sequences f_n and F_n, this is justified by the following result, which follows from the *Downward Löwenheim–Skolem Theorem* of model theory:

Proposition 3.1. *Let G be an oligomorphic permutation group on an infinite set. Then there is an oligomorphic permutation group G' on a countably infinite set such that $F_n(G') = F_n(G)$ and $f_n(G') = f_n(G)$ for all $n \in \mathbb{N}$.*

There is a natural topology on the symmetric group of countable degree. This is the topology of pointwise convergence, where a sequence (g_n) converges to g if $\alpha_i g_n = \alpha_i g$ for all sufficiently large n, where $(\alpha_1, \alpha_2, \ldots)$ is an enumeration of Ω. This topology can be derived from a complete metric. The topology is specified by the first part of the proposition below.

Proposition 3.2.

(a) *G is open in* $\mathrm{Sym}(\Omega)$ *if and only if it contains the pointwise stabiliser of a finite set.*

(b) *G is closed in Ω if and only if G is the automorphism group of a relational structure on Ω, that is, a family of relations (of various arities) on Ω.*

The closure of a permutation group G consists of all permutations which preserve the G-orbits on Ω^n for all n. We remarked earlier that, as far as orbit-counting goes, we cannot distinguish between groups with the same orbits. The largest such group is necessarily closed in $\mathrm{Sym}(\Omega)$. So we may restrict our attention to closed groups of countable degree (that is, automorphism groups of countable relational structures) if we are interested in orbit-counting.

The fact that the topology on the symmetric group is derived from a complete metric means that the same is true for any closed subgroup. This permits use of the *Baire category theorem.* Recall that a subset of a complete metric space is *residual* if it contains the intersection of countably many open dense subsets. The Baire category theorem asserts that a residual set is non-empty. Indeed, residual sets are "large" (for example, they have non-empty intersection with any open set, and the intersection of countably many residual sets is residual). Often it is possible to give a non-constructive existence proof for some object by showing that objects of the required type form a residual set.

3.1.4. *Cycle index*

An important tool in combinatorial enumeration is the *cycle index* of a finite permutation group G, which is the polynomial in the indeterminates $s_1, s_2, \ldots, s_n$ (where n is the degree) given by

$$Z(G; s_1, s_2, \ldots, s_n) = \frac{1}{|G|} \sum_{g \in G} s_1^{c_1(g)} s_2^{c_2(g)} \cdots s_n^{c_n(g)},$$

where $c_i(g)$ is the number of i-cycles in the decomposition of the permutation g into disjoint cycles.

For its role in enumeration, see [18].

One cannot simply take this definition unchanged for infinite permutation groups, for several reasons: the number $c_i(g)$ may be infinite; there may be infinite cycles; and the denominator $|G|$ is infinite.

The trick to generalising it lies in the following result about finite groups, the *Shift Lemma*. Let $\mathcal{P}\Omega$ be the set of all subsets of the finite set Ω, and let $\mathcal{P}\Omega/G$ denote a set of representatives of the G-orbits on $\mathcal{P}\Omega$. Also, if X is any subset of Ω, we let $G[X]$ denote the permutation group on X induced by its setwise stabiliser in G.

Proposition 3.3. *For a finite permutation group G on Ω,*

$$\sum_{X \in \mathcal{P}\Omega/G} Z(G[X]; s_1, s_2, \ldots) = Z(G; s_1 + 1, s_2 + 1, \ldots).$$

Now let G be any oligomorphic permutation group on the (possibly infinite) set Ω. We define the *modified cycle index* of G to be the left-hand side of the Shift Lemma, with one small modification: we replace $\mathcal{P}\Omega$ by $\mathcal{P}_{\text{fin}}\Omega$, the set of all finite subsets of Ω. Thus, the modified cycle index is

$$\tilde{Z}(G; s_1, s_2, \ldots) = \sum_{X \in \mathcal{P}_{\text{fin}}\Omega/G} Z(G[X]; s_1, s_2, \ldots).$$

Each term in the sum is the cycle index of a finite permutation group. Also since G is oligomorphic, there are only a finite number of terms in the sum which contribute to the coefficient of a fixed monomial $s_1^{c_1} \cdots s_r^{c_r}$, namely those corresponding to the $f_n(G)$ orbits on sets of cardinality $n = c_1 + 2c_2 + \cdots + rc_r$. (We see here that the definition would fail if G were not oligomorphic.) So we have defined a formal power series in $s_1, s_2, \ldots$. We also see that if G happens to be a finite permutation group, then we have the ordinary cycle index with 1 added to each indeterminate.

Our previous counting functions can be extracted from the modified cycle index:

Proposition 3.4. *Let G be an oligomorphic permutation group.*

- *$F_G(x)$ is obtained from $\tilde{Z}(G)$ by substituting $s_1 = x$ and $s_i = 0$ for $i > 1$.*
- *$f_G(x)$ is obtained from $\tilde{Z}(G)$ by substituting $s_i = x^i$ for all i.*

Just as before, it is true that

- for any oligomorphic permutation group G, there is an oligomorphic group G' of countable degree satisfying $\tilde{Z}(G) = \tilde{Z}(G')$;
- an oligomorphic group of countable degree and its closure have the same modified cycle index.

So we may consider closed groups of countable degree.

We conclude this section by displaying the modified cycle index for the groups S (the infinite symmetric group) and A (the group of order-preserving permutations of $\mathbb{Q}$).

Proposition 3.5.

$$\tilde{Z}(S) = \exp\left(\sum_{i\geq 1} \frac{s_i}{i}\right);$$

$$\tilde{Z}(A) = \frac{1}{1 - s_1}.$$

3.2. Connections

Oligomorphic permutation groups are closely connected with two other areas of mathematics: model theory, and combinatorial enumeration. In this section we discuss the connections.

3.2.1. *Model theory*

Model theory describes structures consisting of a set with a collection of constants, relations and functions. Much of mathematics can be fitted into this framework, often in different ways. For example, a group has a binary operation (composition), a unary operation (inversion), and a constant (the identity); but the second and third may be defined in terms of the first. A graph can be regarded as a set of vertices with a binary relation (adjacency), or as a set of vertices and edges with a unary relation (to distinguish the vertices) and a binary operation of incidence. The latter is appropriate for multigraphs.

Logic describes such structures by means of collections of formulae. The language includes symbols for the relations, functions, and constants, connectives and quantifiers, equality, brackets, and a supply of variables. We work in *first-order logic*: we are allowed to combine finitely many formulae with connectives, and to quantify over variables which range over the underlying set. I will assume that the language is countable. A *sentence* is a formula with no free variables (all variables are quantified). Thus for example the group axioms can be expressed as a single sentence (the conjunction of the associative, identity and inverse laws, each universally quantified over all variables).

A *theory* is a set of sentences; a structure is a *model* for a theory if every sentence in the theory is true in the model.

Sometimes we want a theory to have a wide range of models (this is the case with group theory). At other times, we have a particular model in mind, and want to capture as much of its essence as possible in a theory (this is the case with the Peano axioms for the natural numbers). It is known that, as long as a theory has infinite models, it cannot have a unique model; there will be models of arbitrarily large cardinality. The best we can do is ask that the theory is α*-categorical*, where α is an infinite cardinal, meaning that there is a unique model of cardinality α up to isomorphism. By a theorem of Vaught, there are only two types of categoricity, countable and uncountable: if a theory is α-categorical for some uncountable cardinal α, then it is α-categorical for all such.

Uncountable categoricity gives rise to a powerful structure theory, extending that for vector spaces over a fixed field or algebraically closed fields of fixed characteristic (where a single invariant, the rank, determines the model). By contrast, countable categoricity is related to symmetry, by the following remarkable theorem due independently to Engeler,

Ryll-Nardzewski and Svenonius:

Theorem 3.2. *Let M be a countable model of a theory T over a countable language. Then T is $\aleph_0$-categorical if and only if the automorphism group of M is oligomorphic.*

In fact, if M is the unique countable model of T, and $G = \mathrm{Aut}(M)$, then $F_n^*(T)$ is the number of n-types over T (maximal consistent sets of formulae in n free variables). So our results are applicable to counting types in such theories.

Note in passing that the automorphism group of $\mathbb{N}$ is trivial; so the Peano axioms have countable "non-standard" models.

3.2.2. *Combinatorial enumeration*

The set-up is similar, but we restrict ourselves to relational structures (no function or constant symbols). We will see that the counting problems for orbits on sets and tuples of oligomorphic permutation groups are identical with those for unlabelled and labelled structures in so-called *oligomorphic Fraïssé classes* of relational structures. These include large numbers of combinatorially important classes of structures, so we have a rather general paradigm for interesting counting problems.

Take a fixed relational language **L** (this means, each set carries named relations of prescribed arities). For example, we could consider graphs, directed graphs, tournaments, partially ordered sets, k-edge-coloured graphs, graphs with a fixed bipartition, or much more recondite examples.

A *Fraïssé class* over **L** is a class $\mathcal{C}$ of finite relational structures over **L** satisfying the following four conditions:

(a) $\mathcal{C}$ is closed under isomorphism;
(b) $\mathcal{C}$ is closed under taking induced substructures (this means, take a subset of the domain, and all instances of all relations which are contained within this subset);
(c) $\mathcal{C}$ has only countably many members up to isomorphism;
(d) $\mathcal{C}$ has the *amalgamation property*: this means that, if $B_1, B_2 \in \mathcal{C}$ have isomorphic substructures, they can be "glued together" along these substructures (or possibly more) to form a structure in $\mathcal{C}$. Formally, if $A, B_1, B_2 \in \mathcal{C}$ and $f_i : A \to B_i$ is an embedding for $i = 1, 2$, then there is a structure $C \in \mathcal{C}$ and embeddings $h_i : B_i \to C$ for $i = 1, 2$ so that $h_1 f_1 = h_2 f_2$.

(The reader is warned that I allow the structure A to be empty here. Some authors exclude this, state the special case where $A = \emptyset$ separately, and call it the *joint embedding property.*)

For a simple example, finite graphs form a Fraïssé class: take the union of B_1 and B_2 identifying the isomorphic substructures, and put any or no edges between $B_1 \setminus A$ and $B_2 \setminus A$. It is just a little more difficult to show that finite total or partial orders form a Fraïssé class.

Let M be a structure over $\mathbf{L}$. We make two definitions:

- The *age* of M is the class of all finite $\mathbf{L}$-structures which are embeddable in M.
- M is *homogeneous* if any isomorphism between finite induced substructures of M can be extended to an automorphism of M.

For example, the totally ordered set $\mathbb{Q}$ is homogeneous, and its age consists of all finite totally ordered sets.

Fraïssé's Theorem asserts:

Theorem 3.3. *A class $\mathcal{C}$ of finite $\mathbf{L}$-structures is the age of a countable homogeneous $\mathbf{L}$-structure M if and only if it is a Fraïssé class. If these conditions hold, then M is unique up to isomorphism.*

We call the countable homogeneous structure M the *Fraïssé limit* of the class $\mathcal{C}$. Thus, $(\mathbb{Q}, <)$ is the Fraïssé limit of the class of finite totally ordered sets. The Fraïssé limit of the class of finite graphs is the celebrated *random graph*, or *Rado graph* R, which is extensively discussed in [4].

Now suppose that a Fraïssé class satisfies the following stronger version of condition (c):

(c′) For any $n \in \mathbb{N}$, the class $\mathcal{C}$ has only finitely many n-element members up to isomorphism.

We call such a class an *oligomorphic Fraïssé class.* All examples mentioned so far are oligomorphic; the condition certainly holds if $\mathbf{L}$ contains only finitely many relations. Let M be its Fraïssé limit, and let $G = \mathrm{Aut}(M)$ be the automorphism group of M. By homogeneity, it follows that

- $F_n(G)$ is the number of labelled n-element structures in $\mathcal{C}$ (that is, structures on the set $\{1, 2, \ldots, n\}$;
- $f_n(G)$ is the number of unlabelled n-element structures in $\mathcal{C}$ (that is, isomorphism classes of n-element structures).

Some Fraïssé classes satisfy a stronger version of the amalgamation property, called *strong amalgamation.* This is said to hold if the amalgam of any two structures can be produced without identifying any points not in the common substructure. More formally, if $h_1(b_1) = h_2(b_2)$, then $b_1 = f_1(a)$ and $b_2 = f_2(a)$ for some $a \in A$.

This holds in the above examples. An example where it fails is given by the class of graphs consisting of isolated vertices and edges. If B_1 and B_2 are edges and A a single vertex, then in the amalgam we are forced to identify the other ends of the two edges as well.

Proposition 3.6. *A Fraïssé class has the strong amalgamation property if and only if the automorphism group of its Fraïssé limit has the property that the stabiliser of any finite set of points fixes no additional points.*

Later we will see a still stronger version.

3.3. Constructions

There are two main sources for examples of (closed) oligomorphic groups: building new examples from old, or constructing Fraïssé classes (the group is then the automorphism group of the Fraïssé limit of the class).

3.3.1. *Direct and wreath products*

Suppose that G_1 and G_2 are oligomorphic permutation groups on Ω_1 and Ω_2 respectively. We can form their direct or wreath product, and each of these has two actions, which we now discuss.

Direct product, intransitive action. The direct product $G = G_1 \times G_2$ acts on the disjoint union of Ω_1 and Ω_2, say Ω. An n-subset of Ω consists of k points of Ω_1 and $n-k$ points of Ω_2, for some k with $0 \leq k \leq n$; two n-sets are in the same G-orbit if and only if their intersections with Ω_i are in the same G_i-orbit for $i = 1, 2$. Thus $(f_n(G))$ is the convolution of $(f_n(G_1))$ and $(f_n(G_2))$:

$$f_n(G) = \sum_{k=0}^{n} f_k(G_1) f_{n-k}(G_2),$$

from which it follows that the generating functions multiply:

$$f_G(x) = f_{G_1}(x) f_{G_2}(x).$$

Similarly, for F_n, we have an exponential convolution, so that the exponential generating functions also multiply:

$$F_G(x) = F_{G_1}(x)F_{G_2}(x).$$

The modified cycle index is also multiplicative:

$$\tilde{Z}(G) = \tilde{Z}(G_1)\tilde{Z}(G_2).$$

Direct product, product action. The direct product $G_1 \times G_2$ also has a product action on the cartesian product $\Omega_1 \times \Omega_2$. This is more difficult to analyse; the recent paper [9] describes the situation.

First, there is a multiplicative property: it is easy to see that

$$F_n^*(G) = F_n^*(G_1)F_n^*(G_2).$$

The modified cycle index of the product can be computed as follows. Define an operation on the indeterminates by

$$s_i \bullet s_j = (s_{\mathrm{lcm}(i,j)})^{\gcd(i,j)}.$$

Extend this operation multiplicatively to monomials and then additively to sums of monomials. Then we have

$$\tilde{Z}(G) = \tilde{Z}(G_1) \bullet \tilde{Z}(G_2).$$

It should clearly be possible to deduce the first of these relations from the second; but this is surprisingly difficult (see the cited paper).

Here is an example. Consider first the group $G = A$, the order-preserving permutations of $\mathbb{Q}$. We have $F_n(G) = n!$, and hence

$$F_n^*(G) = \sum_{k=1}^{n} S(n,k)k! = P(n),$$

the number of *preorders* of an n-set (reflexive and transitive relations).

Now let $G = A \times A$ with the product action. We have $F_n^*(G) = P(n)^2$, and so

$$F_n(G) = \sum_{k=1}^{n} s(n,k)P(k)^2.$$

Moreover, since G (like A) has the property that the setwise stabiliser of an n-set fixes it pointwise, we have

$$f_n(G) = F_n(G)/n!.$$

Now the exponential generating function for $P(n)$ is $1/(2-\mathrm{e}^x)$, with the nearest singularity to the origin at $\log 2$; so $P(n)$ is roughly $n!/(\log 2)^n$. So $F_n(G)$ is about $(n!)^2/(\log 2)^{2n}$. Since the Stirling numbers alternate in sign, it is not completely trivial to find the asymptotics of $f_n(G)$. This was achieved in [10], using three entirely different methods; it turns out that

$$f_n(G) \sim \frac{n!}{4}\mathrm{e}^{-\frac{1}{2}(\log 2)^2}\frac{1}{(\log 2)^{2n+2}}.$$

In broad-brush terms, "factorial times exponential".

Wreath product, imprimitive action. The wreath product $G_1\mathrm{Wr}G_2$ has its imprimitive action on $\Omega_1 \times \Omega_2$; as we have seen, this should be thought of as a covering of Ω_2 with fibres isomorphic to Ω_1. The function $F_G(x)$ is found by substitution:

$$F_G(x) = F_{G_2}(F_{G_1}(x) - 1).$$

In particular, for any oligomorphic group G, we have

$$F_{S\mathrm{Wr}G}(x) = F_G(\mathrm{e}^x - 1) = F_G^*(x),$$

so that $F_n(S\mathrm{Wr}G) = F_n^*(G)$ for all n, as is easily seen directly.

For another example, we note that $F_{G\mathrm{Wr}S}(x) = \exp(F_G(x) - 1)$. Thus, the substitution rule for the wreath product can be regarded as the prototype of a wide generalisation of the *exponential principle* in combinatorics (cf. [16, 33]). If G is the automorphism group of the Fraïssé limit of a Fraïssé class $\mathcal{C}$, then $G\mathrm{Wr}S$ is similarly associated with the class of disjoint unions of $\mathcal{C}$-structures; the exponential principle applies to this situation. In general, if H is associated with a Fraïssé class $\mathcal{D}$, then $G\mathrm{Wr}H$ is associated with the class of disjoint unions of $\mathcal{C}$-structures with a $\mathcal{D}$-structure on the set of parts. For example, if $H = A$, then we have ordered sequences of $\mathcal{C}$-structures; if H is the automorphism group of the random graph, we have a graph whose vertices are $\mathcal{C}$-structures; and so on.

The numbers $f_n(G)$ of orbits on unordered sets cannot be obtained from the sequences $(f_n(G_1))$ and $f_n(G_2))$ alone; we need the modified cycle index of the top group G_2. The generating function $f_G(x)$ is obtained from $\tilde{Z}(G_2)$ by substituting $f_{G_1}(x^i) - 1$ for the variable s_i, for $i = 1, 2, \ldots$.

Wreath product, power action. The wreath product also has a power action on $\Omega_1^{\Omega_2}$, the set of functions from Ω_2 to Ω_1 (the set of transversals to the fibres in the imprimitive action). This action is not in general oligomorphic; it is so if G_2 is a finite group.

It is shown in [9] that $F_n^*(G)$ is obtained from the (ordinary) cycle index of the finite group G_2 by substituting $F_n^*(G_1)$ for all of the variables s_i.

3.3.2. *Other examples*

Further examples are most easily described as automorphism groups of relational structures. As we have seen, any closed oligomorphic group of countable degree is the automorphism group of a homogeneous relational structure, which is the Fraïssé limit of a Fraïssé class of finite structures. Sometimes the easiest way to specify the group is to give the Fraïssé class.

The random graph and some of its relations. Let R be the countable random graph. This graph is the Fraïssé limit of the class of finite graphs. There are direct constructions for it, see [4]: for example, the vertices are the primes congruent to 1 mod 4, and p and q are joined if and only if p is a quadratic residue mod q. Alternatively, as Erdős and Rényi showed, if we form a countable random graph by choosing edges independently with probability 1/2 from the 2-subsets of a countable set, the resulting graph is isomorphic to R with probability 1.

The graph R is homogeneous, and contains all finite (and indeed all countable) graphs as induced subgraphs. So, if $G = \mathrm{Aut}(R)$, then $F_n(G)$ is the number of labelled graphs on n vertices (which is $2^{n(n-1)/2}$), while $f_n(G)$ is the number of unlabelled graphs on n vertices (which is asymptotically $F_n(G)/n!$). Note that these sequences grow so fast that the generating functions converge only at the origin.

There are several related Fraïssé classes. Here are a few examples.

- In place of graphs, we can take directed graphs, or oriented graphs, or tournaments, or k-uniform hypergraphs. Note that the automorphism group of the random k-uniform hypergraph is $(k-1)$-transitive but not k-transitive; so every degree of transitivity is possible for infinite permutation groups.
- In [34] it is shown that there are only five closed supergroups of $\mathrm{Aut}(R)$, namely
  - $\mathrm{Aut}(R)$;
  - The group $D(R)$ of dualities (automorphisms and anti-automorphisms) of R, which is 2-transitive and contains $\mathrm{Aut}(R)$ as a normal subgroup of index 2;

  - The group $S(R)$ of switching automorphisms of R (see below), which is 2-transitive;
  - The group $B(R)$ of switching automorphisms and anti-automorphisms of R, which is 3-transitive and contains $S(R)$ as a normal subgroup of index 2;
  - the symmetric group.

- In [24] the countable homogeneous graphs were determined. There are some trivial ones (disjoint unions of complete graphs of the same size, and their complements); some non-trivial ones, the Henson graphs and their complements; and the random graph. The *Henson graph* H_n is the Fraïssé limit of the class of all finite graphs which contain no complete graph of order n, for $n \geq 3$. They are not very well understood.

The operation σ_X of *switching* a graph with respect to a set X of vertices consists in replacing edges between X and its complement by non-edges, and non-edges by edges, while keeping things inside or outside X the same as before. A permutation g is a *switching automorphism* of Γ if $\Gamma^g = \sigma_X(\Gamma)$ for some set X. Switching anti-automorphisms are defined similarly.

One curious fact is that $F_n(S(R))$ and $f_n(S(R))$ are equal to the numbers of labelled and unlabelled *even graphs* on n vertices (graphs with all vertex degrees even), although the even graphs do not form a Fraïssé class.

Ordered structures. An old theorem of Skolem says that, if two subsets X and Y of $\mathbb{Q}$ are dense and have dense complements, then there is an order-preserving permutation of $\mathbb{Q}$ carrying X to Y. More generally, consider colourings of $\mathbb{Q}$ with m colours so that each colour class is dense. It is not hard to show that there is a unique such colouring up to order-preserving permutations. The automorpism group A_m of such a colouring (the group of permutations preserving the order and the colours) is oligomorphic, and $f_n(A_m) = m^n$. For, if $\{c_1, \ldots, c_m\}$ is the set of colours, then an n-set $\{q_1, \ldots, q_n\}$, with $q_1 < \cdots < q_m$, can be described by a word of length n in the alphabet $\{c_1, \ldots, c_m\}$, whose ith letter is the colour of q_i; two sets lie in the same orbit if and only if they are coded by the same word, and every word arises as the code of some subset.

We can modify this example in the same way we did for $\mathbb{Q}$ itself, allowing ourselves to preserve or reverse the order, or turning it into a circular order.

Treelike structures. There are vast numbers of treelike structures; I cannot give even a brief overview, so I will concentrate on a couple of examples.

Consider the class of *boron trees*, that is, finite trees in which each vertex has degree 1 or 3. Consider any four leaves in such a tree. They are connected by a graph of the following shape:

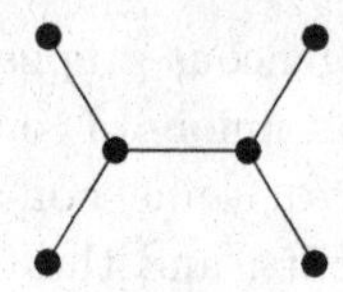

We see that one of the three partitions of the four leaves into two sets of two is distinguished by the fact that the paths joining vertices in the same pair do not intersect. Thus, there is a quaternary relation on the set of leaves. It is possible to show that this relation uniquely determines the boron tree, and that the class of finite structures carrying quaternary relations which arise in this way is a Fraïssé class. So there is a corresponding oligomorphic group G, which is 3-transitive but not 4-transitive, and satisfies $f_n(G) \sim An^{-5/2}c^n$, where $c = 2.483\ldots$.

This construction can be modified in many ways, of which a few are given below.

- We may consider trees with degrees $1, 3, 4, \ldots, m$, or with any possible degree except 2. (Divalent vertices are invisible in this construction.)
- We may consider internal vertices as well as leaves: there will be a ternary betweenness relation saying that one vertex is on the path joining the other two.
- By embedding the trees in the plane, we may impose a circular order on the set of leaves.

Many of these constructions give examples with exponential growth (roughly c^n for some $c > 1$).

Other examples. The symmetric group S on a countable set Ω has an induced action on the set of k-element subsets of Ω, for any k. These groups are oligomorphic, but except in the case $k = 2$, not much is known about the asymptotics of the orbit-counting sequences (see [11] for the case $k = 2$).

Other examples are linear and affine groups on infinite-dimensional vector spaces over finite fields. See [12] for the asymptotics.

3.4. Growth Rates

In this section we survey some known results about the rate of growth of the sequences $(f_n(G))$ and $(F_n(G))$ for an oligomorphic group G.

First, note that there is no upper bound for growth rates. Given any sequence (a_n) of natural numbers, let $\mathbf{L}$ be a relational language containing a_n n-ary relations $\mathcal{C}$ be the Fraïssé class consisting of all structures in which an n-ary relation holds only if all its arguments are distinct. If G is the automorphism group of the Fraïssé limit, then $f_n(G) \geq 2^{a_n}$ for all n. Indeed, it is slow growth which is most interesting!

For intransitive or imprimitive groups, we can have polynomial growth for (f_n). For a simple example, if G is the direct product of r symmetric groups (with the intransitive action), then $f_n(G) = \binom{n+r-1}{r-1}$, with generating function $f_G(x) = (1-x)^{-r}$.

For the same group, $F_n(G) = r^n$, since the orbit of an n-tuple is determined by the orbit containing each of its points. Recently it has been shown (cf. [1]) that, if G is transitive, and the point stabiliser has m orbits on the remaining points, then $F_n(G) \geq m^{n-1}$. Equality is possible here for any m.

There is a gap above polynomial growth for (f_n). The next possible growth rate is fractional exponential, about $\exp(\sqrt{n})$. This is realised by the group $G = S \operatorname{Wr} S$, for which we have $f_n(G) = p(n)$, the number of partitions of the integer n, with growth asymptotically

$$\frac{1}{4n\sqrt{3}} \exp(\pi\sqrt{2n/3});$$

and $F_n(G) = B(n)$, the number of partitions of an n-set (the Bell number), whose growth is faster than exponential but slower than factorial.

However, if we insist that the group is primitive, there is a dramatic change ([25, 29]):

Theorem 3.4. *There is a number $c > 1$ such that the following holds. Suppose that G is primitive but not highly set-transitive. Then*

- $f_n(G) \geq c^n/q(n)$, *for some polynomial q;*
- $F_n(G) \geq n!\, c^n/q(n)$, *for some polynomial q.*

In other words, for (f_n), there is a gap between constant and exponential growth! Merola's proof gives the result with $c = 1.324\ldots$; the best-known examples have $c = 2$ (these are circular versions of the rationals partitioned into two dense subsets, that we met earlier).

The main problems for exponential growth are:

- prove that the "exponential constant" $\lim_{n\to\infty}(f_n)^{1/n}$ always exists;
- find the possible values it can take;
- find a structural description of the examples where the growth is no faster than exponential.

What happens just above exponential growth? Here are some examples.

- Let $\mathcal{C}$ be the Fraïssé class each of whose members is a set with a pair of total orders, and G the automorphism group of its Fraïssé limit. Then $f_n(G) = n!$ and $F_n(G) = (n!)^2$.
- The group $S\mathrm{Wr}S_2$ with the power action is primitive. I do not know what the asymptotic behaviour of $(f_n(G))$ is. As explained earlier, we find that

$$F_n(G) \sim \tfrac{1}{2}B(n)(B(n)+1),$$

  where $B(n)$ is the Bell number.
- For the permutation group induced by S on the set of 2-element subsets of the domain, we have

$$F_n(G) \sim B(2n)2^{-n}n^{-1/2}\exp\left(-[\tfrac{1}{2}\log(2n/\log n)]^2\right).$$

No clear evidence of a gap emerges from this limited data. In [26] there are some theorems connecting growth just faster than exponential with model-theoretic properties such as stability and the strict order property.

For the automorphism group of the random graph, the growth rate is about $\exp(cn^2)$. For such growth, it doesn't make a lot of difference whether we consider F_n or f_n. We make one brief observation about this case.

Proposition 3.7. *Let G be the automorphism group of a homogeneous structure over a finite relational language. Then $F_n(G)$ is bounded above by the exponential of a polynomial.*

The converse of this is not true. For example, the general linear group on a vector space of countable dimension over the field of two elements has $F_n(G)$ roughly $2^{n^2/16}$, but is not the automorphism group of a homogeneous structure over a finite relational language. (Take two n-tuples of vectors, the first linearly independent and the second satisfying the single linear relation that the sum of the vectors is zero. These two n-tuples cannot be distinguished by relations of arity less than n.)

An open problem here is to understand what the implications for G are of being the automorphism group of a homogeneous structure over a finite relational language.

3.5. Graded Algebras

Another part of mathematics where sequences of positive integers occur is the theory of *graded algebras*. Such an algebra is a direct sum

$$A = \bigoplus_{n \in \mathbb{N}} A_n,$$

where the A_n are vector spaces over a field F, and, if $v \in A_m$ and $w \in A_m$, then the algebra product vw belongs to A_{m+n}. The subspaces A_n are the *homogeneous components*. If they are all finite-dimensional, then the sequence of their dimensions (or its ordinary generating function) is the *Hilbert series* of the algebra. If a graded algebra is finitely generated, then the dimensions grow no faster than a polynomial in n, and so the Hilbert series converges inside the unit circle.

We construct graded algebras as follows. First, let Ω be an infinite set, and F a field (for our purposes, always of characteristic zero). Let A_n be the vector space of functions from $\binom{\Omega}{n}$ (the set of n-element subsets of Ω) to F, and define a multiplication on the homogeneous components by the rule

$$fg(X) = \sum_{Y \in \binom{X}{n}} f(Y)g(X \setminus Y)$$

for $f \in A_n$, $g \in A_m$, and $X \in \binom{\Omega}{n+m}$. Extended linearly to $A = \bigoplus A_n$, this multiplication is commutative and associative, and makes A a graded algebra. The constant function 1 on $\binom{\Omega}{0} = \{\emptyset\}$ is the identity. (In fact A_0 is 1-dimensional, spanned by the identity.) This algebra is the *reduced incidence algebra* of the poset of finite subsets of Ω, but we make no use of this fact.

The algebra A is much too large: its homogeneous components have infinite dimension (except for A_0 which has dimension 1), and there are many nilpotent elements.

But it has one important feature: if $e \in A_1$ is the constant function on Ω with value 1, then e *is not a zero-divisor*; that is, multiplication by e is a monomorphism from A_n to A_{n+1}. (This algebraic statement is really a fact about finite combinatorics: the incidence matrix of n-sets and $(n+1)$-sets of a set X of cardinality at least $2n+1$ has full rank.)

Now let G be a permutation group on Ω. There is a natural action of G on A_n, for all n; we let A_n^G be the set of functions invariant under G (that is, constant on the G-orbits), and $A^G = \bigoplus A_n^G$. Then A^G is a graded subalgebra of A. Moreover, if G is oligomorphic, then $\dim(A_n^G) = f_n(G)$, so that *the Hilbert series of* A^G *is* $f_G(x)$.

There are various interesting examples of groups for which A^G is finitely generated. For example, if $G = G_1 \times G_2$ in its intransitive action on $\Omega_1 \cup \Omega_2$, where G_i acts on Ω_i for $i = 1, 2$, we have

$$A^G \cong A^{G_1} \otimes_F A^{G_2},$$

for any field of characteristic zero. In particular, $A^{S \times S}$ is freely generated by (that is, a polynomial ring in) two generators of degree 1, and more generally, A^{S^r} freely generated by r generators of degree 1.

If H is a finite permutation group of degree n, and $G = S \mathrm{Wr} H$ (in its imprimitive action), then A^G is isomorphic to the ring of invariants of H (acting as a linear group via permutation matrices). In particular, if $H = S_n$, then A^G is the ring of symmetric polynomials in n variables, and is freely generated by the elementary symmetric polynomials $e_1, \ldots, e_n$ of degrees $1, 2, \ldots, n$, by Newton's Theorem.

However, it follows from Macpherson's Theorem that, if G is primitive but not highly homogeneous, then A^G cannot be finitely generated.

In the light of this, we are forced to look for other kinds of algebraic properties of A^G. An example is a remarkable recent theorem of [30], confirming a 30-year-old conjecture:

Theorem 3.5. *Let G be a permutation group on an infinite set Ω. Then A^G is an integral domain if and only if G has no finite orbits on Ω.*

One direction is trivial. Suppose that Δ is a G-orbit of size n, and let $f \in A_n$ be the characteristic function of $\{\Delta\}$. Then $f \in A_n^G$ and $f^2 = 0$. The converse requires a new type of Ramsey-type theorem which is likely to have further applications. The result itself is applicable to growth rates, in view of the following result by [6], whose proof uses some easy dimension theory from algebraic geometry:

Theorem 3.6. *If G is oligomorphic and A^G is an integral domain then $f_{m+n} \geq f_m + f_n - 1$.*

Another situation in which we can describe the structure of A^G is when G is associated with a Fraïssé class $\mathcal{C}$ (that is, G is the automorphism group of the Fraïssé limit of $\mathcal{C}$). In [5] it is shown that, if the class $\mathcal{C}$ has notions of connected components, disjoint union, and spanning substructures satisfying a few simple axioms, then A^G is a polynomial algebra whose generators are the characteristic functions of the connected structures in $\mathcal{C}$. (More precisely, for each isomorphism type of n-element connected structure in $\mathcal{C}$, let f be the function in A_n which takes the value 1 on n-sets whose induced substructure is of the given type and 0 elsewhere; the collection of all such functions is a set of free generators for A^G.) Here are a few, reasonably typical, examples.

- Let $\mathcal{C}$ be the class of finite graphs, so that G is the automorphism group of the random graph R. Then a basis for the nth homogeneous component of A^G consists the characteristic functions of all n-vertex graphs (so that $\dim(A^G_n) = f_n(G)$ is the number of unlabelled graphs on n vertices). Now the characteristic functions of connected graphs on n vertices are algebraically independent, and the set of all such elements forms a free generating set for A^G, which is a polynomial algebra (infinitely generated).
- We saw earlier that, if G is associated with a Fraïssé class $\mathcal{C}$, then $G\mathrm{Wr}S$ is associated with the class of disjoint unions of $\mathcal{C}$-structures. There is an obvious notion of connectedness: structures with a single part are connected! Thus, $A^{G\mathrm{Wr}S}$ is a polynomial algebra, with $f_n(G)$ generators of degree n for all n. (Thus, the structure of $A^{G\mathrm{Wr}S}$ is independent of the structure of A^G except for numerical information about dimensions of components.)
- Consider the set $\mathbb{Q}$ with two kinds of structure: the order, and m subsets forming a partition of $\mathbb{Q}$, each one dense in $\mathbb{Q}$. (Think of a colouring of $\mathbb{Q}$ with m colours $c_1, c_2, \ldots, c_m$, so that each colour class is dense.) Let G be the group of permutations preserving the order and the colours. An orbit of G on n-sets is described uniquely by a word of length n in the alphabet $\{c_1, c_2, \ldots, c_m\}$, recording the colours of the elements of the set in increasing order. Now the algebra A^G is the *shuffle algebra* over an alphabet of size m, which occurs in the theory of free Lie algebras (cf. [32]). One can develop an appropriate notion of connectedness, so that the connected words are the so-called *Lyndon words.* The fact that the shuffle algebra is a polynomial algebra generated by the Lyndon

words was first proved in [31], but emerges as a special case of the theory presented here.

3.6. Group Structure

Relatively little is known about the structure of closed oligomorphic permutation groups, but the picture is rapidly changing. We look first at the normal structure in a couple of classical examples.

The countable symmetric group $S = \mathrm{Sym}(\Omega)$ has a normal subgroup $\mathrm{FSym}(\Omega)$, the *finitary symmetric group*, consisting of all permutations moving only finitely many points; this has a normal subgroup of index 2, the *alternating group* $\mathrm{Alt}(\Omega)$, consisting of finitary permutations which are even permutations of their supports. These are the only non-trivial proper normal subgroups; in particular, the quotients are simple.

The group A of order-preserving permutations of $\mathbb{Q}$ has two normal subgroups L and R; L consists of all permutations fixing all sufficiently large positive rationals, and R consists of permutations fixing all sufficiently large negative rationals. Their intersection consists of the order-preserving permutations of bounded support. These are the only non-trivial normal subgroups. In particular, $A/(L \cap R) \cong L/(L \cap R) \times R/(L \cap R)$, and the two factors are isomorphic.

The automorphism group of the countable random graph is simple (cf. [35]). Indeed, given any two elements g, h of this group with $g \neq 1$, it is possible to write h as the product of three copies of g or g^{-1}.

An important property which has had a lot of attention is the small index property. Let G be a permutation group of countable degree. A subgroup H has *small index* if $|G : H| < 2^{\aleph_0}$. (If the Continuum Hypothesis holds, this just says that H has finite or countable index.) The stabiliser of any finite set has small index. We say that G has the *small index property* if any subgroup of G of small index contains the pointwise stabiliser of a finite set; it has the *strong small index property* if every subgroup of small index lies between the pointwise and setwise stabilisers of a finite set. If G is a closed oligomorphic group with the small index property, then the topology of G is determined by the group structure: a subgroup is open if and only if it has small index, so the subgroups of small index form a neighbourhood basis of the identity.

The symmetric group S, the group A of order-preserving permutations of $\mathbb{Q}$, and the automorphism group of the random graph all have the strong small index property ([8, 15, 19]). A typical example of a permutation group which does not have the strong small index property is $S\mathrm{Wr}S$, in its imprimitive action; the stabiliser of a block of imprimitivity is not contained in the setwise stabiliser of any finite set. Examples which do not have the small index property can be constructed by producing automorphism groups of Fraïssé limits which have infinite elementary abelian 2-groups as quotients; they have "too many" subgroups of small index!

I mentioned earlier that the notion of Baire category is important for closed oligomorphic groups. Such a group G is said to have *generics* if there is a conjugacy class which is residual in G. In each of the three groups mentioned above, generics exist. Indeed, the nth direct power of the group has generics (in other words, the original group has generic n-tuples) for all n. This property is closely related to the small index property (see [19]).

We can also ask what group is generated by a "typical" n-tuple of elements. John Dixon proved that almost all pairs of elements of the finite symmetric group S_n generate S_n or the alternating group A_n. Later [13], he proved an analogue for the symmetric group of countable degree: almost all pairs of elements (in the sense of Baire category, that is, a residual set) generate a highly transitive free subgroup of S. (The existence of highly transitive free groups was first shown in [28].)

As noted, very recently these results have been extended to wider classes of groups.

Macpherson and Tent define the free amalgamation property, which is a strengthening of the strong amalgamation property, as follows. A Fraïssé class $\mathcal{C}$ has the *free amalgamation property* if, whenever B_1 and B_2 are structures in $\mathcal{C}$ with a common substructure A, there is an amalgam C of B_1 and B_2 such that

- the intersection of B_1 and B_2 in C is precisely A (this is the content of strong amalgamation, see Section 3.2);
- Every instance of a relation in C is contained in either B_1 or B_2.

This holds for graphs: we can choose to make the amalgam so that there are no edges between $B_1 \setminus A$ and $B_2 \setminus A$.

They prove the following theorem:

Theorem 3.7. *Let $\mathcal{C}$ be a nontrivial Fraïssé class (that is, there are some non-trivial relations) with the free amalgamation property, and G the automorphism group of its Fraïssé limit. Then G is simple.*

The trivial case must be excluded, since then G is the symmetric group S, and as we have seen this group is not simple.

A Polish group (a topological group whose topology comes from a complete metric) is *almost free* if a residual subset of the n-tuples of elements of G freely generate a free group. Now [17] gives a number of equivalent characterisations of such groups, and show:

Theorem 3.8. *Closed oligomorphic groups are almost free.*

[23] give conditions for the existence of generic conjugacy classes in a closed oligomorphic group.

In closing I mention a couple of important papers linking oligomorphic permutation groups, dynamical systems, and Ramsey theory: [20, 22].

References

[1] Applegate, R. and Cameron, P. J. (2008). Orbits on n-tuples, *Communications in Algebra*, to appear.

[2] Bailey, R. A. (2004). *Association Schemes: Designed Experiments, Algebra and Combinatorics.* Cambridge University Press, Cambridge.

[3] Cameron, P. J. (1976). Transitivity of permutation groups on unordered sets. *Math. Z.* **48** 127–139.

[4] Cameron, P. J. (1997). The random graph. In *The Mathematics of Paul Erdős* (ed. R. L. Graham and J. Nešetřil), Springer, Berlin, 331–351.

[5] Cameron, P. J. (1997). The algebra of an age. In *Model Theory of Groups and Automorphism Groups* (ed. David M. Evans), London Mathematical Society Lecture Notes. **244**, Cambridge University Press, Cambridge, 126–133.

[6] Cameron, P. J. (1998). On an algebra related to orbit-counting *J. Group Theory.* **1** 173–179.

[7] Cameron, P. J. (1999). *Permutation Groups.* London Math. Soc. Student Texts **45**, Cambridge University Press, Cambridge.

[8] Cameron, P. J. (2005). The random graph has the strong small index property. *Discrete Math.* **291** 41–43.

[9] Cameron, P. J., Gewurz, D. and Merola, F. (2008). Product action. *Discrete Math.* **308** 386–394.

[10] Cameron, P. J., Prellberg, T. and Stark, D. (2006). Asymptotic enumeration of incidence matrices. *Journal of Physics: Conference Series.* **42** 59–70.

[11] Cameron, P. J., Prellberg, T. and Stark, D. (2008). Asymptotic enumeration of 2-covers and line graphs, preprint.

[12] Cameron, P. J. and Taylor, D. E. (1985). Stirling numbers and affine equivalence. *Ars Combinatoria.* **20B** 3–14.

[13] Dixon, J. D. (1990). Most finitely generated permutation groups are free. *Bull. London Math. Soc.* **22** 222–226.

[14] Dixon, J. D. and Mortimer, B. (1996). *Permutation Groups.* Springer, New York.

[15] Dixon, J. D., Neumann, P. M. and Thomas, S. (1986). Subgroups of small index in infinite symmetric groups. *Bull. London Math. Soc.* **18** 580–586.

[16] Dress, A. and Müller, T. W. (1997). Decomposable functors and the exponential principle. *Advances in Math.* **129** 188–221.

[17] Gartside, P. M. and Knight, R. W. (2003). Ubiquity of free subgroups. *Bull. London Math. Soc.* **35** 624–634.

[18] Goulden, I. P. and Jackson, D. M. (2004). *Combinatorial Enumeration.* Dover Publ. (reprint), New York.

[19] Hodges, W. A., Hodkinson, I., Lascar D. and Shelah, S. (1993). The small index property for ω-stable ω-categorical structures and for the random graph. *J. London Math. Soc.* **48** 204–218.

[20] Hubička, J. and Nešetřil, J. (2005). Finite presentation of homogeneous graphs, posets and Ramsey classes. *Israel J. Math.* **149** 21–44.

[21] Joyal, A. (1981). Une théorie combinatoire des séries formelles. *Adv. Math.* **42** 1–82.

[22] Kechris, A. S., Pestov, V. G. and Todorčević, S. B. (2005). Fraïssé limits, Ramsey theory, and topological dynamics of automorphism groups. *Geometric and Functional Analysis.* **15** 106–189.

[23] Kechris, A. S. and Rosendal, C. (2007). Turbulence, amalgamation, and generic automorphisms of homogeneous structures *Proc. London Math. Soc.* **94** 302–350.

[24] Lachlan, A. H. and Woodrow, R. E. (1980). Countable ultrahomogeneous undirected graphs. *Trans. Amer. Math. Soc.* **262** 51–94.

[25] Macpherson, H. D. (1983). The action of an infinite permutation group on the unordered subsets of a set. *Proc. London Math. Soc.* **46** 471–486.

[26] Macpherson, H. D. (1987). Permutation groups of rapid growth. *J. London Math. Soc.* **35** 276–286.

[27] Macpherson, H. D. and Tent, K. (2008). Simplicity of some automorphism groups, preprint.

[28] McDonough, T. P. (1977). A permutation representation of a free group. *Quart. J. Math. Oxford.* **28** 353–356.

[29] Merola, F. (2001). Orbits on n-tuples for infinite permutation groups. *Europ. J. Combinatorics.* **22** 225–241.

[30] Pouzet, M. (2008). When the orbit algebra of a group is an integral domain? Proof of a conjecture of P. J. Cameron, to appear.

[31] Radford, D. E. (1979). A natural ring basis for the shuffle algebra and an application to group schemes *J. Algebra* **58** 432–454.

[32] Reutenauer, C. (1993). *Free Lie Algebras.* London Math. Soc. Monographs (New Series) **7** Oxford University Press, Oxford.

[33] Stanley, R. P. (1978). Generating functions. In *MAA Studies in Mathematics*, vol. 17 (G.-C. Rota ed.) Mathematical Association of America, Washington, 100–141.

[34] Thomas, S. R. (1991). Reducts of the random graph. *J. Symbolic Logic.* **56** 176–181.

[35] Truss, J. K. (1985). The group of the countable universal graph. *Math. Proc. Cambridge Philos. Soc.* **98** 213–245.

Chapter 4

Descriptive Set Theory and the Geometry of Banach Spaces

Gilles Godefroy
Institute of Mathematics of Jussieu,
Bureau 4B3, 175,
rue du Chevaleret, 75013 Paris, France
godefroy@math.jussieu.fr

4.1. Introduction

Mathematicians began to investigate the intrinsic complexity of subsets of the real line and of more general spaces about a century ago. Souslin's discovery that the projection on the line of a Borel subset of the plane is not always Borel was an early but major success, and his separation theorem as well. When put together, his results say that a set which is defined with an existence quantifier on a real number ($\exists\, x \in \mathbb{R}$ such that ...) cannot in general be defined only with a sequence of quantifier on natural numbers, although it is so if this set can *also* be defined with a universal quantifier on a real variable ($\forall\, x \in \mathbb{R}, \ldots$), in which case the real numbers can be eliminated from the definition.

In the above, speaking of "real numbers" is nothing but a convenience and what really matters is to know whether our set can be defined from, say, closed sets by iterating countable intersections and unions countably many times or if we need to index certain operations with an uncountable set (which is usually metrizable, separable and complete, so that we keep some control on what goes on).

In real life, all the sets we actually meet are measurable, but it must be stressed that quite often they are not Borel, and we will actually display in this survey many classes of separable Banach spaces which are not Borel.

What we mean when speaking of a "Borel class" of Banach spaces has been made clear in Bossard's work [3] (motivated by the seminal work [4]) where it was shown for instance that the relation of linear isomorphism between Banach spaces is complicated and that no decent set contains exactly one space in each isomorphism class (see Theorem 4.2 below). This work opened the way to several applications of descriptive set theory to Banach spaces, in particular through the use of (transfinite) uniform boundedness principles. In the last few years, the work of S. Argyros, P. Dodos and of co-authors deepened the theory with the discovery of "amalgamation" methods which tighten the links (and stress the analogy) between set-theoretical and linear operations. These new techniques provide the right approach to universality problems, by showing that the classes $\mathcal{SE}(Y)$ of isomorphic subspaces of a given Banach space Y are the generic hereditary and analytic classes of Banach spaces.

Let us now outline the contents of this paper. Section 4.2 is a survey of the classical theory of analytic sets. The gist of the results from this section is that knowing that a tree $T \subseteq \omega^{<\omega}$ of integers is not well-founded gives no way of finding the infinite branch which testifies that T is ill-founded. Section 4.3 displays Bossard's frame, which allows a measure of the complexity of given classes of Banach spaces. The main results of this section explain the "non-algorithmic" nature of the isomorphic theory of Banach spaces. Section 4.4 presents some natural coanalytic ranks such as the Szlenk index (and related indices) on the coanalytic non-Borel class of spaces with separable dual, and motivations for their use. Finally, Section 4.5 introduces the recent "converse statements" where analytic collections of separable spaces are "glued together" (with a tree space) for being embedded into the set of subspaces of a "small" Banach space. The theory is far from complete and several open problems are mentioned in the text.

Notation: We denote $\omega = \{0, 1, 2, 3, \ldots\}$ the set of natural numbers. If E is a set, E^ω (respectively, $E^{<\omega}$) is the set of sequences (respectively of finite sequences) of elements of E. In particular, ω^ω is the set of sequences of integers, and $\omega^{<\omega}$ is the set of finite sequences of integers. If $s \in \omega^{<\omega}$, we denote $|s|$ the length of s and $t < s$ means that $|t| \leq |s|$ and that s begins with t. We denote as well $s < \sigma$ if $\sigma \in \omega^\omega$ begins with $s \in \omega^{<\omega}$. The Cantor set 2^ω of subsets of ω is denoted Δ. An ordinal is identified with the set of its predecessors. In particular, we denote ω_1 the set of countable ordinals. If A and B are sets, we define $\pi_1 = A \times B \to A$ by $\pi_1(a, b) = a$. The vector space of functions $u : A \to \mathbb{R}$ which are 0 except for finitely many $a \in A$ is denoted $c_{00}(A)$. Other pieces of notation are classical or will be defined before use.

4.2. A Short Survey on Analytic Sets

The space ω^ω of sequences of integers is a metric complete separable space - in other words, a *Polish* space when equipped with the distance

$$d(\sigma, \sigma') = \sum_{i \in \omega} 2^{-i}\ \delta(\sigma(i), \sigma'(i))$$

where $\delta(n, k) = 0$ if $n = k$ and 1 if not. It is easy to check that given any Polish space P, there exists a continuous onto map $S : \omega^\omega \to P$.

The following definition is due to Souslin and goes back to 1917.

Definition 4.1. A metric space M is analytic if there exists a continuous onto map $S : \omega^\omega \to M$.

It follows immediately from the definition that the class of analytic sets is stable under continuous images.

Lemma 4.1. *Let P be a Polish space, and $A \subseteq P$ be a subset of P. The following are equivalent:*
(i) A is analytic.
(ii) There exists a closed subset $F \subseteq P \times \omega^\omega$ such that $A = \pi_1(F)$.

Proof. (ii) $\Rightarrow$ (i): F is closed in the Polish space $P \times \omega^\omega$ hence it is Polish. Hence F is analytic, and A as well since it is a continuous image of F.

(i) $\Rightarrow$ (ii): Let $S = \omega^\omega \to P$ be a continuous map such that $S(\omega^\omega) = A$. Since S is continuous, the set

$$F = \{(S(\sigma), \sigma) \in P \times \omega^\omega;\ \sigma \in \omega^\omega\}$$

is closed in $P \times \omega^\omega$, and $A = \pi_1(F)$. □

We now observe that the space ω^ω enjoys remarkable stability properties. Indeed $\omega^\omega \times \omega \simeq \omega^\omega$ and $(\omega^\omega)^\omega \simeq \omega^\omega$. This implies:

Lemma 4.2. *The class of analytic subsets of a given Polish space P is stable under countable unions and countable intersections.*

Proof. Let $A_n = \pi_1(F_n)$ with F_n a closed subset of $P \times \omega^\omega$. The set

$$F = \{(x, \sigma, n) \in P \times \omega^\omega \times \omega;\ (x, \sigma) \in F_n\}$$

is a closed subset of $P \times \omega^\omega \times \omega$, and thus

$$\bigcup_{n \in \omega} A_n = \pi_1(F)$$

is analytic. On the other hand the set

$$G = \{(x, (\sigma_n)_n) \in P \times (\omega^\omega)^\omega; (x, \sigma_n) \in P_n \text{ for all } n\}$$

is closed in $P \times (\omega^\omega)^\omega$, and thus

$$\bigcap_{n\in\omega} A_n = \pi_1(G)$$

is analytic. □

Since closed subsets of P are clearly analytic, Lemma 4.2 implies that every Borel subset of P is analytic, and thus one has:

Lemma 4.3. *If P_1 and P_2 are Polish spaces and $B \subseteq P_1 \times P_2$ is a Borel subset of $P_1 \times P_2$ then $\pi_2(S)$ is analytic.*

This lemma provides a substitute to an erroneous statement of Lebesgue, who had claimed that the projection on a line of a Borel subset of the plane is Borel. This claim was actually the starting point of Souslin's work, who showed that although such a projection is in general *not* Borel, it retains some properties of Borel sets such as universal measurability.

We now recall the definition of a *tree* on integers: this is a subset T of $\omega^{<\omega}$ such that if $s \in T$ and $t < s$, then $t \in T$. The *boundary* $[T]$ of such a tree T is the set of all $\sigma \in \omega^\omega$ such that $\sigma_{|n} \in T$ for all $n \in \omega$. In other words, $[T]$ consists of all the "infinite branches" of T. Every boundary is a closed subset of ω^ω, and all closed subsets can be obtained in this way.

In view of Definition 4.1, it is therefore not surprising that trees show up in this context. Indeed let

$$A = \pi_1(F)$$

with F closed in $P \times \omega^\omega$. For $s \in \omega^{<\omega}$, let

$$V_s = \{\sigma \in \omega^\omega;\ s < \sigma\}.$$

Let d be a distance which defines the topology of P, and for $x \in P$ and $\delta > 0$, let

$$B(x, \delta) = \{y \in P; d(x, y) < \delta\}.$$

We define a tree $T(x)$ as follows:

$$T(x) = \{s \in \omega^{<\omega}; (B(x, |s|^{-1}) \times V_s) \cap F \neq \phi\}.$$

It is easily seen that

$$(x, \sigma) \in F \Leftrightarrow \sigma \in [T(x)]$$

and thus $x \in A = \pi_1(F)$ if and only if $[T(x)] \neq \phi$.

We now recall a classical piece of notation: a tree T is *well-founded* if $[T] = \phi$. With this notation, $x \in A$ if and only if $T(x)$ is not well-founded.

Let $\mathcal{T} \subseteq 2^{(\omega^{<\omega})}$ be the set of all trees on integers (including the tree $\{\phi\}$). This is a closed subset of $2^{(\omega^{<\omega})}$, and thus a compact set. Let $WF \subseteq \mathcal{T}$ be the subset consisting of well-founded trees. This subset WF is a transfinite union of "constituents" defined as follows: let $\omega_1 = \{\alpha; \alpha < \omega_1\}$ be the set of countable ordinals. Given $T \in \mathcal{T}$, we define

$$T' = \{s \in \omega^{<\omega};\ \text{there is } n \in \omega \text{ with } s^\frown n \in T\}$$

In other words, T' consists of all $s \in T$ which can be extended to a longer sequence in T. We define then by transfinite induction

$$T^{(\alpha+1)} = (T^{(\alpha)})'$$

and for β a limit ordinal

$$T^{(\beta)} = \bigcap_{\alpha<\beta} T^{(\alpha)}.$$

It is clear that if $T \notin WF$, then for all $\alpha < \omega_1$ and all $\sigma \in [T]$

$$b_\sigma = \{s < \sigma\} \subseteq T^{(\alpha)}.$$

Conversely, if $s \in T^{(\alpha)}$ for all $\alpha < \omega_1$, there is $n \in \omega$ such that $s^\frown n \in T^{(\alpha)}$ for all α, and it follows that if

$$\bigcap_{\alpha<\omega_1} T^{(\alpha)} \neq \phi$$

then $T \notin WF$. If $T \in WF$, we denote by $h(T)$ the *height* of T, that is, the smallest $\alpha < \omega_1$ such that $T^{(\alpha)} = \phi$. If $T \notin WF$, we denote $h(T) = \omega_1$.

Returning to $A = \pi_1(F)$ an analytic subset of P, we define

$$r : P \to \omega_1 \cup \{\omega_1\}$$

by $r(x) = h(T(x))$. By the above, $A = r^{-1}(\{\omega_1\})$ and

$$C = P \backslash A = \bigcup_{\alpha<\omega_1} r^{-1}(\{\lambda; \lambda < \alpha\}).$$

This map r is called a *rank*. Such a set $C = P\backslash A$, whose complement is analytic, is called a *coanalytic* set and if we let

$$C_\alpha = r^{-1}(\{\lambda; \lambda < \alpha\})$$

we have $C = \bigcup_{\alpha\in\omega_1} C_\alpha$, and C appears as a transfinite increasing union of "constituents".

The following combinatorial result on trees is an abstract version of Souslin's separation theorem. We state it without proof.

Theorem 4.1. *For any countable ordinal $\alpha < \omega_1$, the set*

$$B_\alpha = \{T \in \mathcal{T}; h(T) < \alpha\}$$

is a Borel subset of $\mathcal{T}$. Moreover, if A is an analytic subset of $WF \subseteq \mathcal{T}$, then

$$\sup_{T \in A} h(T) < \omega_1.$$

It is an instructive exercise to show that for every $\alpha < \omega_1$, there exists $T \in WF$ such that $h(T) = \alpha$ and then, Theorem 4.1 implies that WF is not analytic, and that $\mathcal{T} \backslash WF$ is analytic but *not* Borel.

If now $A = \pi_1(F)$ is an analytic set, the map $r(x) = h(T(x))$ defined above yields to the following applications of Theorem 4.1: the coanalytic set $C = P \backslash A$ is a transfinite union of the Borel sets $C_\alpha = r^{-1}(B_\alpha)$. If A_1 is an analytic subset of C, there is $\alpha < \omega_1$ such that $A_1 \subseteq C_\alpha$. Hence two disjoint analytic sets can be separated by a Borel set: $A_1 \subseteq C_\alpha$ and $A \subseteq P \backslash C_\alpha$.

So far, we displayed the basic facts of Souslin's theory within the frame of metric spaces. However, most of this can be done in *standard Borel spaces*, that is, in sets X equipped with a σ-field $\mathcal{B}$ such that $(X, \mathcal{B})$ is Borel-isomorphic with some Polish space equipped with its Borel σ-field. In other words, $(X, \mathcal{B})$ is a standard Borel space if there is a Polish distance on X whose Borel σ-field coincide with $\mathcal{B}$. However this distance is not canonical and it plays little role. In order to introduce a "Borel way" to describe analytic sets, we return to the above definition of the trees $T(x)$ and with the same notation, we observe that for all $s \in \omega^{<\omega}$, the set

$$U_s = \{x \in P; (B(x, |s|^{-1}) \times V_s) \cap F \neq \phi\}$$

is open in P, and

$$A = \bigcup_{\sigma \in \omega^\omega} \bigcap_{s < \sigma} U_s.$$

This is called the *Souslin operation*, and it is usually denoted

$$A = \mathcal{S}((U_s)).$$

This shows that every analytic set can be obtained from Borel sets $\{B_s; s \in \omega^{<\omega}\}$ through the Souslin operation $\mathcal{S}$. Conversely, the class of analytic sets is stable under $\mathcal{S}$: if the sets $\{A_s; s \in \omega^{<\omega}\}$ are analytic, so is $\mathcal{S}((A_s))$.

We can therefore define a subset A of a standard Borel space $(X, \mathcal{B})$ to be analytic if there exist $\{B_s; s \in \omega^{<\omega}\} \subseteq \mathcal{B}$ with

$$A = \mathcal{S}((B_s)).$$

The above theory (rank, constituents, separation, ...) applies and no reference is needed to a Polish distance on X. If $(X, \mathcal{B})$ and $(Y, \mathcal{B}')$ are standard Borel spaces, and $F : X \to Y$ is a Borel map, then $F(X)$ is an analytic subset of Y, and every analytic set is obtained in this way. We refer to [6], [26], [31] for much more on Souslin theory and descriptive set theory and to [2] for some connections between this theory and geometry of Banach spaces.

4.3. Bossard's Coding of Separable Banach Spaces

This section is devoted to a general frame (published in [3]) which makes it possible to define Borel, analytic or coanalytic families of separable Banach spaces.

In order to do so, we need a proper parametrization of the collection of separable Banach spaces, which turns it into a set.

If P is a Polish space, we denote $\mathcal{F}(P)$ the set of closed subsets of P. Let $\hat{P}$ be a metrizable compactification of P; with the Hausdorff distance, the set $\mathcal{F}(\hat{P})$ is compact metric, and the map $F \to \bar{F}$ maps $\mathcal{F}(P)$ to the subset $\mathcal{F}_0(\hat{P})$ of $\mathcal{F}(\hat{P})$ defined by

$$\mathcal{F}_0(\hat{P}) = \{F \subseteq \hat{P}; F = \overline{F \cap P}\}.$$

It is easily checked that $\mathcal{F}_0(\hat{P})$ is a G_δ subset of $\mathcal{F}(\hat{P})$ and thus it is homeomorphic to a Polish space. The corresponding Borel structure on $\mathcal{F}(P)$ does not depend upon the compactification, and it is called the *Effros-Borel* structure on $\mathcal{F}(P)$. It is generated by the sets

$$B_U = \{F \in \mathcal{F}(P); F \cap U \neq \phi\}$$

where U is an arbitrary open subset of P.

We denote by $\Delta = 2^\omega$ the Cantor set. It is well-known that every separable Banach space is isometric to a subspace of $\mathcal{C}(\Delta)$ equipped with the supremum norm.

Let $SE(\mathcal{C}(\Delta))$ be the set of all closed *linear subspaces* of $\mathcal{C}(\Delta)$. It is easily checked that $SE(\mathcal{C}(\Delta))$ is a Borel subset of $\mathcal{F}(\mathcal{C}(\Delta))$ equipped with the Effros-Borel structure. It is therefore a standard Borel space, as Borel subset of a standard Borel space.

For sake of shortness we now denote $SE(\mathcal{C}(\Delta))$ simply as SE. Let $\simeq$ by the equivalence relation of isomorphism: if $(X,Y) \in SE^2$, $X \simeq Y$ means that X is isomorphic to Y.

Definition 4.2. A coding of separable Banach spaces up to isomorphism is an onto map from a set Z onto the set $SE/\simeq$. The *canonical coding* is the quotient map $\theta : SE \to SE/\simeq$.

It could of course be objected at this stage that the "canonical" coding is not so canonical, since other parametrizations of Banach spaces could be chosen: for instance, any separable Banach space is a quotient of $\ell_1(\omega)$ and the kernels of such quotient map could be used. However, it is shown in [3] that such alternative parametrizations lead to the same complexity results.

The space SE is a standard Borel space, but is there any chance to equip the quotient space $SE/\simeq$ with a usable structure of standard Borel space? The following result from [3] answers this question in the negative.

Theorem 4.2. *The relation $\simeq$ is analytic non-Borel; that is, the set*

$$I = \{(X,Y) \in SE^2;\ X \simeq Y\}$$

is analytic non-Borel. There exists a space $U \in SE$ whose isomorphism class

$$\langle U \rangle = \{X \in SE;\ X \simeq U\}$$

is analytic non-Borel. The relation $\simeq$ has no analytic section.

Outline of Proof:

This result relies on the fact that the set WF of well-bounded trees is coanalytic non Borel (see Theorem 4.1). We denote by U Pelczynski's universal space (cf. [35]) which has a basis (u_n) and contains complemented copies of every Banach space with a basis. Let $\|\cdot\|$ denote the norm on U. We now define a "tree space" as follows: for $y \in c_{00}(\omega^{<\omega})$, set

$$|||\, y\, |||_2 = \sup \left\{ \left(\sum_{j=0}^{k} \left\| \sum_{s \in I_j} y(s)\ u_{|s|} \right\|^2 \right)^{1/2} \right\}$$

where the supremum is taken over all $k \in \omega$ and over all admissible choices of intervals, namely all finite sets $\{I_j; 0 \leq j \leq k\}$ of intervals $\omega^{<\omega}$ such that every branch of $\omega^{<\omega}$ meets at most one of these intervals.

Let $U(\omega^{<\omega})$ be the completion of $c_{00}(\omega^{<\omega})$ under this norm. If $T \in \mathcal{T}$ is any subtree of $\omega^{<\omega}$, let $U(T)$ be the space of all vectors from $U(\omega^{<\omega})$ supported by T or alternatively the completion of $c_{00}(T)$ under $\|| \cdot \||_2$. Since WF is a non-Borel subset of $\mathcal{T}$, the fact that the isomorphism class $\langle U \rangle$ is non-Borel follows from the following three statements:

(a) the map $T \mapsto U(T)$ is Borel from $\mathcal{T}$ to SE.

(b) If $T \in WF, U(T)$ is reflexive and thus $U(T) \notin \langle U \rangle$.

(c) If $T \notin WF$, $U(T)$ is isomorphic to U.

Statement (a) is easy to check. Statement (b) follows from the fact that an ℓ_2-sum of reflexive spaces is reflexive through a transfinite induction. Statement (c) follows from the fact that if $\sigma \in \omega^\omega$ and $b_\sigma = \{s < \sigma\}$ is the corresponding branch, the space $U(b_\sigma)$ is isomorphic to U. Now, if $T \notin WF$, pick σ such that $b_\sigma \subseteq T$. Then $U(b_\sigma) \simeq U$ is a complemented (by restriction) subspace of $U(T)$, and since $U(T)$ has a basis it is isomorphic to a complemented subspace of U. Now an application of Pelczynski's decomposition scheme shows that $U(T) \simeq U$.

It remains to establish that the relation $\simeq$ (that is, the set I) is analytic, and this follows quite easily from its definition by "there exists an isomorphism". Since the intersection

$$\langle U \rangle = I \cap (SE \times \{U\})$$

is non-Borel, the relation $\simeq$ is analytic non-Borel. Finally the existence of a non-Borel class shows that there is no analytic set A which meets every class in exactly one point. Indeed, if A were such a set let $\{X\} = A \cap \langle U \rangle$ and let $A' = A \backslash \{X\}$. Since A' is analytic as well as $\simeq$, its saturation $SE \backslash \langle U \rangle$ would be analytic. But since $\langle U \rangle$ is analytic, Souslin's separation theorem would show that both are Borel, contradicting the fact that $\langle U \rangle$ is not. □

The proof of Theorem 4.2 is typical of which one should expect in this context: given a tree T, one tries to construct a copy of U in $U(T)$. This will be possible exactly when $T \notin WF$. If $T \in WF$ but is very high, it will take "very long" to check that $U(T)$ does not contain U, and this cannot be done in any "algorithmic" way.

We note that condition (b) from the above proof shows that the class of reflexive spaces is coanalytic non-Borel.

The above proof shows that the isomorphism class $\langle U \rangle$ of U is analytic non-Borel. On the other hand, Kwapien's theorem (see [28]) easily implies that $\langle \ell_2 \rangle$ is Borel. It is not known whether this characterizes ℓ_2:

Problem 4.1. *Let X be a separable Banach space whose isomorphism class $\langle X\rangle$ is Borel. Is X isomorphic to ℓ_2?*

It follows from Bourgain's work [5] on $\mathcal{L}_p$ space that if $X = L_p$ with $1 < p < \infty$, $p \neq 2$ then $\langle X\rangle$ is not Borel (see [5], Th. 4.35). The case of $X = c_0$ is particularly interesting: it is not known (see [19], Remark 2 after the Theorem 5.6) whether c_0 is up to isomorphism the only predual of l_1 with a summable Szlenk index. Such a characerization of c_0 would show that its isomorphism class is Borel.

Clearly, the proof of Theorem 4.2 shows as well that the class of spaces which contain an isomorphic copy of U is analytic non-Borel. This is a special case of the following result from [3], whose proof follows similar lines.

Theorem 4.3. *Let Z be an infinite-dimensional separable Banach space. Then the class $\mathcal{C}_Z$ of separable Banach spaces which contain an isomorphic copy of Z is analytic non-Borel.*

Let us mention that Theorem 4.3 remains true when "isomorphic" is replaced by "isometric", in which case it even extends to every space Z of dimension greater than 1 (see [20]).

The article [3] contains much more results similar to the above, whose common feature is to assert that the natural classes of Banach spaces are as complicated as they look at first sight. For instance, the class of separable spaces not containing ℓ_1 with non-separable dual is the difference $A_1\backslash A_2$ of two analytic sets and is not simpler. One natural question seems however to be left open. Before stating it, let us recall that the relation of embedding, in other words the set

$$J = \{(X, Y) \in SE^2;\ \exists\ Z \subseteq Y \text{ with } X \simeq Z\}$$

is analytic (non-Borel), and then for every space Y, the class $\mathcal{SE}(Y)$ of space which are isomorphic to subspaces of Y is analytic. Is this really optimal?

Problem 4.2. *Does there exist a separable space Y such that $\mathcal{SE}(Y)$ is non-Borel?*

Let us note that for some space Y, such as ℓ_2 or $\mathcal{C}(\Delta)$ it is actually Borel. The behaviour of the isomorphism relation $\simeq$ restricted to such classes $\mathcal{SE}(Y)$ is not well-understood despite important recent progress (see [15], [37], [14], [16]). Actually the following improvement of Theorem 4.2 has been shown in [17]: the relation of isomorphism between separable

Banach spaces is a complete analytic equivalence relation. In other words, this isomorphism relation enjoys maximal complexity among all analytic equivalence relations. Thus, separable Banach spaces up to isomorphism provide complete invariants for essentially every isomorphism problem in analysis, with the by-product that classifying Banach spaces up to isomorphism is definitively beyond reach.

Gowers' solution to the homogeneous space problem (see [21]) states that $Y \simeq \ell_2$ if and only if $\{\mathcal{SE}^*(Y)/\simeq\}$ is a singleton, where $\mathcal{SE}^*$ denotes the set of infinite dimensional closed subspaces. However the following problem is still open.

Problem 4.3. *Let Y be an infinite-dimensional separable Banach space which is not isomorphic to ℓ_2. Does Y contain subspaces $\{X_n; n \in \omega\}$ such that X_n is not isomorphic to X_k if $n \neq k$?*

4.4. Coanalytic Ranks

In Section 4.2 we discussed heights of trees and the corresponding rank function r associated with a representation $A = \pi_2(F)$ of an analytic set as projection of a closed set. This yields to a general definition.

Definition 4.3. Let P be a standard Borel space and C be a coanalytic subset of P. A coanalytic rank σ on C is a map $\sigma = P \to \omega_1 \cup \{\omega_1\}$ such that

(i) $C = \{x \in P;\ \sigma(x) < \omega_1\}$
(ii) $\{(x,y) \in C \times P;\ \sigma(x) \leq \sigma(y)\}$ is coanalytic in P^2.
(iii) $\{(x,y) \in C \times P;\ \sigma(x) < \sigma(y)\}$ is coanalytic in P^2.

For every coanalytic set C, there exists such a coanalytic rank, and the classical result below, which extends Theorem 4.1, states that this rank is quite canonically associated with C.

Theorem 4.4. *Let σ be a coanalytic rank of the coanalytic set C. Then:*

(i) For every $\alpha < \omega_1$, the set $B_\alpha = \{x \in C; \sigma(x) \leq \alpha\}$ is Borel.
(ii) If $A \subseteq C$ is analytic, there is $\alpha < \omega_1$ such that $A \subseteq B_\alpha$.
(iii) If σ' is another coanalytic rank on C, there exists $\phi : \omega_1 \to \omega_1$ such that if $\sigma(x) \leq \alpha$, then $\sigma'(x) \leq \phi(\alpha)$.

Condition (iii) above asserts that all coanalytic ranks on C are somewhat equivalent (a special case of the Kunen-Martin uniform boundedness principle). Indeed the set of all α's such that $\sigma^{-1}([0,\alpha[) = \sigma'^{-1}([0,\alpha[)$ is a closed and cofinal subset of ω_1.

Natural coanalytic ranks can be displayed on the coanalytic classes from section 4.3, such as an "embedding rank" on the class $\mathcal{G}_Z = SE \backslash \mathcal{C}_Z$ of Banach spaces which do not contain an isomorphic copy of a given space Z (cf. [3]). However the most interesting examples where (iii) can be applied, come from Cantor-Bendixon derivations.

We now define this concept, and refer to [25] for its applications to harmonic analysis. Let K be a metrizable compact space, and $\mathcal{F}(K)$ be the set of closed subsets of K equipped with the Hausdorff topology. A Borel derivation

$$d : \mathcal{F}(K) \to \mathcal{F}(K)$$

is a Borel map such that

(i) $d(F) \subseteq F$ for all F.

(ii) If $F \subseteq G$ then $d(F) \subseteq d(G)$.

Such a derivation can be iterated: if $\alpha < \omega_1$, define

$$d^{(\alpha+1)}(F) = d(d^{(\alpha)}(F))$$

and for $\beta < \omega_1$ a limit ordinal,

$$d^{(\beta)}(F) = \bigcap_{\alpha<\beta} d^{(\alpha)}(F)$$

let now

$$\sigma_d(F) = \min\{\alpha; d^{(\alpha)}(F) = \phi\}.$$

If such an ordinal exists, and ω_1 otherwise. Then one has:

Theorem 4.5. *Let $\{d_n; n \in \omega\}$ be a countable family of Borel derivations. Then the map $\Sigma(F) = \sup\{\sigma_{d_n}(F); n \in \omega\}$ is a coanalytic rank on the coanalytic set $C = \{F \in \mathcal{F}(K); \sigma_{d_n}(F) < \omega_1$ for all $n\}$.*

Note that the special case of a single derivation (when all d_n's coincide) is very important.

The first example of derivation, due to Cantor, is when $d(F)$ is the set of accumulation points of F. We now define a similar concept: Let F be a w^*-compact subset of the space $\ell_\infty = \ell_1^*$ and $\epsilon > 0$. We define

$$d_\epsilon(F) = \{x \in F; \|\cdot\| - \text{diam } (V \cap F) \geq \epsilon \text{ for all } V\}$$

where V runs through the set of all w^*-open sets containing x, and

$$\delta_\epsilon(F) = \{x \in F, \|\cdot\| - \text{ diam } (H \cap F) \geq \epsilon \text{ for all } H\}$$

where H runs through the set of all w^*-open half-spaces containing x. It is clear that d_ϵ and Δ_ϵ are derivations, and it can be checked that they are Borel on $\mathcal{F}(B_{\ell_\infty})$. Let now

$$\sigma_\epsilon(F) = \sigma_{d_\epsilon}(F)$$

and

$$\sigma'_\epsilon(F) = \sigma_{\delta_\epsilon}(F).$$

By Theorem 4.5 the map

$$\sigma(F) = \sup\{\sigma_{1/k}(F); k \in \omega\}$$

is a coanalytic rank on the set

$$C = \{F \in \mathcal{F}(B_{K^*}); \sigma_\epsilon(F) < \omega_1 \text{ for all } \epsilon > 0\}.$$

Although the obvious inclusion

$$d_\epsilon(F) \subseteq \delta_\epsilon(F)$$

is in general strict, it follows from a result of [32] that C coincide with the set of F's for which $\sigma'_\epsilon(F) < \omega_1$ for all $\epsilon > 0$. Hence if

$$\sigma'(F) = \sup\{\sigma'_{1/k}(F),\ k \in \omega\}$$

it follows from Theorems 4.4 (iii) and Theorem 4.5 that there exists a map $\phi : \omega_1 \to \omega_1$ such that $\sigma(F) \leq \alpha$ implies $\sigma'(F) \leq \phi(\alpha)$. Hence, although the derivation δ which consists removing small slices is slower than d which allows the removal of small w^*-open sets, it is not *much* slower. Actually, as noted after Theorem 4.4, there is a closed cofinal subset C of ω_1 such that if $\alpha \in C$ then $\sigma(F) < \alpha$ if and only if $\sigma'(F) < \alpha$.

This theory yields to natural indices defined on separable Banach spaces. Let X be separable. Since there exists $Q : \ell_1 \to X$ a quotient map, the dual unit ball B_{X^*} embeds (in w^* and norm topologies) into B_{ℓ_∞}, and we can define

$$Sz(X) = \sigma(B_{X^*})$$
$$Dz(X) = \sigma'(B_{X^*}).$$

These indices are called the Szlenk index (following [39] where it is defined in another equivalent way) and the dentability index (see e.g. [19]). It is easily seen that these indices do not depend upon the choice of the quotient map Q, and that Sz (respectively, Dz) can be defined by removing from B_{X^*} those w^*-open sets (respectively, w^*-slices) which are small in norm.

The frame provided by B_{ℓ_∞} shows however that $Dz(X) \leq \psi(Sz(X))$, where $\psi = \omega_1 \cup \{\omega_1\} \to \omega_1 \cup \{\omega_1\}$ is a universal map such that $\psi(\omega_1) = \omega_1$. Indeed $Sz(X) < \omega_1$ if and only if $Dz(X) < \omega_1$, if and only if X^* is separable. A separable reduction argument shows that ψ actually works for every Asplund space (see [29]).

The first value of the map ψ is known: if $Sz(X) = \omega$, then $Dz(X) \leq \omega^2$ and this estimate is optimal ([22]). Actually, $Dz(X) = \omega$ if and only if X is super-reflexive, and if $Sz(X) = \omega$ and X is not super-reflexive then $Dz(X) = \omega^2$. Note that a third index, denoted $Cz(X)$ in [19], can be defined through the removal of w^*-open slices which can be covered by a union of small w^*-open sets. Equivalently, the convex Szlenk index $Cz(X)$ is defined by the derivations

$$\Delta_\epsilon(F) = \overline{\text{conv}}^{w^*}(d_\epsilon(F)).$$

It is clear that

$$Sz(X) \leq Cz(X) \leq Dz(X).$$

It is shown in [22] that $Dz(X) \leq \omega^\omega \cdot Cz(X)$. It is known that the Szlenk index of any space with separable dual has the form ω^α for some $\alpha < \omega_1$. The values of $\psi(\omega^\alpha)$ for $\alpha \geq 2$ are not known. The existence of a cofinal subset S' of ω_1 such that if $\alpha \in S'$ and $Sz(X) = \alpha$ then $Dz(X) = \alpha$ follows from an application of the classical "pressing-down" lemma (see [30], Prop. 6.4).

There are several motivations for studying these indices. First of all, they are invariants for linear isomorphisms and this sometimes lead to classification results: for instance, if K and L are countable compact spaces, then $\mathcal{C}(\mathcal{K}) \simeq \mathcal{C}(\mathcal{L})$ if and only if $Sz(\mathcal{C}(K)) = Sz(\mathcal{C}(\mathcal{L}))$. Even more, the condition $Sz(X) = \omega$ is equivalent to $Cz(X) = \omega$ and is invariant under uniform homeomorphism ([19]) and in particular under Lipchitz isomorphism.

Also, they happen to be useful in renorming theory. This follows from the fact that the derivations δ_ϵ and Δ_ϵ maps convex sets to convex sets, and convex combinations of distances from these convex sets yield to dual norms with good convexity properties and thus to smooth equivalent norms. We provide two examples: using the inequality $Cz(X) \leq \varphi(Sz(X))$ for all X with separable dual, it is shown in [29] that every space Y such that

$$\sup\{Sz(X); X \text{ separable}, X \subseteq Y\} < \omega_1$$

has an equivalent Fréchet-smooth norm. Using the quantitative behaviour of d_ϵ and Δ_ϵ, it is shown in [19] that $Sz(X) = \omega$ if and only if X has an asymptotically uniformly smooth equivalent norm (a result previously shown in [27] with weaker quantitative estimates).

We refer to [12], [10] and [11] for more applications of Szlenk derivations to Lipschitz isomorphisms, to [18] and [30] for surveys on the Szlenk index, and to the Chapter 2 of [23] for a presentation of the whole theory.

To make a link with section 4.3, let us mention that the Szlenk index $Sz(X)$ and the dentability index $Dz(X)$ are coanalytic ranks on the coanalytic subset of SE consisting of spaces with a separable dual ([3], Th. 4.11). However, it is still not known if the index $r(X) = \sup(Sz(X), Sz(X^*))$ is a coanalytic rank on the coanalytic set of reflexive spaces, although the corresponding index on the set of bases of reflexive spaces is known to be a coanalytic rank ([3], Cor. 5.5).

We will see in the next section how to use such ranks for showing the existence of certain universal spaces.

4.5. A New Direction: The Converse Statements

The main argument in the proof of Theorem 4.2 is that the "tree space" $U(\omega^{<\omega})$ satisfies the following conditions: if $\sigma \in \omega^\omega$ and $b_\sigma = \{s < \sigma\}$ is the corresponding branch, then $U(b_\sigma) \simeq U$ while if $T' \in WF$ the space $U(T)$ is not isomorphic to U (and is actually reflexive). Under one form or another, similar arguments are frequently used for proving that a collection of Banach spaces is analytic non Borel.

A remarkable progress due to S. Argyros and P. Dodos is that such constructions are actually generic. Indeed they showed in [1] that a usable analogue of Definition 4.1 is available for analytic families of Banach spaces with Schauder bases.

We need some terminology. If Λ is a non-empty set, a *pruned tree* on Λ is a subset T of $\Lambda^{<\omega}$ such that for every $s \in T$, there is $s < \sigma \in \Lambda^\omega$ such that $b_\sigma = \{s < \sigma\} \subseteq T$. We denote $[T] = \{\sigma \in \Lambda^\omega;\ b_\sigma \subseteq T\}$.

With this notation, let us call a *Schauder tree space on* T a Banach space X equipped with a collection $\{x_t; t \in T\}$ of normalized vectors such that

(i) $X = \overline{span}\{x_t; t \in t\}$

(ii) For every $\sigma \in [T]$, the sequence $\{x_{\sigma_{|n}}; n \geq 1\}$ is a bi-monotone Schauder basic sequence.

We can now state the following fundamental lemma (see [1], Prop. 83):

Lemma 4.4. *Let $A \subseteq SE$ be an analytic set of Banach spaces, such that*

every $Y \in A$ has a Schauder basis. Then there exist a pruned tree T on ω and a Schauder tree space X on T such that
(1) For every $Y \in A$, there is $\sigma \in [T]$ such that

$$Y \simeq X_\sigma = \overline{span}\{x_{\sigma|n}; n \geq 1\}.$$

(2) For every $\sigma \in [T]$, there is $Y \in A$ such that $X_\sigma \simeq Y$.

This lemma turns out to be a very powerful tool for constructing "small" universal spaces for the analytic set A. For instance, given a Schauder tree space X, let us consider like in the proof of Theorem 4.2 (using some ideas from [24]) its "ℓ_2 Baire sum" T_2^X defined by the norm

$$\|z\|_{T_2^X} = \sup \left\{ \left(\sum_{j=0}^{\ell} \left\| \sum_{t \in S_j} z(t)\, x_t \right\|_X^2 \right)^{\frac{1}{2}} \right\}$$

where the supremum is taken over all choices of intervals such that every branch b_σ meets at most one of them. It is clear that for all $\sigma \in [T]$, X_σ is isometric to the subspace of T_2^X of vectors supported by b_σ. And on the other hand, T_2^X is "close enough" to the family A, in the sense where some control is available on subspaces of T_2^X which would not belong to A. This construction, with several technical refinements (since for using [1], one needs to construct spaces with bases containing the elements in A, in a controlled way from the Effros-Borel point of view) allows to show that several natural classes of Banach spaces are strongly bounded, according to the following definition.

Definition 4.4. A subset $\mathcal{C}$ of SE is said to be strongly bounded if for every analytic subset A of $\mathcal{C}$, there exists $Y \in \mathcal{C}$ such that A is contained in the set $\mathcal{SE}(Y)$ of spaces which are isomorphic to a subspace of Y.

Several natural coanalytic classes of separable Banach spaces have been shown to be strongly bounded (and others are conjectured to be). To make a link with the result shown in section 4.2 asserting that coanalytic sets C can be written

$$C = \bigcup_{\alpha < \omega_1} C_\alpha$$

with C_α Borel sets, we note that strongly bounded coanalytic classes $\mathcal{C}$ can be written

$$\mathcal{C} = \bigcup_{\alpha < \omega_1} \mathcal{SE}(Y_\alpha)$$

and thus a transfinite family $(Y_\alpha)_{\alpha<\omega_1}$ of spaces from $\mathcal{C}$ suffices to exhaust the class. Let us now provide examples.

Theorem 4.6. ([8]): *The following classes are strongly bounded:*
1) The class R of reflexive spaces.
2) The class SD of spaces with a separable dual.

It follows from 1) that there exists a reflexive space Z such that $X \in \mathcal{SE}(Z)$ for every uniformly convex space X (a result from [33]). Indeed the existence of an equivalent uniformly convex norm (in other words, super-reflexivity) is a Borel condition.

As said before, it is not known if the functional

$$r(X) = \sup(Sz(X),\ Sz(X^*))$$

is a coanalytic rank on R (a question from [3]). However, it is known ([34]) that for every $\alpha < \omega_1$, there is a reflexive space Z_α such that $\mathcal{SE}(Z_\alpha)$ contains all reflexive spaces X such that $Sz(X) \leq \alpha$ and $Sz(X^*) \leq \alpha$. Note that it is also proved in [34], *using* the universality results shown there, that the class $\mathcal{C}_\alpha$ of reflexive spaces X such that $r(X) \leq \alpha$ is analytic for every countable ordinal α. Hence the main result of [34] appears to be a consequence of Theorem 4.6. So far however, no *direct* proof is known of the fact that $\mathcal{C}_\alpha$ is analytic.

On the class SD, the Szlenk index $Sz(\cdot)$ is a coanalytic rank. Hence 2) shows that for every $\alpha < \omega_1$, there is Y_α such that Y_α^* is separable and $X \in \mathcal{SE}(Y_\alpha)$ if $Sz(X) \leq \alpha$. This answers positively a question from [37].

Theorem 4.7. ([7]): *The class of non-universal spaces is strongly bounded.*

Actually, P. Dodos shows the existence, for every $\lambda > 1$, of a transfinite family $(Y_\alpha)_{\alpha<\omega_1}$ of non-universal $\mathcal{L}_{\infty,\lambda}$ spaces, such that

$$\mathcal{G}_{\mathcal{C}(\Delta)} = \bigcup_{\alpha<\omega_1} \mathcal{SE}(Y_\alpha)$$

where $\mathcal{G}_{\mathcal{C}(\Delta)}$ denotes the collection of spaces which do not contain an isomorphic copy of $\mathcal{C}(\Delta)$; in other words, the collection of non-universal spaces.

The coanalytic set $\mathcal{G}_{\mathcal{C}(\Delta)}$ is equipped with a natural coanalytic rank obtained from a tree (see [3], Th. 4.4), and this is also true for the set $\mathcal{G}_Y$ where Y is infinite-dimensional but arbitrary. However, it is not known if every class $\mathcal{G}_Y$ is strongly bounded; in particular, it is not known for $\mathcal{G}_{\ell_1}$.

Note that there is no space $Z \in \mathcal{G}_Y$ such that $\mathcal{G}_Y = \mathcal{SE}(Z)$. Indeed, $\mathcal{SE}(Z)$ is analytic while $\mathcal{G}_Y$ is coanalytic non Borel ([3]). Hence, every class $\mathcal{G}_Y$ is hereditary with no universal space.

Theorem 4.8. ([9]): *The class of unconditionally saturated Banach spaces is strongly bounded.*

In this statement, "X unconditionally saturated" means that every subspace of X contains an unconditional basic sequence, or equivalently (by [21]) that X does not contain an hereditary indecomposable (in short, H.I) subspace. Hence if we decide once and for all to live in a universe from where H.I. spaces are banned (in other words, if we restrict our attention to unconditionally saturated Banach spaces) then the above theorem asserts that strong boundedness holds in our world.

Acknowledgment

I gratefully thank Professor T.S.S.R.K. Rao who kindly invited me to participate to the Platinum Jubilee of the Indian Statistical Institute. I am grateful to I.S.I. and in particular to Ms. Asha Lata for the kind help in the preparation and typing of this article.

References

[1] Argyros, S. and Dodos, P. (2007). Genericity and amalgamation of classes of Banach spaces. *Adv. Math.* **209** 666-748.

[2] Argyros, S., Godefroy, G. and Rosenthal, H.P. (2003). Descriptive set theory and Banach spaces. Chapter 23, *Handbook of the geometry of Banach spaces.* Vol. 2, Elsevier pp. 1007-1069.

[3] Bossard, B. (2002). A coding of separable Banach spaces. Analytic and coanalytic families of Banach spaces. *Fund. Math.* **172** 117-152.

[4] Bourgain, J. (1979). The Szlenk index and operators on $\mathcal{C}(K)$ spaces. *Bull. Soc. Math. Belg. Serie B.* **31** 87-117.

[5] Bourgain, J. (1981). *New Classes of $\mathcal{L}^p$ Spaces.* Lecture Notes in Math. **889**, Springer.

[6] Christensen, J. P. R. (1974). *Topology and Borel Structure.* North Holland Math. Studies, **10**.

[7] Dodos, P. (2007). On classes of Banach spaces admitting "small" universal spaces. Preprint.

[8] Dodos, P. and Ferenczi, V. (2007). Some strongly bounded classes of Banach spaces. *Fund. Math.* **193** 171-179.

[9] Dodos, P. and Lopez-Abad, J. (2007). On unconditionally saturated Banach spaces. Preprint.

[10] Dutrieux, Y. (2001). Lipschitz-quotients and the Kunen-Martin theorem. *Comment. Math. Univ. Carolinae.* **42** 641-648.

[11] Dutrieux, Y. and Kalton, N. J. (2005). Perturbations of isometries between $\mathcal{C}(K)$ spaces. *Studia Math.* **166** 181-197.

[12] Dutta, S. and Godard, A. (2008). Banach spaces with property (M) and their Szlenk indices. *Mediterranean J. of Maths.* To appear.

[13] Ferenczi, V. (2006). Minimal subspaces and isomorphically homogeneous sequences in a Banach space. *Israel J. Maths.* **156** 125-140.

[14] Ferenczi, V. and Galego, E. M. (2007). Some results on the Schroeder-Bernstein property for separable Banach spaces. *Canad. J. Math.* **59** 63-84.

[15] Ferenczi, V. and Rosendal, C. (2005). On the number of non-isomorphic subspaces of a Banach space. *Studia Math.* **168** 203-216.

[16] Ferenczi, V. and Rosendal, C. (2005). Ergodic Banach spaces. *Advances in Math.* **195** 259-282.

[17] Ferenczi, V., Louveau, A. and Rosendal, C. (2008). The complexity of classifying Banach spaces up to isomorphism. To appear.

[18] Godefroy, G. (2001). The Szlenk index and its applications, general topology in Banach spaces. (T. Banakh ed.) Nova Science Publishers Inc., 71-79.

[19] Godefroy, G., Kalton, N.J. and Lancien, G. (2001). Szlenk indices and uniform homeomorphisms. *Trans. Amer. Math. Soc.* **353** 3895-3918.

[20] Godefroy, G. and Kalton, N.J. (2007). Isometric embeddings and universal spaces. *Extracta Math.* **22** 179-189.

[21] Gowers, W.T. (2002). An infinite Ramsey theorem and some Banach space dichotomies. *Annals of Math.* **156** 797-833.

[22] Hajek, P. and Lancien, G. (2007). Various slicing indices on Banach spaces. *Mediterranean J. Math.* **42** 179-190.

[23] Hajek, P., Montesinos, V., Vanderwerff, J. and Zizler, V. (2008). *Biorthogonal systems in Banach spaces.* CMS Books in Mathematics, Canadian Math. Society.

[24] James, R.C. (1974). A separable somewhat reflexive Banach space with nonseparable dual. *Bull. AMS* **80** 738-743.

[25] Kechris, A. and Louveau, A. (1987). *Descriptive Set Theory and the Structure of Sets of Uniqueness.* London Math. Soc. Lecture Notes Ser. **128**, Cambridge University Press.

[26] Kechris, A. (1995). *Classical Descriptive Set Theory.* Springer, New York.

[27] Knaust, H., Odell, E. and Schlumprecht, T. (1999). On asymptotic structure, the Szlenk index and UKK properties in Banach spaces. *Positivity* **3** 173-199.

[28] Kwapien, S. (1972). Isomorphic characterizations of inner product spaces by orthogonal series with vector coefficients. *Studia Math.* **44** 583-595.

[29] Lancien, G. (1996). On the Szlenk index and the weak*-dentability index. *Quart. J. Math. Oxford* **47** 59-71.

[30] Lancien, G. (2006). A survey on the Szlenk index and some of its applications. *Rev. R. Acad. Cienc. Exactes Fis. Nat. Ser. A Mat.* **100** 209-235.

[31] Moschovakis, Y. (1980). *Descriptive Set Theory.* North Holland, Amsterdam.

[32] Namioka, I. and Phelps, R. R. (1975). Banach spaces which are Asplund spaces. *Duke Math. J.* **42** 735-750.
[33] Odell, E. and Schlumprecht, T. (2006). A separable reflexive space universal for the uniformly convex Banach spaces. *Math. Ann.* **335** 901-916.
[34] Odell, E., Schlumprecht, T. and Zsak, A. (2008). Banach spaces of bounded Szlenk index. *Studia Math.* To appear.
[35] Pelczynski, A. (1969). Universal bases. *Studia Math.* **32** 247-268.
[36] Rosenthal, H. P. (1979). On applications of the boundedness principle to Banach space theory, according to J. Bourgain. *Seminaire d'initiation a l'analyse, Paris 6* 14p.
[37] Rosendal, C. (2004). Incomparable, non-isomorphic and minimal Banach spaces. *Fund. Math.* **183** 253-274.
[38] Souslin, M. M. (1917). Sur une definition des ensembles mesurables B sans nombres transfinis. *Note aux C.R.A.S. Paris* **141** 88-91.
[39] Szlenk, W. (1968). The non-existence of a separable reflexive Banach space universal for all reflexive Banach spaces. *Studia Math.* **30** 53-61.

Chapter 5

Multiplicity-Free Homogeneous Operators in the Cowen-Douglas Class

Adam Korányi and Gadadhar Misra
Lehman College,
*The City University of New York, Bronx, NY 10468, USA**
adam.koranyi@lehman.cuny.edu
Department of Mathematics,
Indian Institute of Science, Bangalore 560 012, India
gm@math.iisc.ernet.in

In a recent paper, the authors have constructed a large class of operators in the Cowen-Douglas class of the unit disc $\mathbb{D}$ which are *homogeneous* with respect to the action of the group Möb – the Möbius group consisting of bi-holomorphic automorphisms of the unit disc $\mathbb{D}$. The *associated representation* for each of these operators is *multiplicity free*. Here we give a different independent construction of all homogeneous operators in the Cowen-Douglas class with multiplicity free associated representation and verify that they are exactly the examples constructed previously.

The homogeneous operators form a class of bounded operators T on a Hilbert space $\mathcal{H}$. The operator T is said to be *homogeneous* if its spectrum is contained in the closed unit disc and for every Möbius transformation g the operator $g(T)$, defined via the usual holomorphic functional calculus, is unitarily equivalent to T. To every homogeneous irreducible operator T there corresponds an *associated* unitary representation π of the universal covering group $\tilde{G}$ of the Möbius group G:

$$\pi(\hat{g})^* \, T \, \pi(\hat{g}) = (p\hat{g}) \, (T), \; \hat{g} \in \tilde{G},$$

where $p : \tilde{G} \to G$ is the natural homomorphism. In the paper [6] (see also [3]), it was shown that each homogeneous operator T, not necessarily irreducible, in $\mathrm{B}_{m+1}(\mathbb{D})$ admits an associated representation. The representations of $\tilde{G}$ are quite well-known, but we are still far from a com-

*This research was supported in part by a DST – NSF S&T Cooperation Programme.

plete description of the homogeneous operators. In the recent paper [6], the following theorem was proved.

Theorem 5.1. *For any positive real number* $\lambda > m/2$, $m \in \mathbb{N}$ *and an* $(m+1)$*-tuple of positive reals* $\boldsymbol{\mu} = (\mu_0, \mu_1, \ldots, \mu_m)$ *with* $\mu_0 = 1$, *there exists a reproducing kernel* $K^{(\lambda,\boldsymbol{\mu})}$ *on the unit disc such that the adjoint of the multiplication operator* $M^{(\lambda,\boldsymbol{\mu})}$ *on the corresponding Hilbert space* $\mathbf{A}^{(\lambda,\boldsymbol{\mu})}(\mathbb{D})$ *is homogeneous. The operators* $(M^{(\lambda,\boldsymbol{\mu})})^*$ *are in the Cowen-Douglas class* $\mathrm{B}_{m+1}(\mathbb{D})$, *irreducible and mutually inequivalent.*

In the paper [6], we have presented the operators $M^{(\lambda,\boldsymbol{\mu})}$ in as elementary a way as possible, but this presentation hides the natural ways in which these operators can be found to begin with. Here we will describe another independent construction of the operators $M^{(\lambda,\boldsymbol{\mu})}$. We will also give an exposition of some of the fundamental background material. Finally, we will prove that if T is an irreducible homogeneous operator in $\mathrm{B}_{m+1}(\mathbb{D})$ whose associated representation is multiplicity free then, up to equivalence, T is the adjoint of of the multiplication operator $M^{(\lambda,\boldsymbol{\mu})}$ for some $\lambda > m/2$ and $\boldsymbol{\mu} \geq 0$.

5.1. Background Material

Although, we intend to discuss homogeneous operators in the Cowen-Douglas class $\mathrm{B}_n(\mathbb{D})$, the material below is presented in somewhat greater generality. Here we discuss commuting tuples of operators in the Cowen-Douglas class $\mathrm{B}_n(\mathcal{D})$ for some bounded open connected set $\mathcal{D} \subseteq \mathbb{C}^m$. The unitary equivalence class of a commuting tuple in $\mathrm{B}_n(\mathcal{D})$ is in one to one correspondence with a certain class of holomorphic Hermitian vector bundles (hHvb) on $\mathcal{D}$ (cf. [4]). These are distinguished by the property, among others, that the Hermitian structure on the fibre at $w \in \mathcal{D}$ is induced by a reproducing kernel K. It is shown in [4] that the corresponding operator can be realized as the adjoint of the commuting tuple multiplication operator $\mathbf{M}$ on the Hilbert space $\mathcal{H}$ of holomorphic functions with reproducing kernel K.

Start with a Hilbert space $\mathcal{H}$ of $\mathbb{C}^n$-valued holomorphic functions on a bounded open connected set $\mathcal{D} \subseteq \mathbb{C}^m$. Assume that the Hilbert space $\mathcal{H}$ contains the set of vector valued polynomials and that these form a dense subset in $\mathcal{H}$. We also assume that there is a reproducing kernel K for $\mathcal{H}$. We use the notation $K_w(z) := K(z, w)$.

Recall that a positive definite kernel $K: \mathcal{D} \times \mathcal{D} \to \mathbb{C}^{n\times n}$ on $\mathcal{D}$ defines an inner product on the linear span of $\{K_w(\cdot)\xi : w \in \mathcal{D}, \xi \in \mathbb{C}^n\} \subseteq \mathrm{Hol}(\mathcal{D}, \mathbb{C}^n)$ by the rule

$$\langle K_w(\cdot)\xi, K_u(\cdot)\eta\rangle = \langle K_w(u)\xi, \eta\rangle,\ \xi, \eta \in \mathbb{C}^n.$$

(On the right hand side $\langle , \rangle$ denotes the inner product of $\mathbb{C}^n$. We denote by $\varepsilon_1, \ldots, \varepsilon_n$ the natural basis of $\mathbb{C}^n$.) The completion of this subspace is then a Hilbert space $\mathcal{H}$ of holomorphic functions on $\mathcal{D}$ (cf. [1]) in which the linear span of the set of vectors $\{K_w(\cdot)\xi : w \in \mathcal{D}\}$ is dense. The kernel K has the reproducing property, that is,

$$\langle f, K_w\xi\rangle = \langle f(w), \xi\rangle,\ f \in \mathcal{H},\ w \in \mathcal{D},\ \xi \in \mathbb{C}^m.$$

Now, for $1 \leq i \leq m$, we have

$$M_i^* K_w\xi = \bar{w}_i K_w\xi,\ w \in \mathcal{D},\ \text{where}\ (M_i f)(z) = z_i f(z),\ f \in \mathcal{H}$$

and $\{K_w\varepsilon_i\}_{i=1}^n$ is a basis for $\cap_{i=1}^m \ker(M_i - w_i)^*$, $w \in \mathcal{D}$.

The joint kernel of the commuting m-tuple $\mathbf{M}^* = (M_1^*, \ldots, M_m^*)$, which we assume to be bounded, then has dimension n. The map $\sigma_i : w \mapsto K_{\bar{w}}\varepsilon_i$, $w \in \bar{\mathcal{D}}$, $1 \leq i \leq n$, provides a trivialization of the corresponding bundle E of Cowen-Douglas (cf. [4]). Here $\bar{\mathcal{D}} := \{z \in \mathbb{C}^m \mid \bar{z} \in \mathcal{D}\}$).

On the other hand, suppose we start with an abstract Hilbert space $\mathcal{H}$ and a m-tuple of commuting operators $\mathbf{T} = (T_1, \ldots, T_m)$ in the Cowen-Douglas class $\mathrm{B}_n(\mathcal{D})$. Then we have a holomorphic Hermitian vector bundle E over $\mathcal{D}$ with the fibre $E_w = \cap_{i=1}^n \ker(T_i - w_i)$ at $w \in \mathcal{D}$. Following [4], one associates to this a reproducing kernel Hilbert space $\hat{\mathcal{H}}$ consisting of holomorphic functions on $\bar{\mathcal{D}}$ as follows. Take a holomorphic trivialization $\sigma_i : \mathcal{D} \to \mathcal{H}$ with $\sigma_i(w)$, $1 \leq i \leq n$, spanning E_w. For $f \in \mathcal{H}$, define $\hat{f}_j(w) := \langle f, \sigma_j(\bar{w})\rangle_{\mathcal{H}}$, $w \in \bar{\mathcal{D}}$. Set $\langle \hat{f}, \hat{g}\rangle_{\hat{\mathcal{H}}} := \langle f, g\rangle_{\mathcal{H}}$. The function $K_w\varepsilon_j := \widehat{\sigma_j(\bar{w})}$ then serves as the reproducing kernel for the Hilbert space $\hat{\mathcal{H}}$. Note that

$$\begin{aligned}\langle K_w(z)\varepsilon_j, \varepsilon_i\rangle_{\mathbb{C}^n} &= \langle K_w\varepsilon_j, K_z\varepsilon_i\rangle_{\hat{\mathcal{H}}}\\ &= \langle \widehat{\sigma_j(\bar{w})}, \widehat{\sigma_i(\bar{z})}\rangle_{\hat{\mathcal{H}}}\\ &= \langle \sigma_j(\bar{w}), \sigma_i(\bar{z})\rangle_{\mathcal{H}},\ z, w \in \bar{\mathcal{D}}.\end{aligned}$$

If one applies this construction to the case where $\mathcal{H}$ is a Hilbert space of holomorphic functions on $\mathcal{D}$, possesses a reproducing kernel, say K, and the operator $\mathbf{M}^*$ is in $\mathrm{B}_n(\bar{\mathcal{D}})$ then using the trivialization $\sigma_i(w) = K_{\bar{w}}\varepsilon_i$,

$w \in \bar{\mathcal{D}}$ for the bundle E defined on $\bar{\mathcal{D}}$, the reproducing kernel for $\hat{\mathcal{H}}$ is

$$\begin{aligned}\langle K_w(z)\varepsilon_j, \varepsilon_i\rangle_{\mathbb{C}^n} &= \langle K_w\varepsilon_j, K_z\varepsilon_j\rangle_{\mathcal{H}} \\ &= \langle \sigma_j(\bar{w}), \sigma_i(\bar{z})\rangle_{\mathcal{H}} \\ &= \langle K_w\varepsilon_j, K_z\varepsilon_i\rangle_{\hat{\mathcal{H}}}, \ z, w \in \mathcal{D}.\end{aligned}$$

Thus $\mathcal{H} = \hat{\mathcal{H}}$.

Let G be a Lie group acting transitively on the domain $\mathcal{D} \subseteq \mathbb{C}^d$. Let $GL(n, \mathbb{C})$ denote the set of non-singular $n \times n$ matrices over the complex field $\mathbb{C}$. We start with a multiplier J, that is, a smooth family of holomorphic maps $J_g : \mathcal{D} \to \mathbb{C}^{n\times n}$ satisfying the cocycle relation

$$J_{gh}(z) = J_h(z)J_g(h \cdot z), \quad \text{for all} \quad g, h \in G, \ z \in \mathcal{D}, \tag{5.1}$$

Let $\mathrm{Hol}(\mathcal{D}, \mathbb{C}^n)$ be the linear space consisting of all holomorphic functions on $\mathcal{D}$ taking values in $\mathbb{C}^n$. We then obtain a natural (left) action U of the group G on $\mathrm{Hol}(\mathcal{D}, \mathbb{C}^n)$:

$$(U_g f)(z) = J_{g^{-1}}(z)f(g^{-1} \cdot z), \ f \in \mathrm{Hol}(\mathcal{D}, \mathbb{C}^n), \ z \in \mathcal{D}. \tag{5.2}$$

Let $\mathbb{K} \subseteq G$ be the compact subgroup which is the stabilizer of 0. For h, k in $\mathbb{K}$, we have $J_{kh}(0) = J_h(0)J_k(0)$ so that $k \mapsto J_k(0)^{-1}$ is a representation of $\mathbb{K}$ on $\mathbb{C}^n$.

As in [6], we say that if a reproducing kernel K transforms according to the rule

$$J(g, z)K(g(z), g(\omega))J(g, \omega)^* = K(z, \omega) \tag{5.3}$$

for all $g \in \tilde{G}$; $z, \omega \in \mathbb{D}$, then K is *quasi-invariant.*

Proposition 5.1 ([6], Proposition 2.1). *Suppose $\mathcal{H}$ has a reproducing kernel K. Then U defined by* (5.2) *is a unitary representation if and only if K is quasi-invariant.*

Let g_z be an element of G which maps 0 to z, that is $g_z \cdot 0 = z$.

For quasi-invariant K we have

$$K(g_z \cdot 0, g_z \cdot 0) = (J_{g_z}(0))^{-1}K(0, 0)(J_{g_z}(0)^*)^{-1}, \tag{5.4}$$

which shows that $K(z, z)$ is uniquely determined by $K(0, 0)$. For each z in $\mathcal{D}$, the positive definite matrix $K(z, z)$ gives the Hermitian structure of our vector bundle.

Given any positive definite matrix $K(0, 0)$ such that

$$J_k(0)^{-1}K(0, 0) = K(0, 0)J_k(0)^* \text{ for all } k \in \mathbb{K}, \tag{5.5}$$

that is, the inner product $\langle K(0,0)\cdot \mid \cdot\rangle$ is invariant under $J_k(0)$, (5.4) defines a Hermitian structure on the homogeneous vector bundle determined by $J_g(z)$. In fact, $K(z,z)$, for any $z \in \mathcal{D}$ is well defined, because if g'_z is another element of G such that $g'_z \cdot 0 = z$ then $g'_z = g_z k$ for some $k \in \mathbb{K}$. Hence

$$\begin{aligned}
K(g'_z \cdot 0, g'_z \cdot 0) &= K(g_z k \cdot 0, g_z k \cdot 0) \\
&= (J_{g_z k}(0))^{-1} K(0,0)(J_{g_z k}(0)^*)^{-1} \\
&= \big(J_k(0) J_{g_z}(k \cdot 0)\big)^{-1} K(0,0)\big(J_{g_z}(k \cdot 0)^* J_k(0)^*\big)^{-1} \\
&= (J_{g_z}(0))^{-1}(J_k(0))^{-1} K(0,0)(J_k(0)^*)^{-1}(J_{g_z}(0)^*)^{-1} \\
&= (J_{g_z}(0))^{-1} K(0,0)(J_{g_z}(0)^*)^{-1} \\
&= K(g_z \cdot 0, g_z \cdot 0).
\end{aligned}$$

This gives a good overview of all the Hermitian structures of a homogeneous holomorphic vector bundle. But not all such bundles arise from a reproducing kernel. Starting with a positive matrix satisfying (5.5), (5.4) gives us $K(z,z)$, but there is no guarantee (and is false in general) that $K(z,z)$ extends to a positive definite kernel on $\mathcal{D} \times \mathcal{D}$. It is, however, true that if there is such an extension then it is uniquely determined by $K(z,z)$ (because $K(z,w)$ is holomorphic in z and antiholomorphic in w).

This leaves us with the following possible strategy for finding the homogeneous operators in the Cowen-Douglas class. Find all multipliers, (i.e., holomorphic homogeneous vector bundles (hhvb)) such that there exists $K(0,0)$ satisfying (5.5) and consider all such $K(0,0)$. Then determine which of the $K(z,z)$ obtained form (5.4) extends to a positive definite kernel on $\mathcal{D}\times\mathcal{D}$. Then check if the multiplication operator is well-defined and bounded on the corresponding Hilbert space.

Let $\mathcal{H}$ be a Hilbert space consisting of $\mathbb{C}^n$-valued holomorphic functions on some domain $\mathcal{D}$ possessing a reproducing kernel K. The sections of the corresponding holomorphic Hermitian vector bundle defined on $\mathcal{D}$ have many different realizations. The connection between two of these is given by a $n \times n$ invertible matrix valued holomorphic function φ on $\mathcal{D}$. For $f \in \mathcal{H}$, consider the map $\Gamma_\varphi : f \mapsto \tilde{f}$, where $\tilde{f}(z) = \varphi(z) f(z)$. Let $\tilde{\mathcal{H}} = \{\tilde{f} : f \in \mathcal{H}\}$. The requirement that the map Γ_φ is unitary, prescribes a Hilbert space structure for the function space $\tilde{\mathcal{H}}$. The reproducing kernel for $\tilde{\mathcal{H}}$ is easily calculated

$$\tilde{K}(z,w) = \varphi(z) K(z,w) \varphi(w)^*. \tag{5.6}$$

It is also easy to verify that $\Gamma_\varphi M \Gamma_\varphi^*$ is the multiplication operator

$M\colon \tilde f \mapsto z\tilde f$ on the Hilbert space $\tilde{\mathcal H}$. Suppose we have a unitary representation U given by a multiplier J acting on $\mathcal H$ according to (5.2). Transplanting this action to $\tilde{\mathcal H}$ under the isometry Γ_φ, it becomes

$$(\tilde U_{g^{-1}}\tilde f)(z) = \tilde J_g(z)\tilde f(g\cdot z),$$

where the new multiplier $\tilde J$ is given in terms of the original multiplier J by

$$\tilde J_g(z) = \varphi(z)J_g(z)\varphi(g\cdot z)^{-1}.$$

Of course, now $\tilde K$ transforms according to (5.3), with the aid of $\tilde J$. If we want, we can now ensure that, by passing from $\mathcal H$ to an appropriate $\tilde{\mathcal H}$, $\tilde K(z,0)\equiv 1$. We merely have to set $\varphi(z) = K(0,0)^{1/2}K(z,0)^{-1}$. Thus the reproducing kernel $\tilde K$ is almost unique. The only freedom left is to multiply $\varphi(z)$ by a constant unitary $n\times n$ matrix. Once the kernel is normalized, we have

$$J_k(z) = J_k(0),\quad z\in\mathbb D,\ k\in\mathbb K.$$

In fact,

$$I = K(z,0) = J_k(z)K(k\cdot z,0)J_k(0)^* = J_k(z)J_k(0)^{-1}$$

and the statement follows. Therefore, once the kernel K is normalized, we have

$$(U_{k^{-1}}f)(z) = J_k(0)f(k\cdot z),\quad k\in\mathbb K.$$

Given a multiplier J, there is always the following method for constructing a Hilbert space with a quasi-invariant Kernel K transforming according to (5.4). We look for a functional Hilbert space possessing this property among the weighted L^2 spaces of holomorphic functions on $\mathcal D$. The norm on such a space is

$$\|f\|^2 = \int_{\mathcal D} f(z)^*Q(z)f(z)dV(z) \tag{5.7}$$

with some positive matrix valued function $Q(z)$. Clearly, this Hilbert space possesses a reproducing kernel K. The condition that $U_{g^{-1}}$ in (5.2) is unitary is

$$\begin{aligned}\int_{\mathcal D} f(g\cdot z)^*J_g^*(z)Q(z)J_g(z)f(g\cdot z)dV(z) &= \int_{\mathcal D} f(w)^*Q(w)f(w)dV(w)\\ &= \int_{\mathcal D} f(g\cdot z)^*Q(g\cdot z)f(g\cdot z)\left|\frac{\partial(g\cdot z)}{\partial(z)}\right|^2 dV(z),\end{aligned}$$

that is,

$$Q(g \cdot z) = J_g(z)^* Q(z) J_g(z) \left| \frac{\partial(g \cdot z)}{\partial(z)} \right|^{-2}, \tag{5.8}$$

which is equation (5.3) with $J_g(z)$ replaced by $\frac{\partial(g\cdot z)}{\partial(z)} J_g(z)^{*-1}$.

Given the multiplier $J_g(z)$, $Q(z)$ is again determined by $Q = Q(0)$, and (just as in the case of $K(0,0) = A$) it must be a positive matrix commuting with all $J_k(0)$, $k \in \mathbb{K}$. (It is assumed that each $J_k(0)$ is unitary).

In this way, we can construct many examples of homogeneous operators in $\mathrm{B}_n(\mathcal{D})$ but not all.

Even, not all the the homogeneous operators in $\mathrm{B}_1(\mathbb{D})$ come from this construction. There is a homogeneous operator in the class $\mathrm{B}_1(\mathbb{D})$ corresponding to the multiplier $J(g,z) = (g'(z))^\lambda$, $\lambda \in \mathbb{R}$ exactly when $\lambda > 0$. The reproducing kernel is $K(z,w) = (1 - z\bar{w})^{-2\lambda}$. But such an operator arises from the construction outlined above only if $\lambda \geq 1/2$.

Never the less, the homogeneous operators constructed in the manner described above are of interest since they happen to be exactly the subnormal homogeneous operators in this class (cf. [2]).

5.2. Computation of the Multipliers for the Unit Disc

In the case of $\mathrm{B}_n(\mathbb{D})$, it is shown in [6] that the bundle corresponding to a homogeneous Cowen-Douglas operator admits an action of the covering group $\tilde{G}$ of the group $G =$ Möb via unitary bundle maps. This suggests the strategy of first finding all the homogeneous holomorphic Hermitian vector bundles (a problem easily solved by known methods) and then determining which of these correspond to an operator in the Cowen-Douglas class.

We are going to use the method of holomorphic induction. For this, first we describe some basic facts and fix our notation. We follow the notation of [7] which we will use as a reference.

The Lie algebra $\mathfrak{g}$ of $\tilde{G}$ is spanned by $X_1 = \frac{1}{2}\begin{pmatrix} 0 & 1 \\ 1 & 0 \end{pmatrix}$, $X_0 = \frac{1}{2}\begin{pmatrix} i & 0 \\ 0 & -i \end{pmatrix}$ and $Y = \frac{1}{2}\begin{pmatrix} 0 & -i \\ i & 0 \end{pmatrix}$. The subalgebra $\mathfrak{k}$ corresponding to $\tilde{\mathbb{K}}$ is spanned by X_0. In the complexified Lie algebra $\mathfrak{g}^{\mathbb{C}}$, we mostly use the complex basis

h, x, y given by

$$h = -iX_0 = \frac{1}{2}\begin{pmatrix} 1 & 0 \\ 0 & -1 \end{pmatrix}$$
$$x = X_1 + iY = \begin{pmatrix} 0 & 1 \\ 0 & 0 \end{pmatrix}$$
$$y = X_1 - iY = \begin{pmatrix} 0 & 0 \\ 1 & 0 \end{pmatrix}.$$

We write $G^{\mathbb{C}}$ for the (simply connected group) $SL(2, \mathbb{C})$. Let $G_0 = SU(1,1)$ be the subgroup corresponding to $\mathfrak{g}$. The group $G^{\mathbb{C}}$ has the closed subgroups $\mathbb{K}^{\mathbb{C}} = \left\{\begin{pmatrix} z & 0 \\ 0 & \frac{1}{z} \end{pmatrix} : z \in \mathbb{C}, z \neq 0\right\}$, $P^+ = \left\{\begin{pmatrix} 1 & z \\ 0 & 1 \end{pmatrix} : z \in \mathbb{C}\right\}$, $P^- = \left\{\begin{pmatrix} 1 & 0 \\ z & 1 \end{pmatrix} : z \in \mathbb{C}\right\}$; the corresponding Lie algebras $\mathfrak{k}^{\mathbb{C}} = \left\{\begin{pmatrix} c & 0 \\ 0 & -c \end{pmatrix} : c \in \mathbb{C}\right\}$, $\mathfrak{p}^+ = \left\{\begin{pmatrix} 0 & c \\ 0 & 0 \end{pmatrix} : c \in \mathbb{C}\right\}$, $\mathfrak{p}^- = \left\{\begin{pmatrix} 0 & 0 \\ c & 0 \end{pmatrix} : c \in \mathbb{C}\right\}$ are spanned by h, x and y, respectively. The product $\mathbb{K}^{\mathbb{C}} P^- = \left\{\begin{pmatrix} a & 0 \\ b & \frac{1}{a} \end{pmatrix} : 0 \neq a \in \mathbb{C}, b \in \mathbb{C}\right\}$ is a closed subgroup to be denoted T; its Lie algebra is $\mathfrak{t} = \mathbb{C}h + \mathbb{C}y$. The product set $P^+\mathbb{K}^{\mathbb{C}}P^- = P^+T$ is dense open in $G^{\mathbb{C}}$, contains G, and the product decomposition of each of its elements is unique. ($G^{\mathbb{C}}/T$ is the Riemann sphere, $g\tilde{\mathbb{K}} \to gT$, $(g \in G)$ is the natural embedding of $\mathbb{D}$ into it.)

According to holomorphic induction ([5, Chap. 13]) the isomorphism classes of homogeneous holomorphic vector bundles are in one to one correspondence with equivalence classes of linear representations ϱ of the pair $(\mathfrak{t}, \tilde{\mathbb{K}})$. Since $\tilde{\mathbb{K}}$ is connected, here this means just the representations of $\mathfrak{t}$. Such a representation is completely determined by the two linear transformations $\varrho(h)$ and $\varrho(y)$ which satisfy the bracket relation of h and y, that is,

$$[\varrho(h), \varrho(y)] = -\varrho(y). \tag{5.9}$$

The $\tilde{G}$-invariant Hermitian structures on the homogeneous holomorphic vector bundle (making it into a homogeneous holomorphic Hermitian vector bundle), if they exist, are given by $\varrho(\tilde{\mathbb{K}})$-invariant inner products on the representation space. An inner product is $\varrho(\tilde{\mathbb{K}})$-invariant if and only if $\varrho(h)$ is diagonal with real diagonal elements in an appropriate basis.

We will be interested only in bundles with a Hermitian structure. So, we will assume without restricting generality, that the representation space of ϱ is $\mathbb{C}^d$ and that $\varrho(h)$ is a real diagonal matrix.

Furthermore, we will be interested only in irreducible homogeneous holomorphic Hermitian vector bundles, this corresponds to ϱ not being the orthogonal direct sum of non-trivial representations. Suppose we have such a ϱ; we write V_α for the eigenspace of $\varrho(h)$ with eigenvalue α. Let $-\eta$ be the largest eigenvalue of $\varrho(h)$ and m be the largest integer such that $-\eta, -(\eta+1), \ldots, -(\eta+m)$ are all eigenvalues. From (5.9) we have $\varrho(y)V_\alpha \subseteq V_{\alpha-1}$; this and orthogonality of the eigenspaces imply that $V = \oplus_{j=0}^{m} V_{-(\eta+j)}$ and its orthocomplement are invariant under ϱ. So, V is the whole space, and have proved that the eigenvalues of $\varrho(h)$ are $-\eta, \ldots, -(\eta+m)$.

From this it is clear that ϱ can be written as the tensor product of the one dimensional representation σ given by $\sigma(h) = -\eta$, $\sigma(y) = 0$, and the representation ϱ^0 given by $\varrho^0(h) = \varrho(h) + \eta I$, $\varrho^0(y) = \varrho(y)$. Correspondingly, the bundle for ϱ is the tensor product of a line bundle L_η and the bundle corresponding to ϱ^0.

The representation ϱ^0 has the great advantage that it lifts to a holomorphic representation of the group T. It follows that the homogeneous holomorphic vector bundle it determines for $\mathbb{D}, \tilde{G}$, can be obtained as the restriction to $\mathbb{D}$ of the homogeneous holomorphic vector bundle over $G^{\mathbb{C}}/T$ obtained by ordinary induction in the complex analytic category. So, (as a convenient choice) take the local holomorphic cross section $z \mapsto s(z) := \begin{pmatrix} 1 & z \\ 0 & 1 \end{pmatrix}$ of $G^{\mathbb{C}}/T$ over $\mathbb{D}$. In the trivialization given by $s(z)$, the multiplier then appears for $g = \begin{pmatrix} a & b \\ c & d \end{pmatrix} \in G^{\mathbb{C}}$ as

$$\begin{aligned} J_g^0(z) &= \varrho^0\big(s(z)^{-1} g^{-1} s(g \cdot z)\big) \\ &= \varrho^0 \begin{pmatrix} cz+d & 0 \\ -c & (cz+d)^{-1} \end{pmatrix} \\ &= \varrho^0\Big(\exp\Big(\frac{-c}{cz+d} y\Big)\Big)\varrho^0\big(\exp(2\log(cz+d)h)\big). \end{aligned} \tag{5.10}$$

The last two equalities are simple computations.

For the line bundle L_η, the multiplier is $g'(z)^\eta$ (we write $g'(z) = \frac{\partial g}{\partial z}(z)$). Consequently, the multiplier corresponding to the original ϱ is

$$J_g(z) = \big(g'(z)\big)^\eta J_g^0(z). \tag{5.11}$$

5.3. Conditions Imposed by the Reproducing Kernel

We now assume that we have a homogeneous holomorphic vector bundle induced by ϱ as in the preceding sections and that it has a reproducing kernel. Then we derive conditions about the action of $\tilde{G}$ that follow from this hypothesis. In the final section, we will show that these conditions are sufficient: they lead directly to the construction of all homogeneous operators the Cowen-Douglas class with multiplicity free representations.

Under our hypothesis there is a Hilbert space structure on our sections in which the action of $\tilde{G}$ given by (5.4) is unitary. We will study this representation through its $\mathbb{K}$-types (i.e., its restriction to $\tilde{\mathbb{K}}$). We first compute the infinitesimal representation.

For $X \in \mathfrak{g}$, and holomorphic f, we have

$$(U_X f)(z) := \left(\frac{d}{dt}\right)_{|t=0} (U_{\exp(tX)} f)(z)$$

$$= \left(\frac{d}{dt}\right)_{|t=0} \left\{ \left(\frac{\partial(\exp(-tX)\cdot z)}{\partial z}\right)^{\eta} J^0_{\exp(-tX)}(z) f(\exp(-tX)\cdot z) \right\}. \quad (5.12)$$

There is a local action of $G^{\mathbb{C}}$, so this formula remains meaningful also for $X \in \mathfrak{g}^{\mathbb{C}}$. There are three factors to differentiate. For the last one, $(\frac{d}{dt})_{|t=0} f(\exp(-tX)\cdot z) = -(Xz)f'(z)$, and we see that $\exp(tx)\cdot z = \begin{pmatrix} 1 & t \\ 0 & 1 \end{pmatrix} \cdot z = z + t$ gives $x \cdot z = 1$; by similar computations, $y \cdot z = -z^2$, $h \cdot z = z$. For the first factor, we interchange the differentiations and get $-\eta \frac{\partial}{\partial z}(X \cdot z)$, i.e., $0, 2\eta z, -\eta$, respectively for x, y and h.

To differentiate the factor in the middle, we use its expression (5.10). First for $X = y$, we have

$$\left.\frac{d}{dt}\right|_{t=0} \varrho^0\big(\exp(-t(tz+1)^{-1}y)\big) = \left.\frac{d}{dt}\right|_{t=0} \big(\exp(-t(tz+1)^{-1}\varrho^0(y)\big)$$

$$= -\varrho^0(y) \quad (5.13)$$

and

$$\left.\frac{d}{dt}\right|_{t=0} \varrho^0(\exp(2\log(tz+1)h)) = \left.\frac{d}{dt}\right|_{t=0} \exp(2\log(tz+1)\varrho^0(h))$$

$$= 2z\varrho^0(h)\,. \quad (5.14)$$

From these, following the conventions of [7] in defining H, E, F, it follows that

$$(Ff)(z) := (U_{-y}f)(z) = \frac{d}{dt}\Big|_{|t=0} J_{\exp(ty)}(z)f(\exp(ty)\cdot z)$$
$$= \left(-2\eta zI + 2z\varrho^0(h) - \varrho^0(y)\right)f(z) - z^2 f'(z). \tag{5.15}$$

Similar, simpler computations give, for $g = \exp(tx) = \begin{pmatrix} 1 & t \\ 0 & 1 \end{pmatrix}$

$$(Ef)(z) := (U_x f)(z) = -f'(z). \tag{5.16}$$

Finally, for $g = \exp(th) = \begin{pmatrix} e^{t/2} & 0 \\ 0 & e^{-t/2} \end{pmatrix}$, we have

$$J_{\exp(th)}(z) = \varrho\begin{pmatrix} e^{-t/2} & 0 \\ 0 & e^{t/2} \end{pmatrix} = \exp(-t)\varrho^0(h).$$

Hence it is not hard to verify that

$$(Hf)(z) := (U_h f)(z) = \left(-\eta I + \varrho^0(h)\right)f(z) - zf'(z). \tag{5.17}$$

Under our hypothesis, we have a reproducing kernel and U is unitary. From our computations above, we can determine how U decomposes into irreducibles. The infinitesimal representation of U acts on the vector valued polynomials; a good basis for this space is $\{\varepsilon_j z^n : n \geq 0\}$; ε_j is the jth natural basis vector in $\mathbb{C}^d$. We have $H(\varepsilon_j z^n) = -(\eta + j + n)(\varepsilon_j z^n)$, so the lowest $\mathbb{K}$-types of the irreducible summands are spanned by the ε_j. This space is also the kernel of E. So, U is direct sum of discrete series representations ($U^{\eta+j}$, in the notation of [7]), each one appearing as many times as $-(\eta + j)$ appears on the diagonal of $\varrho(h)$.

5.4. The Multiplicity-Free Case

In order to be able to use the computations of [6] without confusion, we introduce the parameter $\lambda = \eta + \frac{m}{2}$.

From the last remark of the preceding section, it is clear that if U is multiplicity-free then $\varrho(h)$ is an $(m+1)\times(m+1)$ matrix with eigenvalues $-\lambda + \frac{m}{2}, -\lambda + \frac{m}{2} - 1, \ldots, -\lambda - \frac{m}{2}$. As $\varrho(h)\varepsilon_j = -(\lambda - \frac{m}{2} + j)\,\varepsilon_j$, (5.9) shows that

$$\varrho(h)\big(\varrho(y)\varepsilon_j\big) = -\left(\lambda + \frac{m}{2} + j + 1\right)\varrho(y)\,\varepsilon_j, \text{that is, } \varrho(y)\,\varepsilon_j = \text{const}\,\varepsilon_{j+1}.$$

So, $\varrho(y)$ is a lower triangular matrix (with non-zero entries, otherwise we have a reducible bundle). The homogeneous holomorphic vector bundle determines $\varrho(y)$ only up to a conjugacy by a matrix commuting with $\varrho(h)$, that is, a diagonal matrix. So, we can choose the realization of our bundle by applying an appropriate conjugation such that $\varrho(y) = S_m$, the triangular matrix whose $(j, j-1)$ element is j for $1 \leq j \leq m$.

By standard representation theory of $\mathrm{SL}(2, \mathbb{R})$, the vectors $(-F)^n \varepsilon_j$ are orthogonal and the irreducible subspaces $\mathcal{H}^{(j)}$ for U are $\mathrm{span}\{(-F)^n \varepsilon_j : n \geq 0\}$ for $0 \leq j \leq m$. There is also precise information about the norms.

Using this, we can construct an orthonormal basis for our representation space.

For any $n \geq 0$, we let $u_n^j(z) = (-F)^n \varepsilon_j$.

To proceed further, we need to find the vectors $u_n^j(z)$ explicitly. This is facilitated by the following Lemma.

Lemma 5.1. *Let u be a vector with $u_\ell(z) = u_\ell z^{n-\ell}$, $0 \leq \ell \leq m$ and $n \geq 0$. We then have*

$$(-Fu)_\ell(z) = (2\lambda - m + \ell + n) u_\ell z^{n+1-\ell} + \ell u_{\ell-1} z^{n+1-\ell},\ 0 \leq \ell \leq m.$$

Proof. We recall (5.15) that $-(Ff)(z) = 2\lambda z f(z) + S_m f(z) - 2z D_m f(z) + z^2 f'(z)$ for $f \in \mathcal{H}(n)$, where $D_m = -\varrho^0(h)$ is the diagonal operator with diagonal $\{-\frac{m}{2}, -\frac{m}{2}+1, \ldots, \frac{m}{2}\}$ and S_m is the forward weighted shift with weights $1, 2, \ldots, m$. Therefore we have

$$(-Fu)_\ell(z) = \big(2\lambda u_\ell + \ell u_{\ell-1} - (m - 2\ell) u_\ell + (n - \ell) u_\ell\big) z^{n+1-\ell}$$

completing the proof. □

Lemma 5.2. *For $0 \leq j \leq m$ and $0 \leq \ell \leq m$, we have*

$$u_{n,\ell}^j(z) = \begin{cases} 0, & 0 \leq \ell \leq j-1 \\ \binom{n}{k} (j+1)_k (2\lambda - m + 2j + k)_{n-k} z^{n-k}, & j \leq \ell \leq m,\ k = \ell - j, \end{cases}$$

where $u_{n,\ell}^j(z)$ is the scalar valued function at the position ℓ of the $\mathbb{C}^{m+1}$-valued function $u_n^j(z) := (-F)^n \varepsilon_j$.

Proof. The proof is by induction on n. The vectors u_n^j are in $\mathcal{H}(n)$ for $0 \leq j \leq m$. For a fixed but arbitrary positive integer j, $0 \leq j \leq m$, we see that $u_{n,\ell}^j(z)$ is 0 if $n < \ell - j$. We have to verify that $(-Fu_n^j)(z) = u_{n+1}^j(z)$. From the previous Lemma, we have

$$(-Fu_n^j)_\ell(z) = (2\lambda - m + \ell + n + j) u_{n,\ell}^j z^{n+j+1-\ell} + \ell u_{n,\ell-1}^j z^{n+j+1-\ell},$$

where $(-Fu_n^j)_\ell(z)$ is the scalar function at the position ℓ of the $\mathbb{C}^{m+1}$ - valued function $(-Fu_n^j)(z)$. To complete the proof, we note (using $k = \ell - j$) that

$$\begin{aligned}
&(-Fu_n^j)_{j+k}(z)\\
&= \Big(\tbinom{n}{k}(j+1)_k(2\lambda - m + 2j + k)_{n-k}(2\lambda - m + 2j + k + n) +\\
&\quad \tbinom{n}{k-1}(j+1)_k(2\lambda - m + 2j + k - 1)_{n-k}\Big)z^{n+1-k}\\
&= (j+1)_k(2\lambda - m + 2j + k)_{n-k}\\
&\quad \Big(\tbinom{n}{k}(2\lambda - m + 2j + k + n) + \tbinom{n}{k-1}(2\lambda - m + 2j + k - 1)\Big)z^{n+1-k}\\
&= (j+1)_k(2\lambda - m + 2j + k)_{n-k}\\
&\quad \Big(\big(\tbinom{n}{k} + \tbinom{n}{k-1}\big)(2\lambda - m + 2j + k - 1) + (n+1)\tbinom{n}{k}\Big)z^{n+1-k}\\
&= (j+1)_k(2\lambda - m + 2j + k)_{n-k}\\
&\quad \Big(\tbinom{n+1}{k}(2\lambda - m + 2j + k - 1) + \tbinom{n+1}{k}(n - k + 1)\Big)z^{n+1-k}\\
&= (j+1)_k(2\lambda - m + 2j + k)_{n-k}\Big(\tbinom{n+1}{k}(2\lambda - m + 2j + n)\Big)z^{n+1-k}\\
&= (j+1)_k\Big(\tbinom{n+1}{k}(2\lambda - m + 2j + k)_{n+1-k}\Big)z^{n+1-k}\\
&= u_{n+1,j+k}^j(z)
\end{aligned}$$

for a fixed but arbitrary j, $0 \le j \le m$ and k, $0 \le k \le m-j$. This completes the proof. □

On $\mathcal{H}^{(j)}$, we have the representation U^{λ_j} acting $(0 \le j \le m)$, where $\lambda_j = \lambda - \frac{m}{2} + j$. Its lowest $\mathbb{K}$-type is spanned by ε_j $(= u_0^j)$ and $H\varepsilon_j = \lambda_j\varepsilon_j$. By [7, Prop 6.14] we have $\|(-F)^k\varepsilon_j\|^2 = \sigma_k^j\|(-F)^{k-1}\varepsilon_j\|^2$ with

$$\sigma_k^j = (2\lambda_j + k - 1)k$$

for all $k \ge 1$. (Here we used that the constant q in [7, Eq. (6.33)] equals $\lambda_j(1 - \lambda_j)$ by [7, Theorem 6.2].) We write

$$\boldsymbol{\sigma}_n^j = \prod_{k=1}^{n} \sigma_k^j$$

which can be written in a compact form

$$\boldsymbol{\sigma}_n^j = ((2\lambda_j)_n(1)_n), \tag{5.18}$$

where $(x)_n = (x+1)\cdots(x+n-1)$. We stipulate that the binomial coefficient $\binom{n}{k}$ as well as $(x)_{n-k}$ are both zero if $n < k$.

The positivity of the normalizing constants $\left(\boldsymbol{\sigma}_{n-j}^j\right)^{\frac{1}{2}}$ $(n \ge j)$ is equivalent to the existence of an inner product for which the set of vectors $\mathbf{e}_{n-j}^j$

defined by the formula:

$$\mathbf{e}^j_{n-j} = (\sigma^j_n)^{-\frac{1}{2}} u^j_{n-j}(z),\ n \geq j,\ 0 \leq j \leq m$$

forms an orthonormal set. Of course, the positivity condition is fulfilled if and only if $2\lambda > m$.

In this way, for fixed j, each e^j_{n-j} has the same norm for all $n \geq j$. Hence the only possible choice for an orthonormal system is $\{\mu_j e^j_{n-j} : n \geq j\}$ for some positive real numbers $\mu_j > 0$ $(0 \leq j \leq m)$. However, we may choose the norm of the first vector, that is, the vector $\mathbf{e}^j_0$, $0 \leq j \leq m$, arbitrarily. Therefore, all the possible choices for an orthonormal set are

$$\mu_j \mathbf{e}^j_{n-j}(z) = \frac{\mu_j}{\sqrt{(2\lambda - m + 2j)_{n-j}}\sqrt{(1)_{n-j}}} u^j_{n-j}(z), \tag{5.19}$$

$n \geq j$, $0 \leq j \leq m$, and μ_j, $0 \leq j \leq m$ are $m+1$ arbitrary positive numbers.

Let us fix a positive real number λ and $m \in \mathbb{N}$ satisfying $2\lambda > m$. Let $\mathcal{H}^{(\lambda,\boldsymbol{\mu})}$ denote the closed linear span of the vectors $\{\mu_j e^j_{n-j} : 0 \leq j \leq m,\ n \geq j\}$. Then the Hilbert space $\mathcal{H}^{(\lambda,\boldsymbol{\mu})}$ is the representation space for U defined in (5.2). Since the vectors $u^j_n \perp u^k_p$ as long as $j \neq k$, it follows that the Hilbert space $\mathcal{H}^{(\lambda,\boldsymbol{\mu})}$ is the orthogonal direct sum $\oplus_{j=0}^m \frac{1}{\mu_j}\mathcal{H}^{(j)}$.

We proceed to compute the reproducing kernel by using the orthonormal system $\{\mu_j e^j_{n-j} : n \geq j\}$, $0 \leq j \leq m$. We point out that for $0 \leq \ell \leq m$, the entry $\mathbf{e}^{\ell,j}_{n-j} z^{n-j}$ at the position ℓ of the vector $\mathbf{e}^j_{n-j}(z)$ is 0 for $n < \ell$. Consequently, $\mathbf{e}^j_{n-j}$ is the zero vector unless $n \geq j$. The set of vectors $\{\mu_j \mathbf{e}^j_{n-j} : 0 \leq j \leq m,\ n \geq j\}$ is orthonormal in the Hilbert space $\mathcal{H}^{(\lambda,\boldsymbol{\mu})}$. We note that

$$\mathbf{e}^j_{n-j}(z) = ((e^{\ell,j}_{n-j} z^{n-k}))_{\ell=0}^m,$$

$$(\mathbf{e}^j_{n-j}(z))_\ell = \begin{cases} 0, & 0 \leq \ell \leq j-1 \\ \sqrt{\frac{(2\lambda+2j-m+k)_{n-j-k}}{(1)_{n-j-k}}}\sqrt{\frac{(n-j-k+1)_k}{(2\lambda+2j-m)_k}}\frac{(j+1)_k}{(1)_k} z^{n-k}, & j \leq \ell \leq m,\ k = \ell - j. \end{cases} \tag{5.20}$$

We have under the hypothesis that we have a reproducing kernel Hilbert space on which the representation U is unitary, explicitly determined an orthonormal basis for this space. Now we are able to answer the question of whether this space really exists. For this it is enough to show that $\sum e_n(z)\,\overline{e_n(w)}^{\mathrm{tr}}$ converges pointwise, the sum then represents the reproducing kernel for this Hilbert space. We will sum the series explicitly, and will verify that it gives exactly the kernels constructed in [6]. This will complete the program of this paper by proving that the examples of [6] give all the homogeneous operators in the Cowen-Douglas class whose associated representation is multiplicity free.

To compute the kernel function, it is convenient to set, for any $n \geq 0$,

$$G(\boldsymbol{\mu}, n, z) = \begin{pmatrix} \mu_0 e_n^{0,0} z^n & \dots & 0 & \dots & 0 \\ \vdots & \dots & \vdots & \dots & \vdots \\ \mu_0 e_n^{j,0} z^{n-j} & \dots & \mu_j e_{n-j}^{j,j} z^{n-j} & \dots & 0 \\ \vdots & \dots & \vdots & \dots & \vdots \\ \mu_0 e_n^{m,0} z^{n-m} & \dots & \mu_j e_{n-j}^{m,j} z^{n-m} & \dots & \mu_m e_{n-m}^{m,m} z^{n-m} \end{pmatrix}$$

$$= \begin{pmatrix} z^n & \dots & 0 \\ \vdots & \ddots & \vdots \\ 0 & \dots & z^{n-m} \end{pmatrix} \begin{pmatrix} e_n^{0,0} & \dots & 0 & \dots & 0 \\ \vdots & \dots & \vdots & \dots & \vdots \\ e_n^{j,0} & \dots & e_{n-j}^{j,j} & \dots & 0 \\ \vdots & \dots & \vdots & \dots & \vdots \\ e_n^{m,0} & \dots & e_{n-j}^{m,j} & \dots & e_{n-m}^{m,m} \end{pmatrix} \begin{pmatrix} \mu_0 & \dots & 0 \\ \vdots & \ddots & \vdots \\ 0 & \dots & \mu_m \end{pmatrix}$$

$$= D_n(z) G(n) D(\boldsymbol{\mu}), \tag{5.21}$$

where $D_n(z)$, $D(\boldsymbol{\mu})$ are the two diagonal matrices and $G(n) = (e_{n-j}^{\ell,j})_{\ell,j=0}^m$ with $\mathbf{e}_{n-j}^{\ell,j} = 0$ if $\ell < j$ or if $n < \ell$. The nonzero entries of the lower triangular matrix $G(n)$, using (5.20), are

$$\begin{aligned} G_{j+k,j}(n) &= \frac{\binom{n-j}{k}(j+1)_k(2\lambda - m + 2j + k)_{n-j-k}}{\sqrt{(2\lambda - m + 2j)_{n-j}}\sqrt{(1)_{n-j}}} \\ &= \frac{\sqrt{(2\lambda - m + 2j + k)_{n-j-k}}}{\sqrt{(2\lambda - m + 2j)_k}} \frac{(n-j-k+1)_k}{\sqrt{(1)_{n-j}}} \frac{(j+1)_k}{(1)_k} \\ &= \sqrt{\frac{(2\lambda - m + 2j + k)_{n-j-k}}{(2\lambda - m + 2j)_k}} \sqrt{\frac{(n-j-k+1)_k}{(1)_{n-j-k}}} \frac{(j+1)_k}{(1)_k} \end{aligned} \tag{5.22}$$

for $0 \leq k \leq m - j$.

Now, we are ready to compute the reproducing kernel K_j for the Hilbert space $\mathcal{H}^{(j)} = \text{span}\{e_{n-j}^j : n \geq j\}$, $0 \leq j \leq m$. Recall that $K(z, w) = \sum_{n=0}^{\infty} e_n(z) e_n(w)^*$ for any orthonormal basis e_n, $n \geq 0$. This ensures that K is a *positive definite* kernel. For our computations, we will use the particular orthonormal basis $\mathbf{e}_{n-j}^j$ as described in (5.19). Since there are j zeros at the top of each of these basis vectors, it follows that (ℓ, p)

will be 0 if either $\ell < j$ or $p < j$. We will compute $(K_j(z,w))$, at (ℓ,p) for $j \le \ell, p \le m$. For ℓ, p as above, we have

$$(K_j(z,w))_{\ell,p} = \sum_{n \ge \max(\ell,p)}^{\infty} \mathbf{e}^j_{n-j,\ell}(z)\overline{\mathbf{e}^j_{n-j,p}(w)}$$

$$= \sum_{n \ge \max(\ell,p)}^{\infty} G_{\ell,j}(n) G_{p,j}(n) z^{n-\ell}\bar{w}^{n-p}.$$

We first simplify the co-efficient $G_{\ell,j}(n)G_{p,j}(n)$ of $z^{n-\ell}\bar{w}^{n-p}$. The values of $G_{\ell,j}(n)$ are given in (5.22). Therefore, we have

$$G_{\ell,j}(n)G_{p,j}(n)$$

$$= \Big(\frac{(2\lambda_j+\ell-j)_{n-\ell}}{(2\lambda_j)_{\ell-j}}\frac{(n-\ell+1)_{\ell-j}}{(1)_{n-\ell}}\frac{(2\lambda_j+p-j)_{n-p}}{(2\lambda_j)_{p-j}}\frac{(n-\ell+1)_{\ell-j}}{(1)_{n-p}}\Big)^{1/2}$$

$$\times \frac{(j+1)_{\ell-j}}{(1)_{\ell-j}}\frac{(j+1)_{p-j}}{(1)_{p-j}}$$

$$= \frac{(2\lambda_j+p-j)_{n-p}(n-\ell+1)_{\ell-j}}{(2\lambda_j)_{\ell-j}(1)_{n-p}}\Big(\frac{(2\lambda_j+\ell-j)_{p-\ell}(n-p+1)_{p-\ell}}{(2\lambda_j+\ell-j)_{p-\ell}(n-p+1)_{p-\ell}}\Big)^{1/2}$$

$$\times \frac{(j+1)_{\ell-j}}{(1)_{\ell-j}}\frac{(j+1)_{p-j}}{(1)_{p-j}}$$

$$= \frac{(2\lambda_j)_{p-j}(2\lambda_j+p-j)_{n-p}(n-\ell+1)_{\ell-j}(n-p+1)_{p-j}}{(2\lambda_j)_{p-j}(2\lambda_j)_{\ell-j}(1)_{n-p}(n-p+1)_{p-j}}\frac{(j+1)_{\ell-j}}{(1)_{\ell-j}}\frac{(j+1)_{p-j}}{(1)_{p-j}}$$

$$= \frac{(2\lambda_j)_{n-j}(n-\ell+1)_{\ell-j}(n-p+1)_{p-j}}{(2\lambda_j)_{p-j}(2\lambda_j)_{\ell-j}(1)_{n-j}}\frac{(j+1)_{\ell-j}}{(1)_{\ell-j}}\frac{(j+1)_{p-j}}{(1)_{p-j}}.$$

Theorem 5.2. *Given an arbitrary set $\mu_0, \ldots, \mu_m$ of positive numbers, and $2\lambda > m$, we have*

$$K^{(\lambda,\boldsymbol{\mu})}(z,w) = \sum_{j=0}^{m} \mu_j^2 K_j(z,w) = \mathbf{B}^{(\lambda,\boldsymbol{\mu})}(z,w).$$

As a result, the two Hilbert spaces $\mathcal{H}^{(\lambda,\boldsymbol{\mu})}$ and $\mathbf{A}^{(\lambda,\boldsymbol{\mu})}$ of [6] are equal.

Proof. We now compare the co-efficients $(K_j(z,w))_{\ell,p}$ with that of a known Kernel. Let $B^{\lambda_j}(z,w) = (1-z\bar{w})^{-2\lambda_j}$, where $B(z,w) = (1-z\bar{w})^{-2}$ is the Bergman kernel on the unit disc. We let ∂ and $\bar{\partial}$ denote differentiation with respect to z and $\bar{w}$ respectively. Put

$$\tilde{\mathbf{B}}^{(\lambda_j)}(z,w) = (\partial^{\ell-j}\bar{\partial}^{p-j}(1-z\bar{w})^{-2\lambda_j})_{j\le\ell,p\le m}.$$

We expand the entry at the position (ℓ, p) of $\tilde{\mathbf{B}}^{(\lambda_j)}(z, w)$ to see that

$$
\begin{aligned}
&(\tilde{\mathbf{B}}^{(\lambda_j)}(z,w))_{\ell,p} \\
&= \sum_{\nu \geq \max(\ell-j,p-j)} \frac{(2\lambda_j)_\nu}{(1)_\nu}(\nu-\ell+j+1)_{\ell-j}(\nu+j-p+1)_{p-j} z^{\nu-(\ell-j)}\bar{w}^{\nu-(p-j)} \\
&= \sum_{n \geq \max(\ell,p)} \frac{(2\lambda_j)_{n-j}}{(1)_{n-j}}(n-\ell+1)_{\ell-j}(n-p+1)_{p-j} z^{n-\ell}\bar{w}^{n-p},
\end{aligned}
$$

where we have set $n = m + j$. Comparing these coefficients with that of $G_{\ell,j}(n)G_{p,j}(n)$, we find that

$$
K_j(z, w) = D_j \tilde{\mathbf{B}}^{(\lambda_j)}(z, w) D_j, \tag{5.23}
$$

where D_j is a diagonal matrix with $\frac{1}{(2\lambda_j)_{\ell-j}} \frac{(j+1)_{\ell-j}}{(1)_{\ell-j}}$ at the (ℓ, ℓ) position with $j \leq \ell \leq m$. Hence $K_j(z, w) = \mathbf{B}^{(\lambda_j)}(z, w)$ which was defined in the equation ([6, Eq. (4.3)]).

Clearly, we can add up the kernels K_j to obtain the kernel $K^{(\lambda,\boldsymbol{\mu})}$ for the Hilbert space $\mathcal{H}^{(\lambda,\boldsymbol{\mu})} = \oplus_{j=0}^{m} \frac{1}{\mu_j} \mathcal{H}^{(j)}$. Hence the proof of the theorem is complete. □

Corollary 5.1. *The irreducible homogeneous operators in the Cowen-Douglas class whose associated representation is multiplicity free are exactly the adjoints of $M^{(\lambda,\boldsymbol{\mu})}$ constructed in* [6].

Proof. In our discussion up to here we proved that the Hilbert space $\mathcal{H}^{(\lambda,\boldsymbol{\mu})}$ corresponding to a homogeneous operator in the Cowen-Douglas class has a reproducing kernel given by $K^{(\lambda,\boldsymbol{\mu})} = \sum_0^m \mu_j^2 K_j$, $2\lambda > 1$, $\mu_1, \ldots, \mu_m > 0$. It follows from the Theorem that the kernels obtained this way are the same as (are equivalent to) the kernels constructed in [6]. These operators were shown to be irreducible ([6]). □

We now consider the action of the multiplication operator $M^{(\lambda,\boldsymbol{\mu})}$ on the Hilbert space $\mathcal{H}^{(\lambda,\boldsymbol{\mu})}$. Let $\mathcal{H}(n)$ be the linear span of the vectors

$$
\{\mathbf{e}_n^0(z), \ldots, \mathbf{e}_{n-j}^j(z), \ldots, \mathbf{e}_{n-m}^m(z)\},
$$

where as before, for $0 \leq \ell \leq m$, $\mathbf{e}_{n-\ell}^j(z)$ is zero if $n - \ell < 0$. Clearly, $\mathcal{H}^{(\lambda,\boldsymbol{\mu})} = \oplus_{n=0}^{\infty} \mathcal{H}(n)$. We have

$$
\begin{aligned}
zG(\mu, n, z) &= D_n(z)G(n)D(\boldsymbol{\mu}) \\
&= D_{n+1}(z)G(n)D(\boldsymbol{\mu}) \\
&= D_{n+1}(z)G(n+1)D(\boldsymbol{\mu})\big(D(\boldsymbol{\mu})^{-1}G(n+1)^{-1}G(n)D(\boldsymbol{\mu})\big).
\end{aligned}
$$

If we let $W(n) = D(\boldsymbol{\mu})^{-1}G(n+1)^{-1}G(n)D(\boldsymbol{\mu})$, then we see that $z\mathbf{e}^j_{n-j}(z) = G(\boldsymbol{\mu}, n+1, z)W_j(n)$, where $W_j(n)$ is the jth column of the matrix $W(n)$. It follows that the operator $M^{(\lambda,\boldsymbol{\mu})}$ defines a block shift W on the representation space $\mathcal{H}^{(\lambda,\boldsymbol{\mu})}$. The block shift W is defined by the requirement that $W : \mathcal{H}(n) \to \mathcal{H}(n+1)$ and $W_{|\mathcal{H}(n)} = W_n^{\mathrm{tr}}$.

Here, we have a construction of the representation space $\mathcal{H}^{(\lambda,\boldsymbol{\mu})}$ along with the matrix representation of the operator $M^{(\lambda,\boldsymbol{\mu})}$ which is independent of the corresponding results from [6].

5.5. Examples

Recall that $G(\boldsymbol{\mu}, n, z) = D_n(z)G(n)D(\boldsymbol{\mu})$. Once we determine the matrix $G(n)$ explicitly, we can calculate both the block weighted shift and the kernel function.

We discuss these calculations in the particular case of $m = 1$. First, it is easily seen that

$$G(n) = \begin{pmatrix} \left(\frac{(2\lambda-1)_n}{(1)_n}\right)^{1/2} & 0 \\ \left(\frac{n}{2\lambda-1}\right)^{1/2}\left(\frac{(2\lambda)_{n-1}}{(1)_{n-1}}\right)^{1/2} & \left(\frac{(2\lambda+1)_{n-1}}{(1)_{n-1}}\right)^{1/2} \end{pmatrix}. \tag{5.24}$$

The block W_n of the weighted shift W is

$$W_n = \begin{pmatrix} \left(\frac{n+1}{2\lambda+n-1}\right)^{1/2} & 0 \\ -\frac{1}{\mu_1}\left(\frac{2\lambda}{2\lambda-1}\right)^{1/2}\left(\frac{1}{(2\lambda+n-1)(2\lambda+n)}\right)^{1/2} & \left(\frac{n}{2\lambda+n}\right)^{1/2} \end{pmatrix}. \tag{5.25}$$

Finally, the reproducing kernel $K^{(\lambda,\boldsymbol{\mu})}$ with $m = 1$ is easily calculated:

$$K^{(\lambda,\boldsymbol{\mu})}(z,w) = \begin{pmatrix} \frac{1}{(1-\bar{w}z)^{2\lambda-1}} & \frac{z}{(1-\bar{w}z)^{2\lambda}} \\ \frac{\bar{w}}{(1-\bar{w}z)^{2\lambda}} & \frac{1}{2\lambda-1}\frac{1+(2\lambda-1)\bar{w}z}{(1-\bar{w}z)^{2\lambda+1}} \end{pmatrix} + \mu_1^2\begin{pmatrix} 0 & 0 \\ 0 & \frac{1}{(1-\bar{w}z)^{2\lambda+1}} \end{pmatrix}. \tag{5.26}$$

One might continue the explicit calculations, as above, in the particular case of $m = 2$ as well. We begin with the matrix

$$G(n) = \begin{pmatrix} \sqrt{\frac{(2\lambda-2)_n}{(1)_n}} & 0 & 0 \\ \sqrt{\left(\frac{n}{2\lambda-2}\right)\left(\frac{(2\lambda-1)_{n-1}}{(1)_{n-1}}\right)} & \sqrt{\frac{(2\lambda)_{n-1}}{(1)_{n-1}}} & 0 \\ \sqrt{\left(\frac{n(n-1)}{(2\lambda-2)(2\lambda-1)}\right)\left(\frac{(2\lambda)_{n-2}}{(1)_{n-2}}\right)} & 2\sqrt{\left(\frac{n-1}{2\lambda}\right)\left(\frac{(2\lambda+1)_{n-2}}{(1)_{n-2}}\right)} & \sqrt{\frac{(2\lambda+2)_{n-2}}{(1)_{n-2}}} \end{pmatrix}. \tag{5.27}$$

The block W_n of the weighted shift W, in this case, is

$$\begin{pmatrix} \sqrt{\frac{n+1}{2\lambda+n-2}} & 0 & 0 \\ \frac{-1}{\mu_1}\sqrt{\left(\frac{2\lambda-1}{2\lambda-2}\right)\left(\frac{1}{(2\lambda+n-1)(2\lambda+n-2)}\right)} & \sqrt{\frac{n}{2\lambda+n-1}} & 0 \\ \frac{-2}{\mu_2}\sqrt{\left(\frac{2\lambda+1}{(2\lambda-2)_3}\right)\left(\frac{n}{(2\lambda+n-2)_3}\right)} & \frac{-2\mu_1}{\mu_2}\sqrt{\left(\frac{2\lambda+1}{2\lambda}\right)\left(\frac{1}{(2\lambda+n-1)(2\lambda+n)}\right)} & \sqrt{\frac{n-1}{2\lambda+n}} \end{pmatrix}. \tag{5.28}$$

Finally, the reproducing kernel $K^{(\lambda,\boldsymbol{\mu})}$ with $m = 2$ has the form:

$$\begin{aligned} &K^{(\lambda,\boldsymbol{\mu})}(z,w) \\ &= \begin{pmatrix} \frac{1}{(1-\bar{w}z)^{2\lambda-2}} & \frac{z}{(1-\bar{w}z)^{2\lambda-1}} & \frac{z^2}{(1-\bar{w}z)^{2\lambda}} \\ \frac{\bar{w}}{(1-\bar{w}z)^{2\lambda-1}} & \frac{1+(2\lambda-2)\bar{w}z}{(2\lambda-2)(1-\bar{w}z)^{2\lambda}} & \frac{z(2+(2\lambda-2)\bar{w}z)}{(2\lambda-2)(1-\bar{w}z)^{2\lambda+1}} \\ \frac{\bar{w}^2}{(1-\bar{w}z)^{2\lambda}} & \frac{\bar{w}(2+(2\lambda-2)\bar{w}z)}{(2\lambda-2)(1-\bar{w}z)^{2\lambda+1}} & \frac{2+4(2\lambda-1)\bar{w}z+(2\lambda-1)(2\lambda-2)z^2\bar{w}^2}{(2\lambda-1)(2\lambda-2)(1-\bar{w}z)^{2\lambda+2}} \end{pmatrix} \\ &\quad + \mu_1^2 \begin{pmatrix} 0 & 0 & 0 \\ 0 & \frac{1}{(1-\bar{w}z)^{2\lambda}} & 2\frac{z}{(1-\bar{w}z)^{2\lambda+1}} \\ 0 & 2\frac{\bar{w}}{(1-\bar{w}z)^{2\lambda+1}} & 2\frac{2}{2\lambda}\frac{1+2\lambda\bar{w}z}{(1-\bar{w}z)^{2\lambda+2}} \end{pmatrix} \\ &\quad + \mu_2^2 \begin{pmatrix} 0 & 0 & 0 \\ 0 & 0 & 0 \\ 0 & 0 & \frac{1}{(1-\bar{w}z)^{2\lambda+2}} \end{pmatrix}. \end{aligned} \tag{5.29}$$

References

[1] Aronszajn, N. (1950). Theory of reproducing kernels, *Trans. Amer.Math. Soc.* **68** 337-404.

[2] Bagchi, B. and Misra, G. (1996). Homogeneous tuples of multiplication operators on twisted Bergman spaces. *J. Funct. Anal.* **136** 171-213.

[3] Biswas, I. and Misra, G. (2008). $\widetilde{SL}(2,\mathbb{R})$-homogeneous vector bundles. *Int. J. Math.* **19** 1-19.

[4] Cowen, M. J. and Douglas, R. G. (1978). Complex geometry and operator theory. *Acta Math.* **141** 187-261.

[5] Kirilov, A. (1976). *Elements of the theory of representations.* Springer-Verlag.

[6] Korányi, A. and Misra, G. (2008). Homogeneous operators on Hilbert spaces of holomorphic functions. *J. Func. Anal.* **254** 2419-2436.

[7] Sugiura, M. (1975). *Unitary representations and harmonic analysis. An introduction.* Kodansha Ltd., Tokyo; Halstead Press [John Wiley & Sons], New York-London-Sydney.

Chapter 6

The Standard Conjectures on Algebraic Cycles

M. S. Narasimhan

Tata Institute of Fundamental Research
and Indian Institute of Science, Bangalore, India
narasim@math.tifrbng.res.in

In his talk at the International Colloquium on Algebraic Geometry held in January 1968 at the Tata Institute of Fundamental Research, Grothendieck formulated two "standard conjectures" on algebraic cycles, which arose "from an attempt at understanding the conjectures of Weil on the ζ functions of algebraic varieties" ([4]). At the end of his talk he says "alongside the problem of resolution of singularities, the proof of the standard conjectures seems to be the most urgent task in algebraic geometry". The conjectures also form the basis of his theory of "motives". (See [5], also [1].) Unfortunately, the conjectures are not yet proved.

These conjectures are of two types:

1) The standard conjecture of Lefschetz type.
2) The standard conjecture of Hodge type (not to be confused with the Hodge conjecture on algebraic cycles, which will be mentioned later).

We will state the conjectures first in the case of complex non-singular projective varieties and later in the case of abstract algebraic varieties. (The standard conjecture of Hodge type is known to be true in the complex case, from the work of Hodge. The conjecture of Lefschetz type is still unknown even in the complex case).

6.1. The Case of Complex Projective Varieties

Let X be a compact connected complex manifold of (complex) dimension n, embedded in a complex projective space $\mathbb{P}^m(\mathbb{C})$. An algebraic cycle (of codimension i) in X is a finite formal linear combination $Z = \sum a_j Z_j$ where

Z_j are irreducible analytic (algebraic) subvarieties of X of codimension i and a_j are rational numbers (or integers, depending on the context). Such a cycle defines an element of the singular cohomology group $H^{2i}(X, \mathbb{Q})$, by associating to an irreducible algebraic variety Z its Poincaré dual $[Z]$, and extending this to cycles in the obvious way.

To define $[Z]$ one associates to Z its fundamental homology class in $H_{2(n-i)}(Z, \mathbb{Z})$ either by using a triangulation or as in Theorem 3.2 and Proposition in 3.3 of [2]; the inclusion of Z in X yields an element of $H_{2(n-i)}(X, \mathbb{Z})$ and $[Z]$ is defined to be the image of this element under the Poincaré duality isomorphism $H_{2(n-i)}(X, \mathbb{Z}) \to H^{2i}(X, \mathbb{Z})$.

(Another way to define $[Z]$ is to use the theorem of Hironaka on resolution of singularities. First, for Z nonsingular, define $[Z]$ to be $i_*(1)$ where i_* is the Gysin homomorphism defined by the inclusion $i : Z \to X$. If Z is singular, let $f : \tilde{Z} \to Z$ be a resolution of singularities, with $\tilde{Z}$ nonsigular and projective and f an isomorphism on the smooth locus of Z. Define $[Z]$ to be $(i \circ f)_*[1]$; the class thus obtained is independent of the resolution.)

The real cohomology class defined by $[Z]$ is the class of the (closed) current obtained by appropriately integrating forms of degree $2(n - i)$ on Z ([2], Proposition in §3.4).

We denote by A^i the $\mathbb{Q}$-subspace of elements of $H^{2i}(X, \mathbb{Q})$ represented by algebraic cycles, and we call A^i the space of algebraic cohomology classes.

Let us denote by H^p the space $H^p(X, \mathbb{Q})$. Let $\xi \in H^2$ be the cohomology class defined by a hyperplane section, and $L : H^p \to H^{p+2}$ the operator $L(u) = \xi \cup u$ for $u \in H^p$, where $\cup$ denotes the cup product in cohomology. We then have the Hard Lefschetz theorem ([10], Corollaire on p. 75): for $0 \leq p \leq n$,

$$L^{n-p} : H^p \to H^{2n-p}$$

is an isomorphism. We now state

The standard conjecture of Lefschetz type. *For $0 \leq 2i \leq n$,*

$$L^{n-2i} : H^{2i} \to H^{2n-2i}$$

maps A^i **onto** *A^{n-i}.* (Note that L^{n-2i} maps A^i into A^{n-i}.)

Remark: The Hodge conjecture is the following: *let $u \in H^{2i}(X, \mathbb{Q})$; if u considered as an element of $H^{2i}(X, \mathbb{C})$ is represented under the de Rham isomorphism by a differential form of type (i, i), then u is an algebraic class.* Let $H^{i,i} \subset H^{2i}(X, \mathbb{C})$ denote the space represented by forms of type (i, i).

By the Hard Lefschetz Theorem, L^{n-2i} maps $H^{2i}(X,\mathbb{Q}) \cap H^{i,i}$ isomorphically onto $H^{2n-2i}(\mathbb{Q}) \cap H^{n-i,n-i}$. So we see that the Hodge conjecture on algebraic cycles implies standard conjecture of Lefschetz type.

An equivalent form of the standard conjecture of Lefschetz type (for all smooth projective varieties) is: *the operator Λ of Hodge theory is algebraic* ([8] Th. 4-1, p. 14).

For the definition of Λ see §4 of [8] or p. 76, [9]. Using the Künneth formula and Poincaré duality, the operator $\Lambda : H^*(X,\mathbb{Q}) \to H^*(X,\mathbb{Q})$ can be considered as an element of $H^*(X \times X,\mathbb{Q})$, and to say that Λ is algebraic means that this element is an algebraic cohomology class in $X \times X$. Grothendieck says that the conjecture seems to be most amenable in this form.

Let X be as above. The kernel of the map

$$L^{n-p+1} : H^p \to H^{2n-p+2}$$

is called the space of *primitive forms* and is denoted by P^p $(0 \le p \le n)$.

Hodge Index theorem. Let X be a complex projective manifold. Then on $P^{2i} \cap A^i$ (the space of primitive algebraic classes) the $\mathbb{Q}$-valued symmetric form

$$(x,y) = (-1)^i \int_X x \cup y \cup \xi^{n-2i}$$

is positive definite ([10], p. 77).

The "abstract" analogue of the above result is the Standard Conjecture of Hodge type, described in the next section.

Homological and numerical equivalence of algebraic cycles. Two cycles Z_1 and Z_2 of codimension i in X are said to be numerically equivalent if the intersection numbers $(Z_1 \cdot Z)$ and $(Z_2 \cdot Z)$ are the same for all irreducible subvarieties of codimension $(n-i)$. They are said to be homologically equivalent if they have the same image in $H^{2i}(X,\mathbb{Q})$. It has been a long standing conjecture that *numerical and homological equivalence of algebraic cycles (with $\mathbb{Q}$-coefficients) coincide.* This result can be seen to be a consequence of the standard conjecture of Lefschetz type as follows. Since the cycle map, which associates to a cycle its cohomology class, takes intersection product into cup product, it is enough to show that the canonical pairing

$$A^i \otimes_{\mathbb{Q}} A^{n-i} \to \mathbb{Q}$$

is non-degenerate. To prove this we define an operator

$$* : H^p \to H^{2n-p}$$

which is an isomorphism and which maps algebraic classes to algebraic classes such that $(a, b) \to \int a \cup (*b)$ is positive definite on A^i. If $a \in H^i$ and

$$a = \sum L^j a_j, \quad a_j \in P^{i-2j}$$

is the primitive decomposition (cf. [5], Th. 5 on p. 75) then $*a$ is defined by

$$*a = \sum (-1)^{\frac{(i-2j)(i-2j+1)}{2}} L^{n-i+j} a_j.$$

That $*$ maps algebraic classes to algebraic classes follows from the standard conjecture of Lefschetz type and positive definiteness follows from Hodge index theorem.

6.2. Standard Conjectures in Abstract Algebraic Geometry

Let X be a smooth irreducible projective algebraic variety of dimension n over an algebraically closed field k, perhaps of positive characteristic p. Let ℓ be a prime number $\ell \neq p$. We denote by H^i the ℓ-adic étale cohomology group $H^i(X, \mathbb{Q}_\ell)$ with coefficients in the field $\mathbb{Q}_\ell$ of ℓ-adic numbers, which has been defined by Grothendieck ([6]). Let μ_{l^m} denote the group of roots of unity of order l^m in k, and set $\mu_{l^\infty} = \bigcup_{m \geq 0} \mu_{l^m}$. For simplicity, we choose an isomorphism between μ_{l^∞} and $\mathbb{Q}_\ell/\mathbb{Z}_\ell$. (Making such a choice is called "a heresy" by Grothendieck!) We then have a cycle map

$$Z^i(X) \otimes_{\mathbb{Z}} \mathbb{Q} \to H^{2i}$$

where $Z^i(X)$ denotes the group of i-codimensional (integral) algebraic cycles ([6], pp. 23-24). We denote by A^i the image and an element of A^i will be referred to as an algebraic cohomology class. Let $\xi \in H^2(X)$ be the class of a hyperplane section. From the work of Deligne it follows that for $m \leq n$

$$\cup \xi^{n-m} : H^m \to H^{2n-m}$$

is an isomorphism (Hard Lefschetz, [3], Theorem 5.5). We can now state

The standard conjecture of Lefschetz type. *For $2i \leq n$,*

$$\cup \xi^{n-2i} : H^{2i} \to H^{2n-2i}$$

maps A^i isomorphically onto A^{n-i}.

For $m \leq n$ we denote by P^m the primitive part of H^m, defined as the kernel of the map

$$\cup \xi^{n-m+1} : H^m \to H^{2n-m+2}.$$

The standard conjecture of Hodge type. *On $P^{2i} \cap A^i$ the $\mathbb{Q}$-valued symmetric form*

$$(a, b) = (-1)^i K(a \cup b \cup \xi^{n-2i})$$

is positive definite, where K denotes the isomorphism $H^{2n} \to \mathbb{Q}_\ell$.

Remark: If both these conjectures are true then numerical and homological equivalence of algebraic cycles would coincide, as indicted above in the case of complex projective varieties.

Remark: A priori it is not clear that A^i are finite dimensional vector spaces over $\mathbb{Q}$. This would be the case if numerical and homological equivalence coincide, since the space of algebraic cycles modulo numerical equivalence is a finite dimensional vector space ([1], 3.5 and [8], Lemma 5.2).

Remark: As in the complex case, the standard conjecture of Lefschetz type (for all varieties X) is equivalent to: *the operator Λ is algebraic.*

Weil conjectures. As mentioned earlier, a motivation for standard conjectures was to understand Weil conjectures (on rational points of varieties defined over a finite field).

Let X be an irreducible smooth projective variety of dimension n, defined by polynomials with coefficients in a finite field $\mathbb{F}_q$. For each $m \geq 1$, let ν_m denote the number of points with coordinates in the extension field $\mathbb{F}_{q^m}$. Define the function $Z(t)$ by:

$$\log Z(t) = \sum_{m \geq 1} \nu_m t^m / m.$$

The function $Z(t)$ is called the zeta function of X (The function $Z(q^{-s})$, denoted by $\zeta(s)$, is also sometimes considered.)

Among the conjectures of A. Weil on the zeta function are the following.

1) $Z(t)$ is a rational function.
2) $Z(t)$ is of the form

$$Z(t) = \frac{P_1(t) \cdots P_{2n-1}(t)}{P_0(t) \cdots P_{2n}(t)}$$

where $P_i(t)$ a polynomial with integer coefficients with $P_0(t) = 1 - t$ and $P_{2n}(t) = (1 - q^n t)$. The absolute value of the roots of $P_i(t)$ is $q^{-i/2}$; i.e., $P_i(t)$ is of the form $\prod_j (1 - \omega_{ij} t)$ with $| \omega_{ij} |= q^{i/2}$: (Riemann hypothesis).

Remark: Moreover, if X is a non-singular algebraic variety defined over a number field K, then for almost all prime ideals $\mathfrak{p}$ of K, the degree of the polynomial P_i associated with $X_\mathfrak{p}$ (the reduction of X mod $\mathfrak{p}$) coincides with the (usual) ith Betti number of the complex variety X. This gives a method of computing Betti numbers of complex projective varieties by computing rational points on varieties defined over finite fields.

We will now indicate how the Riemann hypothesis (for varieties over a finite field), which has been proved by Deligne ([3]), is related to standard conjectures. Arguments due to Weil and Serre ([4]) yield (compare [9], Theorem 5.6 of [8], and [1], 5.4.3, p. 58):

Proposition 6.1. *Assume that the standard conjecture of Lefschetz type holds for X and the standard conjecture of Hodge type holds for the product $X \times X$. (X smooth projective). Let $f : X \to X$ be a morphism such that $f^*(\xi) = q\xi$ where ξ is the cohomology class of a hyperplane section and $q \in \mathbb{Q}, q > 0$. Then the eigenvalues of the endomorphism $f|_{H^i}$ induced on the ith ℓ-adic cohomology group $H^i(X, \mathbb{Q}_\ell)$ are algebraic integers of absolute value $q^{i/2}$.*

Now let X be defined over a finite field $\mathbb{F}_q$ and let $f : X \to X$ be the Frobenius morphism

$$(x_1, \cdots x_N) \to (x_1^q, \cdots, x_N^q)$$

raising the coordinates to q^{th} powers. Note that the number of rational points in $\mathbb{F}_{q^m}$ is the number of points left fixed by the mth iterate f^m, and the latter can be computed in terms of the trace of the action of f^m on H^i by using the Lefschetz fixed points theorem (extended to ℓ-adic étale cohomology). One then finds that

$$P_i(t) = det\ (1 - tf)|_{H^i}.$$

On applying the previous proposition to f, we see that the absolute value of an eigenvalue of $f|_{H^i}$ is $q^{i/2}$ (if we assume that the standard conjectures have affirmative answers).

References

[1] André, Y. (2004). Une introduction aux motives, Panorama et Synthésis. *Société Mathématiques de France.*

[2] Borel, A. and Haefliger, A. (1961). La classe d'homologie fondamentale d'un éspace analytique. *Bull. Soc. Math. France.* **89** 461-513 (Also no. 56 in the collected papers of A. Borel.)

[3] Freitag, E. and Kiehl, R. (1988). *'Etale cohomology and the Weil conjecture.* Springer.

[4] Grothendieck, A. (1969). Standard conjectures on algebraic cycles, Algebraic Geometry. Bombay Colloquium 1968, Oxford, 193-199.

[5] Jannsen, U. (1994). Motivic sheaves and filtrations on Chow groups. *Motives, Symp. in Pure Math.* **55** A.M.S. 245-302.

[6] Katz, N. M. (1994). Review of l-adic cohomology. In *Motives, Symp. in Pure Math.* **55** Part I 21-30.

[7] Kleiman, S. L. (1968). Algebraic cycles and Weil conjectures. *Dix exposes su la cohomologie des Schémas.* North Holland, Amsterdam. 359-386.

[8] Kleiman, S. L. (1994). The standard conjectures. In *Motives, Symp. in Pure Math.* **55** A.M.S. 3-30.

[9] Serre, J. P. (1960). Analogues kahleriens de certaines conjectures de Weil. *Ann of Math.* **71** 392-394.

[10] Weil, A. (1968). *Variétés kähleriennes.* Hermann, Paris.

Chapter 7

On the Classification of Binary Shifts on the Hyperfinite II_1 Factor

Geoffrey L. Price

Department of Mathematics,
9E, United States Naval Academy,
*Annapolis, MD 21402, USA**
glp@usna.edu

We provide a complete classification up to conjugacy of the binary shifts of finite commutant index on the hyperfinite II_1 factor. There is a natural correspondence between the conjugacy classes of these shifts and polynomials over $GF(2)$ satisfying a certain duality condition.

7.1. Introduction

Let R denote the hyperfinite II_1 factor. A pair of *-automorphisms σ, ρ on R are said to be conjugate if there exists a *-automorphism γ on R which satisfies $\gamma \circ \sigma(A) = \rho \circ \gamma(A)$ for all A in R. The notion of conjugacy carries over to the setting of unital *-endomorphisms on R. In this situation it turns out that the Jones index $[R : \sigma(R)]$ of the subfactor $\sigma(R)$ in R is a numerical conjugacy invariant, as is the commutant (or relative commutant) index: this is the first positive integer k (or ∞) for which the relative commutant algebra $\sigma^k(R)' \cap R$ is nontrivial.

In [12] R. T. Powers introduced a family of unital *-endomorphisms on R known as binary shifts. The range $\sigma(R)$ of each binary shift σ is a subfactor of index 2. As a result the minimal possible commutant index for a binary shift is 2 (cf. [7]). Powers has shown [12] that there exist binary shifts of any specified commutant index $k \in \{\infty, 2, 3, \ldots\}$. In particular he showed that for any finite commutant index there are at most countably many conjugacy classes, but that there are uncountably many conjugacy classes having infinite commutant index (see Theorem 7.4).

*Supported in part by a research grant from the National Science Foundation.

In [15] we gave a complete classification of the conjugacy classes of binary shifts of commutant index 2. We showed that there is a natural correspondence between the conjugacy classes of these shifts and polynomials with coefficients in $GF(2)$ which satisfy a certain duality condition (see Theorem 7.6). These are the polynomials which have constant coefficient 1 and which have no self-reciprocal factors (see Definition 7.2) of degree greater than 1. In this paper we extend the results of [15] by establishing a correspondence between binary shifts of finite commutant index $k \geq 2$ and polynomials over $GF(2)$ with constant coefficient 1 which have no self-reciprocal factors of degree exceeding $2k-3$. Unlike the case for $k=2$, this correspondence for higher commutant indices is not one-to-one, but we produce a recursion formula (4.1) which relates the number of conjugacy classes of binary shifts of commutant index k which are associated with each of the polynomials of the form described above. As a consequence we can provide precise information, for example, on the number of binary shifts of fixed finite commutant index which are associated with any irreducible polynomial $p(x)$ over $GF(2)$.

There is another notion of equivalence among unital $*$-endomorphisms on R, in addition to conjugacy, which is known as cocycle conjugacy. A pair α and β of unital $*$-endomorphisms are said to be cocycle conjugate if there exists a unitary $u \in R$ for which $\alpha \circ Ad(u)$ and β are conjugate. In [20] it was shown that all binary shifts of commutant index 2 are cocycle conjugate and partial results were obtained in [21] for binary shifts of higher finite commutant index.

Since their introduction binary shifts have been used by several authors to illustrate various phenomena. In [10] the authors constructed a binary shift σ on the hyperfinite II_1 factor R such that σ has a unique invariant state ϕ and for which the Connes-Størmer entropy (cf. [4]) $h_\phi(\sigma)$ is 0. On the other hand $h_{\phi\otimes\phi}(\sigma \otimes \sigma) = \log(2)$, which shows that the additive tensor product formula for the Connes-Størmer entropy fails in general (see also [6] and [9]). In [18] it was shown that if σ is a binary shift with finite commutant index then its Connes-Størmer entropy $h_\tau(\sigma)$ is $\frac{1}{2}\log(2)$ where τ is the unique tracial state on R.

It is a pleasure to thank Alexis Alevras and Robert T. Powers for helpful conversations. We are also grateful to Professor Powers for writing a very enlightening computer program related to the classification of binary shifts.

This paper appeared in an abbreviated form in the Proceedings of the National Academy of Sciences, see [19].

7.2. Preliminaries

In this section we present Powers' construction of the binary shifts on R. We also state some of the results which are known about binary shifts and which are relevant to the classification of their conjugacy classes. In particular we shall exploit the close connections that have been made between the theory of recurring linear sequences and binary shifts. See [8], Chapter 6, for an extensive bibliography on the subject of recurring linear sequences.

A *spin system* is a sequence of self-adjoint unitary operators $\{u_i : i \in \mathbb{Z}^+\}$ that commute pairwise up to a phase in the sense that

$$u_i u_j = \lambda_{ij} u_j u_i$$

for complex numbers λ_{ij}, where $i, j \in \mathbb{Z}^+$. Since the u_i are self-adjoint, $u_i^2 = I$ and so it follows that $\lambda ij = \lambda_{ji} = \pm 1$ for all i and j.

Now given a subset X (finite or infinite) subset of $\mathbb{N}$, and its characteristic function $g : \mathbb{N} \to \{0, 1\}$, we can use g to define a spin system as follows. Let $\{u_0, u_1, \dots\}$ be a sequence of hermitian unitary operators, or generators, which satisfy the commutation relations

$$u_i u_{i+j} = (-1)^{g(j)} u_{i+j} u_i, i, j \in \mathbb{Z}^+. \tag{1.1}$$

Note that the commutation relations in (1.1) are translation invariant, i.e., the equations above are independent of the choice of subscript i.

One may define words in the generators by setting, for finite ordered subsets $J = \{j_0, j_1, \dots, j_m\}$ of distinct nonnegative integers, $u(J) = u_{j_0} u_{j_1} \cdots u_{j_m}$, and $u(\emptyset) = I$, the identity. In fact, since the $u_j's$ are hermitian and satisfy (1.1), any product of the generators may be rewritten as either $+u(J)$ or $-u(J)$ for some finite ordered subset $J \subset \mathbb{Z}^+$. For $n \in \mathbb{N}$, let $\mathfrak{A}_n$ be the finite-dimensional group algebra over $\mathbb{C}$ consisting of linear combinations of the words in the generators $\{u_0, u_1, \dots, u_{n-1}\}$. Note that $\mathfrak{A}_n$ has dimension 2^n, the number of words in the generators. Since $\mathfrak{A}_n \subset \mathfrak{A}_{n+1}$ for all $n \in \mathbb{N}$, one may obtain an $AF-$algebra by taking the uniform closure of the union $\cup_{n=1}^{\infty} \mathfrak{A}_n$. Following the terminology of [12], Definition 3.2, we refer to $\mathfrak{A}$ as the binary shift algebra associated with X. The binary shift itself is the unital *-homomorphism on $\mathfrak{A}$ defined uniquely by the mappings $\sigma(u_j) = u_{j+1}$ on the generators. We shall refer to X as the *anticommutation set*, to $\mathbf{a} = \{a_0, a_1, \dots\}$, where $a_j = g(j)$, as the *bitstream*, and to $\check{\mathbf{a}} = \{\dots, a_2, a_1, a_0, a_1, a_2, \dots\}$ as the *reflected bitstream* associated with σ.

If the reflected bitstream is periodic it is easy to see that $\mathfrak{A}$ has a nontrivial center. In fact, if $\breve{a}$ has period length p the word $u_0 u_p$ (as well as its shifts) lies in the center. On the other hand, $\mathfrak{A}$ has trivial center if $\breve{a}$ is not periodic.

Theorem 7.1. *(cf. [12], Theorem 3.9, [16], Theorem 2.3, [13], Corollary 5.5) Let $\mathfrak{A}$ be the binary shift algebra with anticommutation set X and corresponding bitstream given by $a_j = g(j), j \in \mathbb{Z}^+$. Then the folloing conditions are equivalent.*

(i) *$\breve{a}$ is not periodic.*
(ii) *The center of $\mathfrak{A}$ consists of scalar multiples of the identity.*
(iii) *For any nontrivial word w, $wu_j = -u_j w$ for some u_j.*
(iv) *$\mathfrak{A}$ has a unique normalized trace.*
(v) *$\mathfrak{A}$ is isomorphic to the UHF algebra of type 2^∞.*

Hence if any of the above conditions hold then with respect to the unique trace on $\mathfrak{A}$, the weak operator closure of $\mathfrak{A}$ is isomorphic to the hyperfinite II_1 algebra R.

(For the structure of the algebras generated by more general spin systems see [1].)

Assuming any of the conditions of the theorem hold the mapping $\sigma : u_i \to u_{i+1}$ for all $i \in \mathbb{Z}^+$ uniquely defines a binary shift on R.

Theorem 7.2. *(cf. [12], Theorem 3.6) A pair of binary shifts on R are conjugate if and only if they are defined via the same bitstream.*

Observe that the trace τ satisfies $\tau(u(J)) = 0$ for any nontrivial word $w = u(J)$: for by condition (iii) there is for any word w a generator u_j such that $\tau(w) = \tau(u_j{}^2 w) = \tau(u_j w u_j) = -\tau(w)$.

It is straightforward to see that the binary shift σ on $\mathfrak{A}$ extends to a unital *-homomorphism on R which we shall also denote by σ. Then we have the following result for binary shifts on R.

Theorem 7.3. *(cf. [12], [7], Example 2.3.2) For any binary shift on R the subfactor $\sigma(R)$ has Jones index $[R : \sigma(R)] = 2$.*

As a consequence of the previous theorem the minimal commutant index for a binary shift is 2, [7], Corollary 2.2.4. On the other hand, if $X = \{k-1\}$, then $u_0 \in \sigma^k(R)' \cap R, \sigma^{k-1}(R)' \cap R = \mathbb{C}I$, so the associated binary shift has commutant index k. Hence there exist examples of binary shifts for any finite commutant index $k \geq 2$. The following theorem characterizes those binary shifts with finite commutant index.

Theorem 7.4. *(cf. [2], Theorem 2.1, [17], Theorem 5.8) A binary shift on R has finite commutant index if and only if its bitstream* **a** *is eventually periodic. For a binary shift σ of finite commutant index k and generators $u_j, j \in \mathbb{Z}^+$, there is a word $w = u(j_0, j_1, \ldots, j_n)$ such that*

(i) *$j_0 = 0$, i.e., w "starts" with u_0, and*
(ii) *for any $s \geq 0$, $\sigma^{k+s}(R)' \cap R$ is a 2^s-dimensional algebra generated by the words $w, \sigma(w), \ldots, \sigma^r(w)$.*

Definition 7.1. Let σ be a binary shift on R with generators $u_j, j \in \mathbb{Z}^+$. Let z be a word in the generators, and let $p(x) = c_0 + c_1x + \cdots + c_nx^n$ be a polynomial with coefficients in $GF(2)$, then $< z, p >$ is the word $z^{c_0}\sigma(z)^{c_1} \cdots \sigma^n(z)^{c_n}$ in R.

Suppose $p(x)$ is the polynomial for which $w =< u_0, p >$ in the preceding theorem. In the study of binary shifts of commutant index 2 in [15] we described connections among polynomials $p(x)$ words $w =< u_0, p >$ generating relative commutant algebras, and conjugacy classes of shifts. Theorem 7.6 describes this connection. In order to state the theorem we need to identify polynomials which possess a special symmetry.

Definition 7.2. A polynomial $p(x)$ with constant term 1 is called *reciprocal* or *self-reciprocal* if its coefficients are flip-symmetric (see [3], [8]), i.e., $p(x) = c_0 + c_1x + \cdots + c_nx^n = c_n + c_{n-1}x + \cdots + c_0x^n$.

Remark 7.1. Note that if a polynomial $p(x) = c_0 + c_1x + \cdots + c_nx^n$ with constant coefficient 1 then $p^*(x) = x^n p(\frac{1}{x}) = c_n + c_{n-1}x + \cdots + c_0x^n$. Hence $p(x)$ is reciprocal if and only if $p(x) = p^*(x)$.

The following results about reciprocal polynomials will be used in the next section.

Lemma 7.1. *If $h(x)$ is a polynomial with constant coefficient 1 then $h^*(x)h(x)$ is reciprocal. The product of reciprocal polynomials is reciprocal.*

Proof. Obvious. □

Theorem 7.5. *(see [15], Theorem 4.3) Any polynomial $p(x)$ has a unique reciprocal divisor of maximal degree.*

Theorem 7.6. *Let $p(x)$ be a polynomial over $GF(2)$ with constant coefficient 1. Then there exists a binary shift of commutant 2 and generators $u_j, j \in \mathbb{Z}^+$ such that $< u_0, p >$ generates $\sigma^2(R)' \cap R$ if and only if $p(x)$ has no reciprocal factors of degree exceeding 1. Moreover there is a one-to-one correspondence between such polynomials and the family of binary shifts of commutant index 2.*

It is possible to show that among all polynomials of degree $n \geq 3$ and constant coefficient 1 there are 2^{n-2} which satisfy the hypotheses of the theorem, (see [15], Theorem 4.4). Hence there are countably many conjugacy classes of binary shifts of commutant index 2. Corollary 7.7 establishes the same conclusion for any finite commutant index.

Although we are interested in analyzing binary shifts on R it is useful for that purpose to understand the structure of the $AF-$algebras $[\cup_{n=1}^{\infty} \mathfrak{A}_n]^-$ which have nontrivial centers (see the proof of Theorem 7.10).

Theorem 7.7. *Let $\mathbf{a}$ be a bitstream for which $\breve{\mathbf{a}}$ is periodic. Let $\mathfrak{A}$ be the corresponding $AF-$ algebra, and let τ be the trace on $\mathfrak{A}$ which vanishes on nontrivial words. Let M be the von Neumann algebra obtained by completing $\mathfrak{A}$ in the weak operator closure with respect to τ. Then there exists a word $w =< u_0, p >= u_0^{c_0} u_1^{c_1} \cdots u_n^{c_n}$ such that*

(i) $c_0 = 1$,
(ii) *$p(x)$ is reciprocal, and*
(iii) *the center of M is generated by w and its shifts.*

Hence the center of M is isomorphic to the algebra of continuous functions on the Cantor set.

Suppose $\mathbf{a}$ is a bitstream in $GF(2)$ with $a_0 = 0$. Then for each $n \in \mathbb{N}$ one may construct an $n \times n$ Toeplitz matrix A_n whose first row consists of the first n elements of $\mathbf{a}$, viz.,

$$A_n = \begin{bmatrix} a_0 & a_1 & a_2 & a_3 & \dots & a_{n-1} \\ a_1 & a_0 & a_1 & a_2 & \dots & a_{n-2} \\ a_2 & a_1 & a_0 & a_1 & \dots & a_{n-3} \\ \vdots & \vdots & \vdots & \vdots & \ddots & \vdots \\ a_{n-1} & a_{n-2} & a_{n-3} & \dots & \dots & a_0 \end{bmatrix} \tag{1.2}$$

Since A_n is a skew-symmetric matrix it has even rank (this holds true even for matrices over $GF(2)$, (see [11], Theorem IV.11). Considered as a sequence of matrices, the Toeplitz matrices associated with a bitstream exhibit a remarkable property.

Theorem 7.8. *(cf. [13], Theorem 5.4) Let $\nu(A_n)$ denote the nullity of the $n \times n$ Toeplitz matrix above. If $\breve{\mathbf{a}}$ is not periodic the sequence $\{\nu(A_n) : n \in \mathbb{N}\}$ consists of the concatenation of strings of non-negative integers of the form $1, 2, \ldots, m-1, m, m-1, \ldots, 2, 1, 0$. If $\breve{\mathbf{a}}$ is periodic then the nullity sequence consists of finitely many strings of the above mentioned form followed by the sequence $1, 2, \ldots$.*

Theorem 7.9. *(cf. [5], Corollary 2.10, see also [22]) For any even positive integer n there are 2^{n-2} $n \times n$ invertible Toeplitz matrices of the form above.*

Finally it will be helpful to use the following properties of the operations $< z, p >$ for a word z in the generators of a binary shift on R and for polynomials $p(x)$ with coefficients in $GF(2)$ (see [16], Section 4):

$$< z, p >< z, q > = \pm < w, p + q > \tag{1.3.1}$$

$$<< w, p >, q > = \pm < w, pq > \tag{1.3.2}$$

7.3. Bitstreams and Polynomials

In this section we prove some elementary results about bitstreams over finite fields which are based on some well-known results from the theory of linearly recurring sequences, (see [8], Chapter 6). Our results stem from important connections which exist between eventually periodic bitstreams with entries in a finite field and polynomials with coefficients in the same field. Here we deal exclusively with the finite field $GF(2)$. We shall say that a polynomial $p(x) = c_0 + c_1 x + \cdots + c_n x^n$ *annihilates* a bitstream $\mathbf{a}$ if for any $j \in \mathbb{Z}^+$, $\sum_{l=0}^{n} c_l a_{j+l} = 0$. If a bitstream is eventually periodic, i.e., if there exists a positive integer s such that $a_k = a_{k+s}$ for all $k \geq N$, some N, then the polynomial $x^N + x^{N+s}$ annihilates $\mathbf{a}$. In particular one has the following result.

Lemma 7.2. *(cf. [8], Theorem 6.11) A bitstream $\mathbf{a}$ over $GF(2)$ is eventually periodic if and only if it is annihilated by some polynomial $p(x)$ with coefficients in $GF(2)$. $\mathbf{a}$ is periodic if and only if it is annihilated by a polynomial with constant coefficient 1.*

If $\mathbf{a}$ and $\mathbf{b}$ are bitstreams then define addition by $\mathbf{a} + \mathbf{b} = \mathbf{c}$ where $c_j = a_j + b_j$ for all $j \in \mathbb{Z}^+$. Defining scalar multiplication in the obvious way, one sees that the set of periodic (respectively, eventually periodic bitstreams) forms a vector space over $GF(2)$: for if $\mathbf{a}$ is periodic (respectively, eventually periodic) with period s and $\mathbf{b}$ is periodic (respectively, eventually periodic) with period t, then $\mathbf{c}$ is periodic (resp., eventually periodic) with period a divisor of st, (see [8]). The same is true for doubly-infinite periodic sequences, and also for the subspace which consists of the reflected bitstreams $\breve{\mathbf{a}} = \{\ldots, a_2, a_1, a_0, a_1, a_2, \ldots\}, a_0 = 0$, which happen to be periodic. Thus we have:

Lemma 7.3. *The set of periodic reflected bitstreams forms a vector space over $GF(2)$.*

Definition 7.3. (cf. [8], Section 6.5) For a polynomial $p(x)$ with coefficients in $GF(2)$ let $S(p)$ be the vector space of periodic reflected bitstreams annihilated by $p(x)$.

Proposition 7.1. *Suppose $\breve{\mathbf{a}}$ is a periodic reflected bitstream in $S(p)$. If $r(x)$ is the maximal reciprocal factor of $p(x)$ then $\breve{\mathbf{a}} \in S(r)$.*

Proof. Since the reflected bitstream $\breve{\mathbf{a}}$ is symmetric about its entry $a_0 = 0$, it is clear that $p(x)$ annihilates $\breve{\mathbf{a}}$ if and only if $p^*(x)$ does, too (see Remark 7.1). Hence $\breve{\mathbf{a}} \in S(p) \cap S(p^*)$. But for any pair p, q of polynomials, $S(p) \cap S(q) = S(\gcd(p, q))$, by [8], Theorem 6.54. Clearly r divides the polynomial $d = \gcd(p, p^*)$. Suppose $p(x)/r(x) = \prod h_i(x)$, where the h_i's are irreducible, then clearly $d(x) = r(x) \cdot \gcd(\prod h_i(x), \prod {h_i}^*(x))$. But if $h(x)$ is an irreducible factor of both $\prod h_i(x)$ and $\prod {h_i}^*(x)$ then $h(x) = h_i(x) = {h_j}^*(x)$ for some i and j. But then by Lemma 7.1, $h_i h_j = h_i(x){h_i}^*(x)$ is reciprocal, which contradicts the maximality of the degree of r. Hence $r = \gcd(p, p^*) = S(p) \cap S(p^*)$ annihilates $\breve{\mathbf{a}}$. □

Proposition 7.2. *Let $r(x)$ be a reciprocal polynomial with constant coefficient 1 and degree either $2l$ or $2l + 1, l \in \mathbb{Z}^+$, The vector space of periodic reflected bitstreams annihilated by $r(x)$ has dimension l.*

Proof. Let $\breve{\mathbf{c}} = \{\ldots, c_2, c_1, c_0, c_1, c_2, \ldots\}$ be a periodic reflected bitstream. Since r is reciprocal of degree $2l$ or $2l+1$, and $\breve{\mathbf{c}}$ is symmetric about c_0, r annihilates $\breve{\mathbf{c}}$ if and only if it annihilates $\{c_l, c_{l-1}, \ldots, c_1, c_0, c_1, \ldots\}$. Since degree$(r) \in \{2l, 2l + 1\}$, it is clear that $c_1, c_2, \ldots, c_l$ may be chosen arbitrarily but that $c_{l+1}, c_{l+2}, \ldots$ depend on the choice of c_1 through c_l. □

Corollary 7.1. *If $r(x)$ is a reciprocal polynomial with constant coefficient* 1 *and degree either $2l$ or $2l+1, l \in \mathbb{Z}^+$, there are exactly 2^l periodic reflected bitstreams annihilated by $r(x)$.*

Definition 7.4. Let $k \geq 2$ be a fixed positive integer. Let $w = u_0^{c_0} u_1^{c_1} \cdots u_n^{c_n}$, with $c_0 = 1$, be a word in the generators $u_j, j \in \mathbb{Z}^+$ of a binary shift σ. Then w is called a qkword if $w \in \sigma^k(R)' \cap R$ but $w \notin \sigma^{k-1}(R)' \cap R$.

Remark 7.2. Suppose **a** is a bitstream and w is a word of the form above. Using (1.1) repeatedly it follows that for $j \in \mathbb{Z}^+$, $wu_j = (-1)^{a_j c_0 + a_{j-1} c_1 + \cdots + a_{|n-j|} c_n} u_j w$. Hence w is a qkword if and only if there exists a bitstream **a** satisfying the following linear system.

$$
\begin{aligned}
a_{k-1}c_0 + a_{k-2}c_1 + a_{k-3}c_2 + \cdots + a_{|n-k-1|}c_n &= 1 \\
a_k c_0 + a_{k-1}c_1 + a_{k-2}c_2 + \cdots + a_{|n-k|}c_n &= 0 \\
a_{k+1}c_0 + a_k c_1 + a_{k-1}c_2 + \cdots + a_{|n-k+1|}c_n &= 0 \\
&\vdots
\end{aligned}
\tag{2.1}
$$

The first equation holds since w must anticommute with u_{k-1} and the remaining equations hold since w commutes with $u_k, u_{k+1}, \ldots$.

Remark 7.3. As $[R : \sigma(R)] = 2, \sigma(R)' \cap R = \mathbb{C}I$ ([7], Corollary 2.2.4), so there are no q1words.

We shall see below that for fixed values $c_0, c_1, \ldots, c_n$ it is possible to have more than one bitstream **a** which satisfies (2.1). For that reason we shall need the following terminology.

Definition 7.5. Let $p(x) = c_0 + c_1 x + \cdots + c_n x^n$ be a polynomial with constant coefficient 1. If the system (2.1) is satisfied then $p(x)$ is said to *meet* **a** *at the integer* k. If there is an integer k for which $p(x)$ meets **a** then we say that $(p, \mathbf{a})$ are a *binary pair*.

Remark 7.4. Note that if $(p, \mathbf{a})$ meet at k then if σ is the binary shift on R with generators $u_j, j \in \mathbb{Z}^+$ and bitstream **a**, $w =< u_0, p >\in \sigma^k(R)' \cap R$, $w \notin \sigma^{k-1}(R)' \cap R$. Hence w is a qkword and σ is a binary shift of commutant index $\leq k$. On the other hand, since 2 is the minimal possible commutant index for a binary shift, if a binary pair $(p, \mathbf{a})$ meet at 2 then σ has commutant index 2.

Theorem 7.10. *Let $k \geq 2$ and $n > 2k-2$ be fixed integers. Then for any polynomial $p(x)$ of degree n and constant coefficient 1 there are at most 2^{k-2} distinct bitstreams $\mathbf{a} = \{a_0, a_1, \dots\}$ which meet $p(x)$ at k.*

Proof. Induction on k. Using the remark above, a restatement of Theorem 7.6 shows that any polynomial of any degree n with constant coefficient 1 can meet at most one bitstream $\mathbf{a}$ at the integer 2. Suppose the assertion holds for $j = 2, \dots, k-1$. Suppose there is a polynomial $p(x)$ of degree $n > 2k-2$ with constant coefficient 1 which meets some bitstream $\mathbf{a}$ at k. Let σ be the corresponding binary shift which, by the preceding remark, has commutant index $\leq k$. Let $u_j, j \in \mathbb{Z}^+$ be the generators for σ.

Now suppose ρ is any other binary shift, with generators $\{v_j : j \in \mathbb{Z}^+\}$ and bitstream $\mathbf{b}$, such that $< v_0, p >$ is a qkword. Let $\mathbf{s} = \mathbf{a} + \mathbf{b}$. Note that since $p(x)$ meets $\mathbf{a}$ at k, $< u_0, p >$ is a qkword for σ. Hence the coefficients $c_0, \dots, c_n$ of $p(x)$ satisfy the infinite linear system (2.1), as well as system obtained from (2.1) by replacing each a_j with b_j. Hence $[c_0, \dots, c_n]$ satisfies the linear system of equations.

$$\begin{aligned} s_{k-1}c_0 + s_{k-2}c_1 + s_{k-3}c_2 + \cdots + s_{|n-k-1|}c_n &= 0 \\ s_k c_0 + s_{k-1}c_1 + s_{k-2}c_2 + \cdots + s_{|n-k|}c_n &= 0 \\ s_{k+1}c_0 + s_k c_1 + s_{k-1}c_2 + \cdots + s_{|n-k+1|}c_n &= 0 \\ &\vdots \end{aligned} \tag{2.2}$$

It follows that if $z_0, z_1, \dots$ is a family of hermitian generators satisfying the commutation relations given by $\mathbf{s}$ and if η is the shift on the von Neumann algebra M generated by the z_j's, then either $z =< z_0, p >$ commutes with all of the z_j's and lies in the center of M (in which case the reflected bitstream $\check{\mathbf{a}}$ of $\mathbf{a}$ associated with η is periodic, Theorem 7.7), or M is the hyperfinite II_1 factor R, η is a binary shift on R, and there is a $j \in \{2, 3, \dots, k-1\}$ such that $z \in \eta^j(M)' \cap M, z \notin \eta^{j-1}(M)' \cap M$, i.e., that z is a qjword.

Now let $\mathbf{s}^{(1)}, \mathbf{s}^{(2)}, \dots, \mathbf{s}^{(m)}$ be the list of all bitstreams $\mathbf{s}$ satisfying one of the following two conditions with respect to the fixed polynomial $p(x)$: either (i) the reflected bitstream $\check{\mathbf{s}}$ of $\mathbf{s}$ is periodic and annihilated by p, or (ii) $p(x)$ meets s at j for some $j \in \{2, \dots, k-1\}$. Then by the preceding paragraph it follows that the bitstreams $\mathbf{b}$ which meet $p(x)$ at the integer k are $\mathbf{a}, \mathbf{a} + \mathbf{s}^{(1)}, \dots, \mathbf{a} + \mathbf{s}^{(m)}$.

Let $r(x)$ be the maximal reciprocal factor of $p(x)$. If $r(x)$ has degree either 0 or 1 then by the preceding lemmas there are no periodic reflected bitstreams $\breve{\mathbf{a}}$ annihilated by $r(x)$ (and hence by $p(x)$, Proposition 7.1, Proposition 7.2) other than $\breve{\mathbf{a}} = \{\ldots, 0, 0, 0, \ldots\}$. Hence the nontrivial bitstreams $\mathbf{s}$ in the previous paragraph meet the polynomial $p(x)$ at some integer j for $2 \leq j \leq k-1$. By the induction assumption there are no more than 2^{j-2} bitstreams which meet $p(x)$ at each such j. Then from the preceding paragraph, the number of bitstreams $\mathbf{b}$ which meet $p(x)$ at k is no greater than $1 + \sum_{j=2}^{k-1} 2^{j-2} = 2^{k-2}$.

Next suppose that $r(x)$ satisfies $2 \leq \deg(r(x)) \leq 2k-2$. Write $\deg(r(x)) = 2l$ or $2l+1$. By Corollary 7.1 the number of periodic reflected bitstreams $\breve{\mathbf{a}}$ annihilated by $r(x)$ is 2^l. By Proposition 7.1 these are exactly the periodic reflected bitstreams which are annihilated by p. By Lemma 7.5 below there are no bitstreams which meet $p(x)$ at j for $1 \leq j \leq l+1$. On the other hand, by the induction assumption there are, for $l+2 \leq j \leq n$, no more than 2^{j-2} bitstreams which meet $p(x)$ at j. Then the number of binary shifts which meet $p(x)$ at k cannot exceed $2^l + \sum_{j=l+2}^{k-1} 2^{j-2} = 2^{k-2}$. □

7.4. Counting Polynomials with Symmetry

In this section we complete the analysis necessary to enumerate the binary shifts of finite commutant index on the hyperfinite II_1 factor R. As in [15], where a classification was made of the binary shifts of commutant index 2, we establish a natural connection between binary shifts of finite commutant index and polynomials over $GF(2)$ which satisfy a certain symmetry condition. In the course of making this connection, it is convenient first to study the family of binary pairs which meet at a fixed integer $k \geq 2$ (see Definition 7.5) and subsequently to match these pairs with binary shifts. In Lemmas 7.4 and 7.5 we show that a polynomial $p(x)$ with constant coefficient 1 meets a bitstream at k if and only if $p(x)$ has no reciprocal factors of degree $\geq 2k-2$. Using these results we are able to provide a recursion formula which counts, for each polynomial $p(x)$ above, the number of binary shifts σ of commutant index k, with generating family $u_j, j \in \mathbb{Z}^+$ of hermitian unitaries for which $< u_0, p >$ generates the first nontrivial relative commutant algebra $\sigma^k(R)' \cap R$.

Recall that a polynomial $p(x) = c_0 + c_1 x + \cdots + c_n x^n$ with constant coefficient $c_0 = 1$ meets a bitstream $\mathbf{a}$ at an integer k if and only if the coefficients of $p(x)$ satisfy the following infinite system (3.1) of linear equations over F. The first equation holds since $w =< u_0, p >$ must anticommute with u_{k-1}, and the remaining equations are satisfied since w commutes with the generators $u_k, u_{k+1}, \ldots$. If $n < 2k-2$ then $|n-k-1| < k-1$ and we observe, since $c_0 = 1$, that for any choice of $a_1, a_2, \ldots, a_{k-2}$, there exists one and only one choice of $a_k, a_{k+1}, \ldots$ such that the system holds. Hence w is a qkword which corresponds to exactly 2^{k-2} distinct bitstreams $\mathbf{a}$.

For $n = 2k - 2$ the system above becomes

$$\begin{aligned} a_{k-1}c_0 + a_{k-2}c_1 + a_{k-3}c_2 + \cdots + a_{k-1}c_{2k-2} &= 1 \\ a_k c_0 + a_{k-1}c_1 + a_{k-2}c_2 + \cdots + a_{k-2}c_{2k-2} &= 0 \\ a_{k+1}c_0 + a_k c_1 + a_{k-1}c_2 + \cdots + a_{k-3}c_{2k-2} &= 0 \\ &\vdots \end{aligned} \tag{3.1}$$

If the first equation holds then clearly there is one and only one choice for each of the remaining terms $a_k, a_{k+1}, \ldots$ to satisfy the system of equations. Note that the first equation may be rewritten as $a_0c_{k-1} + a_1(c_k + c_{k-2}) + a_2(c_{k+1} + c_{k-3}) + a_3(c_{k+2} + c_{k-4}) + \cdots + a_{k-1}(c_{2k-2} + c_0) = 1$, or (since $a_0 = 0$ and $c_{2k-2} + c_0 = 0$), as $a_1(c_k + c_{k-2}) + a_2(c_{k+1} + c_{k-3}) + a_3(c_{k+2} + c_{k-4}) + \cdots + a_{k-2}(c_{2k-3} + c_1) = 1$. Hence if $a_1, \ldots, a_{k-2}$ satisfy this equation then a_{k-1} may be chosen arbitrarily. If $p(x)$ is reciprocal, however, the left side of this equation is necessarily 0 and so no solution is possible. On the other hand, if $p(x)$ is not reciprocal then $[c_k + c_{k-2}, c_{k+1} + c_{k-3}, c_{k+2} + c_{k-4}, \ldots, c_{2k-3} + c_1] \neq [0, 0, \ldots, 0]$. Hence there is at least one solution to the initial equation above. By viewing the initial equation as a nonhomogeneous linear system of 1 equation in $k - 2$ unknowns $a_1, a_2, \ldots, a_{k-2}$, since there is at least one solution, there are therefore exactly 2^{k-3} solutions $[a_1, a_2, \ldots, a_{k-2}]$ to this first equation. As noted above, the entry a_{k-1} may be chosen arbitrarily but $a_k, a_{k+1}, \ldots$ are determined by their predecessors, hence there are 2^{k-2} bitstreams satisfying the system for each polynomial which is not reciprocal. Note that there are $2^{n-1} - 2^{k-1} = 2^{2k-3} - 2^{k-1} = 2^{n-1} - 2^{n-k+1}$ polynomials of degree n with constant coefficient 1 which are not reciprocal. Hence we have shown the following.

Lemma 7.4. *If $k \geq 2$ and $n < 2k - 2$, any polynomial of degree n with constant coefficient 1 meets at k with 2^{k-2} distinct bitstreams. If $n = 2k-2$ then $2^{n-1} - 2^{n-k+1}$ polynomials of degree n with constant coefficient 1 which meet at k with some bitstream. These are the polynomials which are not reciprocal. Each of these polynomials meets at k with 2^{k-2} distinct bitstreams.*

The analysis pertaining to polynomials of degree exceeding $2k - 2$ is considerably more difficult, and we devote the remainder of this section to studying this case. As we shall see below, whether a polynomial of degree $n \geq 2k-2$ meets a bitstream at the integer k is determined by the degree of its maximal reciprocal factor. Polynomials with reciprocal factors of high degree will not correspond to qkwords, and therefore we will be led to counting the number of polynomial of fixed degree $n \geq 2k-2$ having maximal reciprocal factors exceeding a certain degree (see Theorem 7.12).

Lemma 7.5. *Let $k \geq 2$. No polynomial $p(x)$ with coefficients in $GF(2)$, constant coefficient 1, and a reciprocal factor of degree $\geq 2k-2$ meets at k with any bitstream.*

Proof. Suppose $p(x) = q(x)r(x)$ where $r(x)$ is a reciprocal polynomial of degree $m \geq 2k-2$, and suppose $p(x)$ meets at k with some bitstream $\mathbf{a}$. Then there is a binary shift σ with generators $u_j, j \in \mathbb{Z}^+$, say, such that $w =< u_0, p >$ is a qkword for σ. Let $z =< u_0, q >$, then $w =< u_0, p >=< u_0, qr >= \pm << u_0, q >, r >= \pm < z, r >$. It is straightforward to see that if $z_j = \sigma^j(z), j \in \mathbb{Z}^+$, then σ restricts to a binary shift on the von Neumann algebra M generated by the z_j's and $w \in \sigma^k(M)' \cap M$ but $w \notin \sigma^{k-1}(M)' \cap M$. Hence w anticommutes with z_{k-1} so that if $\mathbf{b}$ is the bitstream for the restricted binary shift on M and $r(x) = l_0 + l_1 x + \cdots + l_m x^m$,

$$b_{k-1}l_0 + b_{k-2}l_1 + \cdots + b_{m-k+1}l_m = 1. \qquad (3.2.1)$$

If $m = 2k-2$ then since $r(x)$ is reciprocal (3.2.1) becomes

$$b_{k-1}l_0 + b_{k-2}l_1 + \cdots + b_1 l_{k-2} + b_0 l_{k-1} + b_1 l_{k-2} + b_{k-1}l_0 = 0, \qquad (3.2.2)$$

a contradiction. If $m > 2k-2$ then since $r(x)$ is reciprocal (3.2.1) may be rewritten

$$b_{k-1}l_m + b_{k-2}l_{m-1} + \cdots + b_{m-k+1}l_0 = 1, \qquad (3.2.3)$$

which implies that w anticommutes with z_{m-k+1}, also a contradiction (since $m-k+1 \geq k$ and $w \in \sigma^k(M)' \cap M$). By contradiction w cannot be a qkword. Equivalently, $p(x)$ does not meet at k with any bitstream. □

Remark 7.5. We shall see below (Corollary 7.4) that for $n \geq 2k-2$ all other polynomials with constant coefficient 1 meet at k with at least one bitstream. In fact, they meet at k with exactly 2^{k-2} bitstreams.

Below we shall count the number of polynomials of fixed degree $n \geq 2k-2$ which have maximal reciprocal factors of degree $\geq 2k-2$. In order to make this calculation we require both a definition and a result from [15].

Definition 7.6. (cf. [15], Definition 4.1.) A polynomial $f(x) \in F[x]$ with constant coefficient 1 is completely free if $f(x)$ has no reciprocal factors except for the constant polynomial 1. $z(n)$ denotes the number of completely free polynomials of degree n with constant coefficient 1.

Theorem 7.11. *(cf. [15], Theorem 4.4.) Let $r \geq 1$ be a fixed integer. If $n = 2r$, then $z(n) = \frac{1}{3}(2 \cdot 4^{r-1} + 4) - 2$ and if $n = 2r+1$ then $z(n) = \frac{1}{3}(4^r - 4) + 2$.*

Theorem 7.12. *With the same notation as above, there are, for $n > 2k-2$, exactly 2^{n-k} polynomials of degree n with constant coefficient 1 whose maximal reciprocal factor has degree $\geq 2k-2$. If $n = 2k-2$ there are $2^{n-(k-1)} = 2^{k-1}$ such polynomials.*

Proof. For $n \geq 2k-2$ let $r(n)$ denote the number of polynomials satisfying the hypotheses of the theorem. For $j \in \mathbb{N}$ let $s(n)$ denote the number of reciprocal polynomials of degree n with constant coefficient 1. Obviously if $n = 2l$ or $2l+1$, $s(n) = 2^l$. From [15], Theorem 4.3, each polynomial with constant coefficient 1 can be decomposed uniquely into a product of a reciprocal and a completely free polynomial. If $n = 2k-2$ then clearly $r(n) = s(n) = s(2k-2) = 2^{k-1}$. If a polynomial $p(x)$ of degree $2k-1$ or $2k$ has a reciprocal factor $g(x)$ of degree at least $2k-2$ then $f(x)/g(x)$, a factor of degree 2 or 1, is itself reciprocal, so by Lemma 7.1 $f(x)$ is reciprocal. Hence $r(2k-1) = s(2k-1) = s(2(k-1)+1) = 2^{k-1} = 2^{n-k}$, and $r(2k) = s(2k) = 2^k = 2^{n-k}$.

Suppose $n > 2k$ then since $z(1) = z(2) = 0$,

$$r(n) = \sum_{j=2k-2}^{n-3} s(j)z(n-j) + s(n)z(0).$$

By direct calculation or by [15], proof of Theorem 4.4,

$$r(n) = \sum_{j=2}^{n-3} s(j)z(n-j) + s(n)z(0) = 2^{n-2},$$

so we calculate, with $d = 2k - 2$

$$\sum_{j=2}^{d-1} s(j)z(n-j) = s(2)z(n-2) + s(3)z(n-3) + \\ s(4)z(n-3) + s(5)z(n-4) + \\ \vdots \\ s(d-2)z(n-d+2) + s(d-1)z(n-d+1).$$

Suppose n is even. Then with $n = 2m$, by the preceding theorem,

$$\begin{aligned} &s(2t)z(n-2t) + s(2t+1)z(n-2t-1) \\ &\quad = 2^t\{z(2(m-t)) + z(2(m-t-1)+1)\} \\ &\quad = 2^t\left\{\frac{1}{3}(2 \cdot 4^{m-t-1} + 4) - 2 + \frac{1}{3}(4^{m-t-1} - 4) + 2\right\} \\ &\quad = 2^t\{4^{m-t-1}\} = 2^{2m-t-2}, \end{aligned}$$

so

$$\begin{aligned} \sum_{j=2}^{d-1} s(j)z(n-j) &= \sum_{j=2}^{2k-3} s(j)z(2m-j) \\ &= \sum_{t=1}^{k-2} 2^{2m-t-2} \\ &= 2^{2m-3} + \cdots + 2^{2m-k} \\ &= 2^{2m-2} - 2^{2m-k} \\ &= 2^{n-2} - 2^{n-k}. \end{aligned}$$

Therefore

$$\begin{aligned} r(n) &= \sum_{j=2k-2}^{n-3} s(j)z(n-j) + s(n)z(0) \\ &= \sum_{j=2}^{n-3} s(j)z(n-j) + s(n)z(0) - \sum_{j=2}^{d-1} s(j)z(n-j) \\ &= 2^{n-2} - \{2^{n-2} - 2^{n-k}\} = 2^{n-k} \end{aligned}$$

We omit the similar argument for the case when n is odd. □

Combining the preceding result with Lemma 7.5 gives the following.

Corollary 7.2. *With the same notation as above, let $n \geq 2k-2$ be a fixed integer. If $n = 2k-2$ then the number of polynomials of degree n with constant coefficient 1 which meet at k is at most $2^{n-1} - 2^{k-1}$ (equivalently, the maximum number of polynomials of coefficient 1 for which $< u_0, p >$ is a qkword is $2^{n-1} - 2^{k-1}$). If $n > 2k-2$ there are at most $2^{n-1} - 2^{n-k}$ such polynomials.*

Definition 7.7. For a fixed nonnegative integer n and an integer $k \geq 2$ we denote by $BP(n,k)$ the set of pairs $(p, \mathbf{a})$ such that (i) $p(x)$ is a polynomial of degree n with constant coefficient 1, (ii) $\mathbf{a}$ is a bitstream, and (iii) $< u_0, p >$ is a qkword for the binary shift σ corresponding to the bitstream $\mathbf{a}$ and $u_0, u_1, \ldots$ are the generators for σ. We shall refer to a pair $(p, \mathbf{a})$ satisfying $(i), (ii), (iii)$ as a binary pair which meeting at k.

For the remainder of the section we assume that $n > 2k-2$. For such n we shall show that there are at least $(2^{n-1} - 2^{n-k})2^{k-2}$ binary pairs in $BP(n,k)$. Note first that if $(p, \mathbf{a}) \in BP(n,k)$ if and only if the coefficients of $p(x) = c_0 + c_1 x + \cdots + c_n x^n$ satisfy the following linear system, for some choice of elements $l_0, l_1, \ldots, l_{k-2}$ in F.

$$
\begin{aligned}
a_1 c_0 + a_0 c_1 + a_1 c_2 \cdots + a_{n-3} c_{n-2} + a_{n-2} c_{n-1} + a_{n-1} c_n &= l_1 \\
a_2 c_0 + a_1 c_1 + a_0 c_2 \cdots + a_{n-4} c_{n-2} + a_{n-3} c_{n-1} + a_{n-2} c_n &= l_2 \\
&\vdots \\
a_{k-2} c_0 + a_{k-3} c_1 + a_{k-4} c_2 \cdots + a_{n-k} c_{n-2} + a_{n-k+1} c_{n-1} + a_{n-k+2} c_n &= l_{k-2} \\
a_{k-1} c_0 + a_{k-2} c_1 + a_{k-3} c_2 \cdots + a_{n-k-1} c_{n-2} + a_{n-k} c_{n-1} + a_{n-k+1} c_n &= 1 \\
a_k c_0 + a_{k-1} c_1 + a_{k-2} c_2 \cdots + a_{n-k-2} c_{n-2} + a_{n-k-1} c_{n-1} + a_{n-k} c_n &= 0 \\
a_{k+1} c_0 + a_k c_1 + a_{k-1} c_2 \cdots + a_{n-k-3} c_{n-2} + a_{n-k-2} c_{n-1} + a_{n-k-1} c_n &= 0 \\
&\vdots
\end{aligned}
\tag{3.3}
$$

Theorem 7.13. *Let $k \geq 2$ be an integer. Then for all integers $n > 2k-2$, $BP(n,k) \geq (2^{n-1} - 2^{n-k}) \cdot 2^{k-2}$.*

Proof. We divide the proof into cases depending on the parity of n. The proof is obtained as an application of the unimodality properties obtained in

[13] (see also [5, 22]) for the nullity sequence $\{\nu(A_m) : m \in \mathbb{N}\}$ corresponding to the $m \times m$ matrices associated with any bitstream $\mathbf{a} = \{a_j : j \in Z^+\}$.

Suppose first that n is odd. Let $\mathbf{a} = a_1, a_2, \ldots$ be a bitstream for which A_{n-1} is invertible. If there is a polynomial p for which (3.3) holds, then since $c_0 = c_n = 1$ the first $n-1$ equations of the system above may be rewritten as

$$\begin{bmatrix} a_0 & a_1 & a_2 & a_3 & \cdots & a_{n-2} \\ a_1 & a_0 & a_1 & a_2 & \cdots & a_{n-3} \\ a_2 & a_1 & a_0 & a_1 & \cdots & a_{n-4} \\ \vdots & \vdots & \vdots & \vdots & \ddots & \vdots \\ a_{n-2} & a_{n-3} & a_{n-4} & \cdots & \cdots & a_0 \end{bmatrix} \cdot \begin{bmatrix} c_1 \\ c_2 \\ c_3 \\ \vdots \\ c_{n-1} \end{bmatrix} = \begin{bmatrix} a_1 + a_{n-1} + l_1 \\ a_2 + a_{n-2} + l_2 \\ a_3 + a_{n-3} + l_3 \\ \vdots \\ a_{k-2} + a_{n-k+2} + l_{k-2} \\ a_{k-1} + a_{n-k+1} + 1 \\ a_k + a_{n-k} \\ \vdots \\ a_{n-1} + a_1 \end{bmatrix}$$

Since A_{n-1} is assumed to be invertible, the matrix equation above has a solution for any choice of elements $a_{n-1}, l_0, l_1, \ldots, l_{k-2}$. Also note that once a_1 through a_{n-1} have been chosen, as well as l_1 through l_{k-2}, there is one and only one choice for each of the remaining entries $a_n, a_{n+1}, \cdots$ in order to satisfy (3.3). Recalling from [5], Corollary 2.10, that there are 2^{n-3} choices of $a_1, \ldots, a_{n-2}$ such that A_{n-1} is invertible, we see that there are $2^{n-3} \cdot 2^{k-1} = 2^{n-2} \cdot 2^{k-2}$ binary pairs $(p, \mathbf{a})$ corresponding to the case when A_{n-1} is invertible.

More generally let s be a fixed integer such that $1 \leq s \leq k-1$. By [5], Theorem 2.9, it is possible to find a string of elements $a_1, a_2, \ldots, a_{n-s}$ of F such that the matrix $A_{n-(2s-1)}$ is invertible and that $\nu(A_{n-(2s-1)+b}) = b$ for $0 \leq b \leq s$. (Note that the case considered in the preceding paragraph corresponds to the case $s = 1$ here.) Momentarily deleting the first $s-1$ equations of the system (3.3) above we see that the next $n-2s+1$ equations may be written in the form

$$\begin{bmatrix} a_0 & a_1 & a_2 & a_3 & \cdots & a_{n-2s} \\ a_1 & a_0 & a_1 & a_2 & \cdots & a_{n-2s-1} \\ a_2 & a_1 & a_0 & a_1 & \cdots & a_{n-2s-2} \\ \vdots & \vdots & \vdots & \vdots & \ddots & \vdots \\ a_{n-2s} & a_{n-2s-1} & a_{n-2s-2} & \cdots & \cdots & a_0 \end{bmatrix} \cdot \begin{bmatrix} c_s \\ c_{s+1} \\ c_{s+2} \\ \vdots \\ c_{n-s} \end{bmatrix}$$

$$
= \begin{bmatrix} a_s c_0 & + & a_{s-1} c_1 & + \cdots + & a_1 c_{s-1} \\ a_{s+1} c_0 & + & a_s c_1 & + \cdots + & a_2 c_{s-1} \\ \vdots & & & & \\ a_{k-2} c_0 & + & a_{k-3} c_1 & + \cdots + & a_{k-s-1} c_{s-1} \\ a_{k-1} c_0 & + & a_{k-2} c_1 & + \cdots + & a_{k-s} c_{s-1} \\ a_k c_0 & + & a_{k-1} c_1 & + \cdots + & a_{k-s+1} c_{s-1} \\ \vdots & & & & \\ a_{n-s} c_0 & + & a_{n-s-1} c_1 & + \cdots + & a_{n-2s-1} c_{s-1} \end{bmatrix}
$$

$$
+ \begin{bmatrix} a_{n-2s+1} c_{n-s+1} & + \cdots + & a_{n-s} c_n \\ a_{n-2s} c_{n-s+1} & + \cdots + & a_{n-s-1} c_n \\ \vdots & & \\ a_{n-s-k+3} c_{n-s+1} & + \cdots + & a_{n-k+2} c_n \\ a_{n-s-k+2} c_{n-s+1} & + \cdots + & a_{n-k+1} c_n \\ a_{n-s-k+1} c_{n-s+1} & + \cdots + & a_{n-k} c_n \\ \vdots & & \\ a_1 c_{n-s+1} & + \cdots + & a_s c_n \end{bmatrix} + \begin{bmatrix} l_s \\ l_{s+1} \\ \vdots \\ l_{k-2} \\ 1 \\ 0 \\ \vdots \\ 0 \end{bmatrix}.
$$

To count the number of binary pairs $(p, \mathbf{a})$ arising from this case we recall the following facts from [5]. By [5], Corollary 2.10, the number of choices of $a_1, a_2, \ldots, a_{n-2s}$ for which $A_{n-(2s-1)}$ is invertible is 2^{n-2s-1}. By [5], Theorem 2.7, since $\nu(A_{n-(2s-1)}) = 0$, then $\nu(A_{n-2s+2}) = 1$ regardless of the choice of a_{n-2s+1}. This latter condition follows from the phenomenon that the sequence $\{\nu(A_j) : j \in \mathbb{N}\}$ is unimodal, i.e., it is the concatenation of strings of positive integers of the form $1, 2, \ldots, m-1, m, m-1, \ldots, 2, 1, 0$. By the proof of Theorem 2.9 of [5] there is one and only one choice for the entries $a_{n-2s+2}, a_{n-2s+3}, \ldots, a_{n-s}$ so that $\nu(A_{n-2s+1+b}) = b$ for $2 \leq b \leq s-1$. Observe from the matrix equation above that the coefficients $c_1, \ldots, c_{s-1}, c_{n-s+1}, \ldots, c_{n-1}$ may be chosen arbitrarily, as may $l_s, \ldots, l_{k-2}$. Once the strings a_1 through a_{n-s}, c_1 through c_{s-1}, c_{n-s+1} through c_{n-1} and l_s through l_{k-2} have been chosen, it follows by considering all equations of (3.3) subsequent to the first $n-s$ equations that the entries $a_{n-s+1}, a_{n-s+2}, \ldots$ are all uniquely determined. It then follows by examining the first $s-1$ equations of the system (3.3) that l_1 through l_{s-1} are all uniquely determined. Counting our choices for the entries above, the number of solutions to the system (3.3) for which $\nu(A_{n-(2s-1)+b}) = b$ for $0 \leq b \leq s$ are $2^{n-2s-1} \cdot 2^1 \cdot 2^{s-1} \cdot 2^{s-1} \cdot 2^{k-1-s}$, or $2^{n-s-1} \cdot 2^{k-2}$.

We show that no binary pair $(p, \mathbf{a})$ can arise from the construction above for more than one choice of $s \in \{1, \ldots, k-1\}$. For suppose $1 \leq s' < s \leq k-1$ and $(p, \mathbf{a})$ is obtained for both indices s and s'. Recall from the construction above corresponding to s that $\nu(A_{n-2s+1}) = 0, \nu(A_{n-2s+2}) = 1, \ldots, \nu(A_{n-s+1}) = s$. It then follows from the unimodality properties of the nullity sequence $\{\nu(A_j) : j \in \mathbb{N}\}$ (see Theorem 7.8) that for $n-2s+1 < t \leq n-1$, $\nu(A_t) > 0$, i.e., that A_t is not invertible. Hence $\nu(A_j) > 0$ for $n-2s+2 \leq j \leq n-1$. Since $(p, \mathbf{a})$ arises from the analysis for the integer s', however, we have $\nu(A_{n-2s'+1}) = 0$. But $n-2s+1 < n-2s'+1 \leq n-1$ so $\nu(A_{n-2s'+1}) > 0$. This contradiction yields the result. Adding all of the binary pairs obtained from each of the cases for s, $1 \leq s \leq n-1$ we get

$$\sum_{s=1}^{k-1} 2^{n-s-1} \cdot 2^{k-2} = (2^{n-1} - 2^{n-k}) \cdot 2^{k-2}$$

distinct binary pairs.

We sketch the proof for the case when n is even. For a fixed positive integer $s \in \{1, 2, \ldots, k-1\}$ consider finite sequences $a_0, a_1, \ldots, a_{n-s-1}$ of elements of F for which the corresponding sequence of Toeplitz matrices $A_j, j \in \{1, 2, \ldots, n-s\}$ satisfies $\nu(A_{n-2s+b}) = b, 0 \leq b \leq s$. An analysis similar to the one above shows that we obtain at least $2^{n-2s} \cdot 2^{k-2}$ binary pairs which meet at k for each s, and therefore we have $(2^{n-1} - 2^{n-k}) \cdot 2^{k-2}$ binary pairs which meet at k. □

Corollary 7.3. *Let $k \geq 2$ be an integer. Then*

(i) $BP(0, k) = 2^{k-2}$.
(ii) $BP(n, k) = 2^{n-1} \cdot 2^{k-2}$ *if* $1 \leq n < 2k-2$.
(iii) $BP(n, k) = (2^{n-1} - 2^{n-k+1}) \cdot 2^{k-2}$ *if* $n = 2k-2$.
(iv) $BP(n, k) = (2^{n-1} - 2^{n-k}) \cdot 2^{k-2}$ *if* $n > 2k-2$.

Proof. The first three equations are obtained by combining the results of Lemmas 7.4 and 7.5. So suppose $n > 2k-2$. By Theorem 7.12 there are 2^{n-k} polynomials of degree $n > 2k-2$ with constant coefficient 1 having a reciprocal factor of degree $\geq 2k-2$. By Lemma 7.5 none of these polynomials meets with any bitstream at the integer k. Therefore, among all polynomials of degree n with constant coefficient 1, there are at most $2^{n-1} - 2^{n-k}$ which meet some bitstream at k. By Theorem 7.10 each such polynomial meets at k with at most 2^{k-2} distinct bitstreams. Hence $BP(n, k) \leq (2^{n-1} - 2^{n-k}) \cdot 2^{k-2}$. But by the preceding theorem, $BP(n, k) \geq (2^{n-1} - 2^{n-k}) \cdot 2^{k-2}$. □

The following result follows as a corollary to the proof of the preceding corollary.

Corollary 7.4. *The following are equivalent for any polynomial $p(x)$ with constant coefficient* 1 *over $GF(2)$, and any integer $k \geq 2$.*

(i) *$p(x)$ meets at least one bitstream at k.*
(ii) *$p(x)$ meets exactly 2^{k-2} bitstreams at k.*
(iii) *$p(x)$ has no reciprocal factors of degree $\geq 2k - 2$.*

7.5. Conjugacy Classes of Binary Shifts

As a consequence of the preceding results we are now in a position to establish a correspondence between the conjugacy classes of binary shifts of finite commutant index and the family of polynomials over $GF(2)$ with constant coefficient 1. Specifically we provide an algorithm which can be used to compute, for any polynomial $p(x)$ over $GF(2)$ with constant coefficient 1, and any integer $k \geq 2$, the number of binary shifts σ of commutant index k associated with $p(x)$ in the sense that $w =< u_0, p >$ generates $\sigma^k(R)' \cap R$, (where $u_j, j \in \mathbb{Z}^+$ are the generators for σ. In Corollary 7.4 it is shown that for a fixed index $k \geq 2$, any polynomial $p(x)$ meets either 2^{k-2} bitstreams at k or it meets no bitstreams at k. In terms of binary shifts, using Remark 7.4, this means that for a polynomial $p(x)$ there are either 2^{k-2} binary shifts σ for which $< u_0, p >$ is a qkword, or no such binary shifts. Note, however, that if $w =< u_0, p >$ is a qkword for some binary shift σ it is not necessarily the case that σ has commutant index k. As an elementary example consider the polynomial $p(x) = x + 1$ and the binary shift σ with bitstream $\{0, 1, 0, 0, 0, \dots\}$. Then $< u_0, p >= u_0 u_1$ is a q3word. On the other hand, σ is a binary shift of commutant index 2, with the word u_0 generating the relative commutant algebra $\sigma^2(R)' \cap R$. What is needed, therefore, is a way to determine how many of the bitstreams $\mathbf{a}$ which meet $p(x)$ at k actually correspond to binary shifts of commutant index k. The following three results provide the key.

Theorem 7.14. *Let* $\mathbf{a}$ *be an eventually periodic but not mirror-periodic bitstream, i.e., the reflected bitstream* $\breve{\mathbf{a}}$ *is not periodic. Let σ, with generators $\{u_j : j \in \mathbb{Z}^+\}$, be the binary shift on R corresponding to* $\mathbf{a}$. *Let $k \in \{2, 3, \dots\}$ be the commutant index of σ. Then if $p(x)$ is such that the word $w =< u_0, p >$ generates $\sigma^k(R)' \cap R$, then*

(i) *$(p, \mathbf{a})$ is a binary pair,*

(ii) $(p, \mathbf{a})$ *meets at the integer* k,

(iii) *if* $f(x)$ *is a polynomial with constant coefficient 1 then* $(pf, \mathbf{a})$ *meets at* $k+deg(f)$.

(iv) *if* $(g, \mathbf{a})$ *is a binary pair for some polynomial* g *with constant coefficient* 1, *then* p *is a factor of* g *and* $(g, \mathbf{a})$ *meets at* $k + deg(g/p)$.

Proof. First note that σ is indeed a binary shift on R, since $\breve{\mathbf{a}}$ is not periodic, Theorem 7.1. Also, σ has finite commutant index, since $\mathbf{a}$ is eventually periodic, Theorem 7.4. Let $p(x) = c_0 + c_1 x + \cdots + c_n x^n$, then $w = u_0^{c_0} u_1^{c_1} \cdots u_n^{c_n}$. Since w anticommutes with u_{k-1} and commutes with $u_k, u_{k+1}, \ldots$, the infinite linear system (2.1) is satisfied. Hence $(p, \mathbf{a})$ is a binary pair meeting at the integer k. This proves (i) and (ii). To see (iii) note that if $f(x) = l_0 + l_1 x + \cdots + l_m x^m$ with $l_0 = 1 = l_m$ then by (1.3), $< u_0, pf >= \pm << u_0, p >, f >= \pm < w, f >= \pm w^{l_0} \sigma(w)^{l_1} \cdots \sigma^m(w)^{l_m}$. It follows that $< u_0, pf >$ anticommutes with u_{k-1+m} and commutes with $u_{k+m}, u_{k+m+1}, \ldots$ whence (iii).

To see (iv) let $y =< u_0, g >$. Since $(g, \mathbf{a})$ is a binary pair there is an integer k_0 where they meet. It follows that the word y anticommutes with u_{k_0-1} and commutes with $u_{k_0}, u_{k_0+1}, \ldots$, i.e., $y \notin \sigma^{k_0-1}(R)' \cap R$ but $y \in \sigma^{k_0}(R)' \cap R$. Since σ has commutant index k then $k \leq k_0$ and $y \notin \sigma^{k_0-1}(R)' \cap R$, $y \in \sigma^{k_0}(R)' \cap R = \{w, \sigma(w), \ldots, \sigma^{k_0-k}(w)\}''$. We conclude that there is a polynomial $h(x)$ of degree $k_0 - k$ and constant coefficient 1 such that $y = \pm < w, h >$. But then $< u_0, g >= y = \pm < w, h >= \pm << u_0, p >, h >= \pm < u_0, ph >$, so $p(x)h(x) = g(x)$. Hence $p(x)$ is a factor of $g(x)$. That $(g, \mathbf{a})$ meet at $k + deg(g/p)$ now follows from (iii). □

As an immediate corollary we have the following.

Corollary 7.5. *Suppose* σ *is a binary shift on the hyperfinite* II_1 *factor* R *with corresponding bitstream* $\mathbf{a} = \{a_0, a_1, \ldots\}$. *Suppose there are an integer* k_0 *and a polynomial* $g(x) \in F[x]$, *with constant coefficient* 1, *such that* g *is paired at* k_0 *with* $\mathbf{a}$. *Then* σ *has finite commutant index. In particular there is a unique polynomial* p *with constant coefficient* 1 *such that* p *and* $\mathbf{a}$ *are paired at* k, *the commutant index of* σ. *Moreover,*

(i) p *is a factor of* g, *and*

(ii) *if the polynomial* g/p *has degree* s, *then* $k + s = k_0$.

Corollary 7.6. *Suppose* σ *is a binary shift on* R *with finite commutant index* k *and corresponding bitstream* $\mathbf{a}$. *Then for* $s \geq 0$ *there are exactly* 2^{s-1}

binary pairs $(g, \mathbf{a})$ *which meet at the integer* $k+s$. *Each such polynomial* g *has the form* $g(x) = p(x)f(x)$ *where* $p(x)$ *is the unique polynomial which meets* $\mathbf{a}$ *at* k, *and* $f(x)$ *is a polynomial with constant coefficient* 1 *of degree* s.

Proof. The result follows from the preceding result and the fact that there are exactly 2^{s-1} distinct polynomials of degree s with constant coefficient 1. □

Definition 7.8. Let $p(x)$ be a polynomial with constant coefficient 1 in $GF(2)$. For any integer $k \geq 2$ let $C(p; k)$ denote the family of binary shifts σ of commutant index k on R for which the word $w =< u_0, p >$ in the generators of σ generates the first nontrivial relative commutant algebra $\sigma^k(R)' \cap R$.

Remark 7.6.

(i) Restating Theorem 7.6 (see also Remark 7.4 in terms of this notation, we have $C(p, 2) = \emptyset$ if $p(x)$ has any reciprocal divisors of degree exceeding 1 and $|C(p, 2)| = 1$, otherwise.
(ii) Since there are no q1words, by Remark 7.3, $C(p, 1) = \emptyset$.
(iii) Let $p(x) = 1$. Then $C(p, k)$, for $k \geq 2$, consists of all binary shifts of commutant index k for which the word $w =< u_0, p >= u_0$ generates $\sigma^k(R)' \cap R$. It is not difficult to show that these are the binary shifts each of whose bitstreams $\mathbf{a} = \{a_0, a_1, \ldots\}$ satisfies $a_{k-1} = 1, a_k = a_{k+1} = \cdots = 0$. Note that, as $a_1, a_2, \ldots, a_{k-2}$ may be chosen arbitrarily, there are 2^{k-2} such binary shifts, i.e., $|C(p, k)| = 2^{k-2}$.

The following result gives a recursive formula for computing the cardinality of $C(p, k)$.

Theorem 7.15. *Let* $k \geq 2$ *be a fixed integer. Let* $p(x)$ *be a polynomial of degree* n. *If* $p(x)$ *has a reciprocal factor* $r(x)$ *with* $deg(r(x)) \geq 2k-2$ *then* $C(p, k) = \emptyset$. *Otherwise, for each* $j = 0, 1, \ldots, n-1$ *let* $q_{j1}, q_{j2}, \ldots, q_{jm_j}$ *be the distinct factors of* $p(x)$ *of degree* j. *Then*

$$|C(p, k)| = 2^{k-2} - \sum_{\substack{\max(\{0, 2+n-k\}) \leq j \leq n-1 \\ 1 \leq i \leq m_j}} |C(q_{ji}, k-(n-j))| \tag{4.1}$$

Proof. If $p(x) = 1$ statement (iii) of the remark indicates that $|C(p, k)| = 2^{k-2}$. It is clear that the summation in this case is 0 and the formula holds in this situation. If $k = 2$ and $p(x)$ has no reciprocal factors of degree

> 1 then by statement (iii) the formula should be 1. Since $C(p, 1) = \emptyset$ by statement (ii) the formula holds in this case. So we may assume that $p(x)$ has degree ≥ 1 and that $k \geq 3$. Let $p(x)$ be a polynomial of degree n with constant coefficient 1. By Lemma 7.5 (see also Remark 7.4) $C(p, k) = \emptyset$ if $p(x)$ has a reciprocal factor of degree $\geq 2k - 2$, so we may assume that the maximal reciprocal factor of $p(x)$ has degree $< 2k - 2$. Suppose $|C(q, l)|$ is known, for all polynomials $q(x)$ of degree $< n$ and all $l \in \{2, 3, \ldots, k - 1\}$. Suppose $\mathbf{a}$ is a bitstream which meets $p(x)$ at k. Let σ be the corresponding binary shift on R. Then either $\sigma \in C(p, k)$ or by Corollary 7.5 there is an $l \in \{2, \ldots, k - 1\}$ such that σ has commutant index $l < k$. In the latter case there is a unique polynomial $q(x)$ for which $\sigma \in C(q, l)$. Suppose $deg(q) = j \leq n - 1$. By Corollary 7.5 $q(x)$ is a proper factor of $p(x)$ and $k = l + deg(p/q) = l + n - j$, so $l = k - (n - j)$. Since $C(q, l) = 0$ unless $l \geq 2$,, we must have $k - (n - j) \geq 2$, or $j \geq 2 + (n - k)$. Of course $j \geq 0$ also. On the other hand, $q(x)$ is a proper factor of $p(x)$, so $j = deg(q) < deg(p) = n$, hence $max(\{0, 2 + (n - k)\}) \leq j \leq n - 1$. Hence every binary shift σ of commutant index less than k, for which $< u_0, p >$ is a qkword (where $\{u_j : j \in \mathbb{Z}^+\}$ are the generators of σ) is accounted for in the summation in the formula above.

Conversely, suppose $q(x)$ is a proper factor of $p(x)$, and suppose $\sigma \in C(q, k - deg(p/q) = C(p, k - n + deg(q))$. Let $\mathbf{a}$ be the bitstream corresponding to σ. Then by Corollary 7.4 and Theorem 7.14, $\mathbf{a}$ is one of the 2^{k-2} bitstreams which meet $p(x)$ at k. Hence the summation in the formula subtracts from the 2^{k-2} bitstreams corresponding to $p(x)$ any bitstream associated with a binary shift σ of commutant index $< k$ for which $< u_0, p >$ is a qkword. Hence the right side of the formula above counts all binary shifts σ of commutant index equal to k for which $< u_0, p >$ generates $\sigma^k(R)' \cap R$. □

Corollary 7.7. *There are countably many conjugacy classes of binary shifts of any finite commutant index.*

Corollary 7.8. *Let $p(x)$ be an irreducible polynomial over $GF(2)$ of degree $n \geq 1$. Let $k \geq 2$ be an integer. If $p(x)$ is reciprocal then*

(i) $|C(p, k)| = 0$ *if* $n \geq 2k - 2$,

(ii) $|C(p, k)| = 2^{k-2}$ *if* $k - 1 \leq n \leq 2k - 2$, *and*

(iii) $|C(p, k)| = 2^{k-2} - 2^{k-n-2}$ *if* $0 < n \leq k - 2$.

If $p(x)$ is not reciprocal then

(iv) $|C(p,k)| = 2^{k-2}$ *if* $n \geq k-1$, *and*

(v) $|C(p,k)| = 2^{k-2} - 2^{k-n-2}$ *if* $0 < n \leq k-2$.

Proof. (i) follows immediately from the first assertion of the theorem. Otherwise, since 1 is the only proper factor of $p(x)$ the formula in the theorem reduces to $|C(p,k)| = 2^{k-2} - |C(1,k-n)|$. If $n \geq k-1$ then $|C(1,k-n)| = 0$, so (ii) and (iv) follow. If $0 < n \leq k-2$ then $|C(1,k-n)| = 2^{k-n-2}$ by (iii) of the remark preceding the theorem, and $|C(p,k)| = 2^{k-2} - |C(1,k-n)| = 2^{k-2} - 2^{k-n-2}$, giving (iii) and (v). □

What follows is an algorithm for determining the bitstreams of those binary shifts which lie in $C(p,k)$, for any polynomial $p(x)$ over $GF(2)$ with constant coefficient 1. If $p(x) = 1$ then $C(p,k)$, by Remark 7.6(iii), consists of binary shifts whose bitstreams are of the form $\mathbf{a} = \{0, a_1, a_2, \ldots, a_{k-2}, 1, 0, 0, \ldots\}$. Suppose $p(x)$ has degree $n > 0$ and suppose $k \geq 2$. Suppose moreover that $p(x)$ has no reciprocal factors of degree $\geq 2k-2$. Assuming the bitstreams for all binary shifts in $C(q,l)$, $\deg(q(x)) < n, l \leq n-1$, have been determined we find the bitstreams associated with the binary shifts in $C(p,k)$.

To do this we first seek any bitstream $\mathbf{a}$ which meets $p(x)$ at k. If $q(x)$ is a factor of $p(x)$ such that $C(q, k - deg(p/q)) \neq \emptyset$ then by Theorem 7.14 the bitstream associated with a binary shift in this set meets $p(x)$ at k. If no such factor $q(x)$ exists we find $\mathbf{a}$ as follows. If $\deg(p(x)) \leq 2k-2$ then we may obtain $\mathbf{a}$ as in the proof of Lemma 7.4. If $\deg(p(x)) > 2k-2$ then we can find a bitstream $\mathbf{a}$ by solving the system consisting of the first $n+1$ equations in the infinite system (3.1). Having done that, the remaining elements $\{a_{n-k}, a_{n-k+1}, \ldots\}$ are obtained from the remaining equations in (3.1) using the fact that $c_0 = 1$. For the next step, let $\mathbf{s}^{(1)}, \mathbf{s}^{(2)}, \ldots, \mathbf{s}^{(m)}$ be the bitstreams as in the proof of Theorem 7.10. For $1 \leq j \leq m$ let $\mathbf{b}^{(j)} = \mathbf{a} + \mathbf{s}^{(m)}$. (From that theorem it was determined that $m \leq 2^{k-2}$. From the computation of $BP(n,k)$ in Corollaries 7.3 and 7.4, however, it turns out that $m = 2^{k-2}$.) By the theorem we have found all of the bitstreams which meet $p(x)$ at k. Note that all of these bitstreams correspond to binary shifts in $C(p,k)$. For if not, the proof of the theorem above implies that one of these bitstreams meets with some factor $q(x)$ of $p(x)$ at the integer $k - deg(p/q)$, a case which we have ruled out.

Since a bitstream is a complete conjugacy invariant for binary shifts on R, the procedure above leads to the following.

Theorem 7.16. *The algorithm above gives a complete classification of the binary shifts of finite commutant index up to conjugacy.*

References

[1] Arveson, W. and Price, G. (2003). The structure of spin systems. *Internat. J. Math.* **14** 119–137.

[2] Bures, D. and Yin, H. (1990). Shifts on the hyperfinite II_1 factor. *Pacific J. Math.* **142** 245–257.

[3] Carlitz, L. (1967). Some theorems on irreducible reciprocal polynomials over a finite field. *J. Reine Angew. Math.* **226** 212–220.

[4] Connes, A. and Størmer, E. (1975). Entropy for automorphisms of II_1 von Neumann algebras. *Acta Math.* **134** 289–306.

[5] Culler, K. and Price, G. L. (1976). On the ranks of skew-centrosymmetric matrices over finite fields. *Linear Algebra Appl.* **248** 317–325.

[6] Golodets, V. and Størmer, E. (1998). Entropy of C^*-dynamical systems defined by bitstreams. *Ergodic Theory Dynam. Systems* **18** 859–874.

[7] Jones, V. F. R. (1983). Index for subfactors. *Invent. Math.* **72** 1–25.

[8] Lidl, R. and Neiderreiter, H. (1994). *Introduction to finite fields and their applications.* Cambridge University Press.

[9] Neshveyev, S. and Størmer, E. (2006). *Dynamical entropy in operator algebras.* Springer-Verlag.

[10] Narnhofer, W., Størmer, E. and Thirring, W. (1995). C^*-dynamical systems for which the tensor product formula for entropy fails. *Ergodic Theory Dynam. Systems.* **15** 961–968.

[11] Newman, M. (1972). *Integral matrices* Academic Press.

[12] Powers, R. T. (1988). An index theory for semigroups of $*$-endomorphisms of $B(H)$ and type II_1 factors. *Canad. J. Math.* **40** 86–114.

[13] Powers, R. T. and Price, G. L. (1994). Cocycle conjugacy classes of shifts on the hyperfinite II_1 factor. *J. Func. Anal.* **121** 275–295.

[14] Price, G. L. (1998). Cocycle conjugacy classes of shifts on the hyperfinite II_1 factor, *II. J. Operator Theory.* **39** 177-195.

[15] Price, G. L. (1999). On the classification of binary shifts of minimal commutant index. *Proc. Nat. Acad. Sci.* **96** 8839–8844.

[16] Price, G. L. (1987). Shifts on finite II_1 factors. *Canad. J. Math.* **39** 492–511.

[17] Price, G. L. (1998). Shifts on the hyperfinite II_1 factor. *J. Func. Anal.* **156** 121–169.

[18] Price, G. L. (1998). The entropy of rational Powers shifts. *Proc. Amer. Math. Soc.* **126** 1715–1720.

[19] Price, G. L. (1999). On the classification of binary shifts of finite commutant index. *Proc. Nat. Acad. Sci.* **96** 14700–14705.

[20] Price, G. L. (1998). Cocycle conjugacy classes of shifts on the hyperfinite II_1 factor. II. *J. Operator Theory.* **39** 177–195.

[21] Price, G. L. (1998). Shifts on the hyperfinite II_1 factor. *J. Funct. Anal.* **156** 121–169.
[22] Price, G. L. and Truitt, G. H. (1999). On the ranks of Toeplitz matrices over finite fields. *Lin. Alg. Appl.* **294** 49–66.

Chapter 8

Symmetric and Quasi-Symmetric Designs and Strongly Regular Graphs

Sharad S. Sane

Department of Mathematics,
University of Mumbai, Vidyanagari,
Santacruz East, Mumbai-400098, India
sane@math.math.mu.ac.in

8.1. Introduction and Preliminaries

Combinatorial Designs first appeared in the medieval literature mainly as puzzles. We will dwell on this theme somewhat later during the course of this article. The medieval forerunner of design theory is thus the famous Euler conjecture, disproved by Bose, Shrikhande and Parker. In the twentieth century, Combinatorial designs, mainly in the form of constructions were first looked at by Statisticians. Specifically a designs or a 2-design has the following definition. By an incidence structure, we mean a triple, consisting of the set of points, the set of blocks and an incidence relation between the two sets. It is common to identify a block with the set of points incident to it; but if there are repeated blocks, then the blocks form a multiset of the sets of points. An incidence structure is called a 2-*design* (or a balanced incomplete block design) if every block has a constant size k, which is strictly less than v, the number of points, and any two points occur together (are commonly contained in) the same number λ of blocks. If b the number of blocks and r the number of blocks containing a given point, then such a configuration is called a (v, b, r, k, λ)-design or sometimes also called a (v, k, λ)-design.

A two-way counting produces following two equations on the parameters of a design **D**.

$$\lambda(v-1) = r(k-1)$$
$$vr = bk.$$

If we look at the (point-block) incidence matrix N of a design **D**, then it is easy to see that the rows of N are linearly independent. Since N is a $v \times b$ matrix, we see that

Theorem 8.1. (**Fisher's inequality**) *We always have $v \leq b$, i.e., the number of points is less than or equal to the number of blocks. Further if D is a design with $v = b$, the incidence structure in which blocks are points and points are blocks is also a design with the same parameter set (v, k, λ).*

Definition 8.1. A design with $v = b$ is called a **symmetric design**.

2-designs were called Balanced Incomplete Block Designs (BIBDs) by statisticians and were used by them in order to get rid of "two-way heterogeneity". For a statistician, the mathematician balance, which just means that every pair of points is contained in exactly λ, a constant number of blocks, translates into statistical balance which means that all the elementary treatment contrasts are estimated with the same variance.

If $k = 3$ and $\lambda = 1$, then the necessary conditions yield $v = 2r + 1$ and hence v is odd. Also the second condition says that 3 divides $r(2r+1)$ and hence v has the form $6m+1$ or $6m+3$. Such designs, with $k = 3$ are called Steiner triple systems. Fano plane is the smallest example of a Steiner triple system. Through recursive constructions, it is not very difficult to show that for all orders $v \equiv 1, 3 \pmod 6$, a Steiner triple system of order v exists. In the year 1847, Kirkman, who was a minister with the Church of England, posed the following question: 15 schoolgirls are to be arranged in batches of three girls each for seven days of the week such that every two of them are together in the same batch in exactly one batch on some day and every day we have 5 batches of girls. This amounts to constructing a Steiner triple system on 15 points and $5 \times 7 = 35$ blocks (or lines) such that the system is "resolvable". The general problem of constructing a resolvable Steiner triple system for every order v of the form $6m + 3$ was solved in 1970s by Ray-Chaudhuri and Wilson.

8.2. Symmetric Designs

Let us have a second look at Fisher inequality and the Symmetric Designs. If N denotes the point-block incidence matrix of such a design D, then the design properties translate into the following matrix equations:

$$NN^t = (r - \lambda)I - \lambda J, \quad JN = kJ, \quad NJ = rJ$$

where I is the identity matrix and J denotes the all 1-matrix (of appropriate order). If $v = b$, then the matrix N is square and non-singular and hence multiplication of the first equation by N^{-1} on the left and N on the right (note that $r = k$, in this case) obtains:

$$N^t N = (k - \lambda)I + \lambda N^{-1} JN$$

But $NJ = kJ$ and hence $N^{-1}J = k^{-1}J$ and we thus get: $N^t N = (k - \lambda)\ \lambda J$. An interpretation of this is the following:

Theorem 8.2. *If D is a symmetric design, the block-point incidence structure D^t called the (combinatorial) dual of D is also a symmetric design with the same parameters v, k and λ.*

Let H denote a Hadamard matrix of order $4t$. This is a matrix of order $4t$ with entries ± 1 such that any two rows of H are orthogonal. Multiplying rows/columns by -1 does not change the Hadamard property and hence we can assume w.l.o.g. that H is in a standard form. That is the first row and the first column of H consist only of $+1$'s. Then deleting the first row and the first column of H and changing -1's to zeroes, we get an incidence matrix of a $(v, k, \lambda) = (4t - 1, 2t - 1, t - 1)$ symmetric design which is called a Hadamard symmetric design. This procedure is reversible and hence the existence of a Hadamard matrix and a Hadamard symmetric design go hand in hand. .

We look at a projective space of dimension 2 over $GF(q)$ and declare its points as points and its lines as blocks of a symmetric design. This is called a projective plane and has parameters $(q^2 + q + 1, q + 1, 1)$. This construction works for every prime power q. More generally, we may take an n-dimensional projective space over $GF(q)$ and declare its points as points and its hyperplanes as blocks of a symmetric design. This is called a projective design and has parameters

$$v = \frac{q^{n+1} - 1}{q - 1},\ k = \frac{q^n - 1}{q - 1},\ \lambda = \frac{q^{n-1} - 1}{q - 1}$$

In the special case, when the projective space has dimension 2, the corresponding symmetric designs have $\lambda = 1$. Symmetric designs of this kind with $\lambda = 1$ (that is, with parameters $v, k, \lambda) = (q^2 + q + 1, q + 1, 1))$ *are defined to be projective planes of order q.*

Construction: Let π be an affine plane of order q with parallel classes of lines $\pi_1, \pi_2, \ldots, \pi_{q+1}$. Let the lines of a-th parallel class π_a be denoted by l_{at}, $t = 1, 2, \ldots, q$. Let the points of π be $c_0, c_1, \ldots, c_{q^2-1}$. For π_a, define a matrix A of order q^2 indexed by the points of π as follows. $A = [a_{ij}]$, $a_{ij} = 1$ iff $i, j \in l_{at}$ for some t and $a_{ij} = 0$ otherwise. If we denote this A by A_a (for the a th parallel class) then,

- A_a is a symmetric matrix.
- $A_a(A_{a'})^t = \begin{cases} \mathrm{q}A_a & \text{if } a' = a \\ \mathrm{J} & \text{if } a' \neq a \end{cases}$
- $\sum_{a=1}^{q+1} A_a = qI + J.$

Now let $L = [L_{\alpha\beta}], \alpha, \beta = 0, 1, 2, \ldots, q+1$ denote a Latin square on the symbol-set $\{0, 1, 2, \ldots, q+1\}$. Replace $L_{\alpha\beta}$ by the matrix A_a if $L_{\alpha\beta} = a$ and replace $L_{\alpha\beta}$ by 0 (a zero matrix of order q^2) if $L_{\alpha\beta} = 0$. Then let the resulting matrix be $M = [M_{\alpha\beta}], \alpha, \beta = 0, 1, 2, \ldots, q+1$. Then (1), (2) and (3) imply that $MM^t = q^2I + qJ$ proving that M is a incidence matrix of a symmetric design with parameters

$$v = q^2(q+2), \quad k = q(q+1), \quad \lambda = q \quad (*)$$

Also, the construction (see [5, 6]) shows that we have a point-partition in which every point class (this corresponds to the juxtaposed $q^2 \times (q^3+2q^2)$ matrix $[M_{\alpha\beta}]$ where β takes all values from 0 to $q+1$) is an affine plane of order q (repeated q times). Hence M is an incidence matrix of a *quasi-affine design (defined later in this exposition).* This family of designs is important for several other reasons. The first construction of such designs was obtained by Ahrens and Szekeres and these designs naturally give rise to strongly regular graphs. Finally, the constructions of Ahrens and Szekeres also connects these objects with generalized quadrangles. We will get back to this theme a bit later in this exposition.

For the reasons of structural symmetry and better connections with group theory, symmetric designs are objects of considerable interest. Many other constructions of symmetric designs are known. The existence question for Symmetric designs is the question of constructing a $(0,1)$-matrix satisfying the matrix equations given above. Algebraic number theory has been employed in order to answer this existence question and the relevant seminal result is called the Bruck-Ryser-Chowla theorem. Unfortunately, it works only in one direction. That is, it provides us with only a necessary condition which, may not be sufficient. For example, it is not known whether there is a projective plane of order twelve but it is known, thanks to the Bruck-Ryser-Chowla theorem that

Theorem 8.3. *If q is the order of a projective plane such that $q \equiv 1, 2 \pmod 4$ then q is a sum of two integer squares. In particular, there are no projective planes of orders q such that $q \equiv 6 \pmod 8$.*

An extensive search ([13]) for almost 200 hours on the fastest CRAY computer available then proved in the late 1980s that there is no projective plane of order ten.

In a recent seminal paper on this difficult problem of constructions of symmetric designs, Ionin ([9]) has obtained an elegant general method that yields about ten different families of symmetric designs not known until three years ago. Improving very strongly on the methods of Dinesh Rajkunlia where the use of weighing matrices and designs was first made, Ionin's constructions have far reaching consequences. Some simplifications of Ionin constructions are published in the last couple of years and the author also expects to write out some simplification of Ionin constructions. We give below one of the theorems of Ionin.

Theorem 8.4. *There exist symmetric designs D with parameters:*

$$v = 1 + 2(q+1)\frac{(q+1)^{2m} - 1}{q+2}, k = (q+1)^{2m}, \quad \lambda = \frac{(q+1)^{2m-1}(q+2)}{2}$$

where $q = 2^p - 1$ is a Mersene prime and m is a positive integer.

About 23 infinite families of symmetric designs are known. However, all the *known examples* of symmetric designs seem to be only of the following types:

- When $\lambda = 1$, we have a projective plane of order q with parameters $(q^2 + q + 1, q + 1, 1)$. These exist for every prime power q. *No other examples are known.*
- When $\lambda = 2$, we have a biplane with parameters $(\binom{k}{2} + 1, k, 2)$. These are known to exist for the following values of k: $3, 4, 5, 6, 9, 11, 13$. *No other examples are known.*
- When $\lambda = 3$, all the *known examples have k bounded by* 15.
- When $\lambda \geq 4$, all the *known examples have k bounded by* $\lambda^2 + \lambda$.

The known situation led to the following informal conjecture attributed to M. Hall.

Conjecture 8.1.

(a) $\forall\lambda \geq 2$, *there exist only finitely many symmetric* (v, k, λ).
(b) Stronger form: $\forall\lambda \geq 4$, *the parameters of a symmetric design satisfy* $k \leq \lambda^2 + \lambda$.

Symmetric designs with parameters (v, k, λ) where k equals $q^2 + q$ and λ equals q are also related to the famous design extension theorem of Cameron. We will refrain from discussing the actual statement of Cameron extension theorem but will discuss the original problem of permutation group extensions from which the design extension problem arose. Let (G, X) denote a permutation group. We call the action of G on X transitive, if given any points x and y in X there is some α in G such that $\alpha(x) = y$. In general, G is said to be t-transitive on X if given any t-tuples of points X say $(x_1, x_2, \ldots, x_t)$ and $(y_1, y_2, \ldots, y_t)$, we have some α in G such that $\alpha(x_i) = y_i$ for all $i = 1, 2, \ldots, t$. Further, this action is sharp (i.e. G is sharp t-transitive) if such an α is unique. Trivial examples of sharp n-transitive actions are the symmetric groups S_n. Jordan proved in the last century that no non-trivial sharp t-transitive group exists for $t \geq 6$ and for $t = 4, 5$ the non-trivial sharp 4 and 5-transitive groups are precisely the Mathieu groups M_{11} and M_{12} that act on sets of orders 11 and 12 respectively. We call a symmetric design $\mathbf{D}$ with parameters (v, k, λ) extendable if there is a 3-design $\mathbf{D}'$ with parameters $v' = v + 1, k' = k + 1, \lambda' = \lambda$ such that if we look at all the blocks containing a single point ∞ of $\mathbf{D}'$, then the resulting incidence structure is $\mathbf{D}$. The symmetric design extension theorem of Cameron is the following.

Theorem 8.5. *Let* $\mathbf{D}$ *be a symmetric* 2*-design that has an extension (to a* 3*-design). Then* $\mathbf{D}$ *has one of the following as its parameters.*

(1) $\mathbf{D}$ *is a Hadamard* 2*-design.*
(2) $v = (\lambda + 2)(\lambda^2 + 4\lambda + 2)$ *and* $k = \lambda^2 + 3\lambda + 1$.
(3) $(v, k, \lambda) = (495, 39, 3)$.
(4) $\mathbf{D}$ *is a projective plane of order ten and hence is a* 2*-design with* $(v, k, \lambda) = (111, 11, 1)$.

The first type in this list contains all the Hadamard symmetric designs and extensions exists for *all the Hadamard symmetric designs.* Other than this infinite class, there is a sporadic 3-design and a parametrically possible projective plane of order ten. Besides these, the Cameron classification theorem also The second type is a putative infinite list of extendable symmetric designs, the first object of which is the projective plane of order four. This object does have an extension, not just to a 3-design but even to 4 and 5-designs. These extensions, called the Witt designs, are intimately connected with the Mathieu groups M_{22}, M_{23} and M_{24}. The next parameter set in this list is a symmetric design with $(v, k, \lambda) = (56, 11, 2)$. As we already saw in the discussion on M. Hall's conjecture, none of the designs of the second type with $\lambda \geq 3$ has been constructed so far and nor has any one shown that designs with such parameters cannot exist. (the question of extension naturally makes sense only after that). The same remark also applies to a design of the third type with $(v, k, \lambda) = (495, 39, 3)$ whose existence is an open question. Fortunately, thanks to the result of Lam cited earlier, the fourth type cannot arise because there is no projective plane of order ten (though this was not known at the time Cameron proved his theorem).

We briefly deal with the question of non-isomorphism. An isomorphism ϕ between two symmetric designs D_1 and D_2 is a pair (ϕ', ϕ'') of two permutations, the first from the point-set of D_1 to the point-set of D_2 and the second from the block-set of D_1 to the block-set of D_2 such that the incidence is preserved. When D_1 equals D_2 this is called an automorphism and it is easily seen that the set of automorphisms of a symmetric design form a group. This is one way in which we can actually construct a symmetric design from a group. This is called the method of difference set. For example developing the initial block $(0, 1, 3)$ modulo 7 gives us all the 7 blocks of the Fano plane. The problem of determining whether or not two given symmetric designs are isomorphic is essentially a uniform hypergraph isomorphism problem (hypergraphs are objects for which edges are subsets of size greater than 2). This problem is evidently at least as hard as the corresponding problem of graph isomorphism which is known to be an NP-complete problem.

It is on this background that producing a large number of non-isomorphic symmetric designs (with the same parameters) is a pertinent question. This question was first handled in the case of Hadamard symmetric designs by Bhat-Nayak, Shrikhande and Singhi. For the case of projective designs (the point-hyperplane incidence structure), this was done first by Kantor and later by Jungnickel. In both the cases, the lower bounds are asymptotically exponential. A recent result establishes the same type of result for designs of Ahrens-Szekeres type, i.e., those designs with $k = \lambda^2+\lambda$. Among various lower bounds proved in that paper of Gharge and Sane ([6]), one is in terms of Catalan numbers and one is in terms of special partitions of integers. Since both these numbers are exponential, the established lower bound is also exponential. We conclude by pointing that there is a marked difference between techniques employed by the earlier authors (Shrikhande, Singhi, Jungnickel) and those in the recent paper. In the former situation, techniques are recursive while in the latter situation they are direct and depend on graph isomorphisms as well as partition functions in number theory. This forms a part of the recent Ph.D. thesis of Gharge. Specifically, call a symmetric design of the Ahrens-Szekeres type, that is one with parmeters

$$v = q^2(q+2), \quad k = q(q+1), \quad \lambda = q \quad (*)$$

quasi-affine if its point-set admits a partition into $q+2$ point classes each with q^2 points such that the induced incidence structure in each point class is an affine plane of order q (repeated q times). We have already come across this special class of symmetric designs with parameters as in $(*)$ in Construction in Section 8.2. It was shown by the author that such a partition (if it exists) is unique (that is, it is uniquely determined by the structure of the given incidence structure. This was exploited in the same paper of the author, to show that, upto isomorphism, there are at least two non-isomorphic quasi-affine designs for any given prime power q. This result was vastly improved recently and two of the main results of Sanjeevani Gharge's Ph.D. thesis ([5]) are the following.

Theorem 8.6. *Let q be a prime power. Then:*

(1) The number of non-isomorphic solutions of quasi-affine designs with parameters $(*)$ is at least $m^{\sqrt{q}}$.

(2) Let C_m denote the m-th Catalan number defined by

$$C_m = \frac{1}{m}\binom{2m-2}{m-1}$$

and let $q \geq 7$. Then the number of non-isomorphic quasi-affine designs with parameters $(*)$ is at least $2mC_m$ (in both the cases m denotes $[\frac{q}{2}]$ and here $[x]$ denotes the largest integer less than or equal to x).

8.3. Strongly Regular Graphs

Recall that a regular graph Γ of degree k is a simple graph in which every vertex has degree k. The graph Γ with n vertices and regular of degree k is called a strongly regular graph with parameters (n, k, a, c) if we have two more constants a and c such that

- If x and y are adjacent vertices then the number of vertices commonly adjacent to both of them is a.
- If x and y are non-adjacent vertices then the number of vertices commonly adjacent to both of them is c.

It is not difficult to see that the complement of a strongly regular graph (srg) Γ is also strongly regular. *In order to make the situation non-trivial and interesting, we insist that both Γ and its complement are connected graphs on at least $n \geq 4$ vertices. In particular, this means that neither Γ nor its complement are complete graphs.* Here are some examples.

(1) *A Triangular Graph T_m* is defined to be the line graph $L(K_m)$ of a complete graph. Equivalently, T_m has a vertex set that consists of all unordered pairs (2-subsets) of the set $M = \{1, 2, \ldots, m\}$ of order m with two vertices adjacent iff the corresponding pairs have an element in common. The parameters of T_m are:

$$\left(\binom{m}{2}, 2m-4, m-2, 4\right)$$

(2) *A Lattice Graph (sometimes also called a Latin Square Graph) X* $L(m)$ is the line graph $L(K_{m,m})$ and has the following description. $V = M \times M$ where M is as in the previous example. Two vertices of X are adjacent iff they have the same first or second co-ordinate. X has the parameters:

$$(m^2, 2m-2, m-2, 2)$$

When $m \neq 4$, such graphs are uniquely determined by their parameters. For $m = 4$, there is an exceptional graph called the Shrikhande graph. This is not isomorphic to $L(4)$ and in fact has an embedding on a torus (which $L(4)$ does not have). Another reason for non-isomorphism is that the full automorphism of the Shrikhande graph has order $16 \times 12 = 192$ (in fact, a point stabilizer is the dihedral group D_6) while the full group of $L(4)$ has order $2 \times (4!)^2$.

(3) The Petersen graph P is a graph on 10 vertices and is the complement of T_5. Its parameters are $(10, 3, 0, 1)$. This graph is also a quotient of the graph of the 1-skeleton of a regular dodecahedron. Its automorphism is the symmetric group S_5.

Strongly regular graphs were first defined by Bose in 1959 in the context of embedding problems of Bruck nets, though, in the form of association schemes these objects were introduced earlier in the paper of Bose and Nair. These objects have connections with group theory. Specifically, let G denote a rank 3 permutation group. Recall that this is a transitive group in which the stabilizer G_x of any point x has precisely two orbits other than the singleton x. It is easily seen that this is independent of the choice of x. Equivalently, rank of a transitive permutation group G acting on a set X is the number of orbits of $X \times X$ (again, one orbit here is the diagonal $\{(x, x)\}$). If these orbits are symmetric (and this happens if and only if G has even order), that is (x, y) and (y, x) are in the same orbit, then we can define a graph Γ whose edges are the pairs in one of the non-diagonal orbits. This graph is strongly regular and the given group G acts as an automorphism group of this graph. A prototypical situation where this holds is the group G that consists of the set of all affine transformations of a finite field K given by: $x \to a^2x + b$ where a and b are elements of K and a is a non-zero quadratic residue. To get symmetric orbits, we require that $|K| \equiv 1(mod4)$. The family of graphs obtained through this construction is called the Paley family of graphs.

An adjacency matrix A of a graph Γ or order n is a binary square matrix of order n with entry 1 at (i, j)-th place iff the i-th and j-th vertices are adjacent and 0 otherwise. A is a real symmetric matrix and hence must have real eigenvalues.

$$\begin{aligned} A^2 &= kI + aA + c(J - I - A) \\ &= (k - c)I + (a - c)A + cJ \end{aligned}$$

and hence we get:

$$A^2 - (a - c)A - (k - c)I \ = cJ$$

Since Γ is connected and regular of degree k, k is a simple eigenvalue of A with a corresponding eigenvector that consists of all 1's. The other eigenvalues can be read off from the above matrix equation and they satisfy the quadratic equation:

$$x^2 - (a - c)x - (k - c) \ = 0 \qquad (*)$$

Let θ and τ denote the other eigenvalues of A. Then Γ connected implies that these eigenvalues have smaller modulus than k. Further, since the product $\theta\tau$ is $-(k-c) < 0$ as shown by $(*)$ we see that one of these numbers is positive and the other is negative, where, by convention, we take θ to be positive. *It was shown by Shrikhande and Bhagwandas that a regular and connected graph Γ is strongly regular iff it has exactly two eigenvalues besides the degree of regularity.* Now let Δ denote the discriminant of $(*)$. Then we have

$$\begin{aligned} \Delta &= (a-c)^2 + 4(k-c) \\ &= (\theta+\tau)^2 - 4\theta\tau \\ \text{2pt]} &= (\theta-\tau)^2 \end{aligned}$$

Let m_θ and m_τ denote the respective multiplicities of the eigenvalues θ and τ. Since the trace of A is 0, we can read off two equations for m_θ and m_τ in terms of θ and τ:

$$m_\theta + m_\tau = n - 1$$

$$m_\theta \theta + m_\tau \tau = -k$$

Theorem 8.7. (Friendship theorem of Erdös and Sös and the (v, k, λ)-graphs) *In a gathering of n people every two persons have exactly one common friend. The Friendship theorem then asserts that there is a person who knows all the rest (and the corresponding graph is the windmill graph whose picture is shown below):*

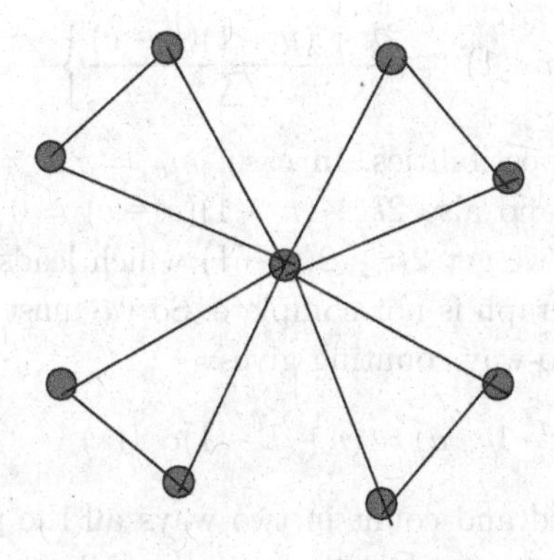

For a proof by contradiction, we assume there is no one who knows all the rest. Then a small argument will prove that every person has the same number of friends and hence the graph Γ is regular and the given conditions imply that it is strongly regular with $a = c = 1$ and plugging these values shows that Δ equals $2\sqrt{k-1}$. This gives the values of θ and τ to be $\pm 2\sqrt{k-1}$ and since m_θ and m_τ must be integers, we see that $(m_\theta - m_\tau)(2\sqrt{k-1})$ equals $-k$ from which it follows that $k-1$ is a perfect square and $\sqrt{k-1}$ divides k which holds only if $k = 2$, a contradiction. This investigation can be generalized to the following: Every two persons have the same number λ of friends (here $\lambda \geq 2$) What are the possibilities? The corresponding strongly regular graphs are called (v, k, λ)-graphs ([19]) and they give rise a class of symmetric designs.

Further Investigations of Strongly Regular Graphs: ([7])

We could, of course, solve the equations explicitly for θ, τ and m_θ and m_τ to get:

$$\theta = \frac{a - c + \sqrt{\Delta}}{2} \; : \; \tau = \frac{a - c - \sqrt{\Delta}}{2}$$

and

$$m_\theta = \frac{(n-1)\tau + k}{\theta - \tau}$$

$$m_\tau = \frac{(n-1)\theta + k}{\theta - \tau}$$

Alternative expressions for m_θ and m_τ are

$$\frac{1}{2}\left\{(n-1) \mp \frac{2k + (n-1)(a-c)}{\sqrt{\Delta}}\right\}$$

There are two distinct possibilities. In case, $m_\theta = m_\tau = m$, clearly, we must have $2m = n - 1$ and also $2k + (n-1)(a-c) = 0$ so that $a < c$ and if $a \leq c - 2$, then we get $2k \geq 2(n-1)$ which leads to $k = n - 1$ a contradiction since the graph is not complete. So we must have $a = c - 1$ and $k = m$. Finally a two-way counting gives:

$$k(k-1-a) = (n-k-1)c \quad (*)$$

(Hold a vertex x fixed and count in two ways all the pairs (y, z) such that y is adjacent to x but z is not adjacent to x while y and z themselves are adjacent.)

Substitution of $n = 2m + 1$ and $k = m$ gives the simplification $a + c = m - 1$ and hence we see that $c = \frac{m}{2}$ and the parameters are:

$$(n, k, a, c) = \left(q, \frac{q-1}{2}, \frac{q-5}{4}, \frac{q-1}{4}\right)$$

where $q \equiv 1 \ (mod\ 4)$. This is customarily called the half-case and the Paley graphs are examples of the half-case. We have already seen the construction of Paley graphs. Paley graphs exhibit a remarkable level of structural symmetry. Their automorphism groups are rank 3 groups. The following two points should be noted about the half-case strongly regular graphs.

(1) A strongly regular graph falls into the half-case if and only if has the same parameter as its complement. In particular, a strongly regular graph which is isomorphic to its complement (a self-complementary graph) belongs to this case.
(2) A necessary condition for the existence of a graph falling in the half-case is that n must be a sum of two squares.

Not the Half-Case; Moore Graphs:

A regular graph of degree k and diameter 2 has at the most $k^2 + 1$ vertices and in case of equality such a graph is called a *Moore graph.* Our investigation will now include these graphs. In case we are not in the half-case, the difference between m_θ and m_τ is non-zero and hence it is easily seen that $2k + (n-1)(a-c)$ is a non-zero integer. In order for m_θ and m_τ to be rational (in fact integers) it is then essential that Δ is a square of an integer and the integer $\sqrt{\Delta}$ must divide $2k + (n-1)(a-c)$. We record this in the following theorem.

Theorem 8.8. *Let Γ be a strongly regular graph which is not a half-case. Then:*

- $\Delta = (a-c)^2 + 4(k-c)$ *is a perfect square say* u^2.
- u *divides* $2k + (n-1)(a-c)$.

Moore graphs are characterized by $a = 0$ and $c = 1$. Here we neither have triangles (because $a = 0$) nor 4-cycles (because $c = 1$) but have a very large number of pentagons. Here we have $u^2 = 4k - 3$. The counting equation produces $n = k^2 + 1$ and hence u divides $k^2 - 2k = k(k-2)$. If $k = 2$ then we have C_5 the cycle with 5 vertices. In other cases, u divides

$(\frac{u^2+3}{4})(\frac{u^2-5}{4})$ and hence it follows that u divides 15 giving $1, 3, 5, 15$ as the only possibilities for u. Here $u = 3$ corresponds to $k = 3$ and we have the Petersen graph. When $u = 5$, we get $k = 7$ and $n = 50$ and we get the Hoffman-Singleton graph. When $u = 15$, we have $k = 57$ and the existence of a graph in this situation is an open question.

The question of existence of a strongly regular graph (n, k, a, c) satisfying the obvious necessary conditions $(*)$ is wide open. There are some other intricate necessary conditions such as the Krein condition. A good source of strongly regular graphs, from the point of view of a design theorist, are the 2-designs that have $\lambda = 1$. For such a design, any two blocks intersect in either 0 or 1 point and we can thus declare two blocks to be adjacent if they are disjoint. This graph turns out to be strongly regular (the class of quasi-symmetric designs studied in the next section is a generalization of such designs). Even good characterizations of such specialized strongly regular graphs are not available. Designs of this type (those with $\lambda = 1$) are a special case of what Bose called partial geometries but we will not go into these things in the present exposition. A special class of partial geometries is the class of Bruck nets, which are equivalent to sets of mutually orthogonal Latin squares. Therefore, sets of mutually orthogonal Latin squares provide prolific source of examples of strongly regular graphs.

8.4. Quasi-Symmetric Designs

Take a design D, not necessarily symmetric. An integer x is called a block intersection number of D if we have two blocks X and Y the cardinality of whose intersection is x. Which numbers occur as block intersection numbers of a design? Thanks to the proof of Fisher's inequality, we see that D has exactly one block intersection number iff it is a symmetric design.

Definition 8.2. A design $\mathbf{D}$ with two block intersection numbers x and y (with $x < y$ by convention) is called a **Quasi-symmetric design**. Such a design is called a *proper quasi-symmetric* design if both the block intersection numbers occur (and hence $\mathbf{D}$ is *not a symmetric design*).

Reasons for studying quasi-symmetric designs are many. A mundane and practical reason is that symmetric designs are more difficult to study (this is not completely true but sometimes believed to be so). On a more serious level quasi-symmetric design allows one to construct its block graph which in most cases of interest can be shown to be strongly regular. Finally quasi-symmetric designs are connected with combinatorial configurations arising out of finite simple groups. Here is a small list of examples.

(1) As was observed in the last section block designs with $\lambda = 1$ are quasi-symmetric with $(x, y) = (0, 1)$.
(2) Every n-dimensional affine geometry (with hyperplanes as blocks) over $GF(q)$ furnishes an example of a quasi-symmetric design. Here $v = q^n$ and $k = q^{n-1}$ while the intersection pair (x, y) is given by $(0, q^{n-2})$.
(3) Let s be a fixed positive integer. Take a symmetric (v, k, λ)-design $\mathbf{D}$ and let $\mathbf{D}'$ be obtained from $\mathbf{D}$ as follows: $\mathbf{D}'$ has the same point-set as that of $\mathbf{D}$. As blocks of $\mathbf{D}'$ we take s copies of all the blocks of $\mathbf{D}$ (customarily, this is expressed by saying that each block of $\mathbf{D}$ is repeated s times). Then $\mathbf{D}'$ has parameters

$$v' = v, b' = sv, r' = sk, k' = k, \lambda' = \lambda \quad (*)$$

This design is easily seen to be quasi-symmetric (there are repeated blocks) and we have $x = \lambda$ and $y = k$. However, a 2-design $\mathbf{D}'$ with parameters as above may actually exist independent of the existence of a symmetric (v, k, λ)-design. Such a 2-design $\mathbf{D}'$ is called an *s-quasi-multiple design* (of a symmetric (v, k, λ)-design), where s is an integer, $s \geq 1$. Observe also that the existence question of an s-quasi-multiple design makes sense irrespective of whether a corresponding symmetric design exists or not. But this question naturally assumes a lot of importance when a corresponding symmetric design does not exist (is known not to exist). It is obvious that quasi-multiple designs with particular parameters can be constructed using suitable subsets of a finite field. For example, a $(7, 3, 1)$ symmetric design is constructed by taking quadratic residues in the field with 7 elements. Similarly, a $(37, 9, 2)$ symmetric design is constructed by taking biquadratic residues in the field $GF(37)$. In general, suppose $v = \frac{k(k-1)}{2} + 1$ is such that a $GF(v)$ exists (necessarily then v is a prime power). Now suppose $u = \frac{k-1}{2}$ is an integer. Now consider the "development" of sets of the following kind in the sense of C. R. Rao or R. C. Bose:

$$A_i = \{\alpha^i, \alpha^{u+i}, \ldots, \alpha^{(k-1)u+i}\}$$

where α is a primitive element of the field. Then by taking sufficiently many A_i's and their developments, one will have constructed a quasi-multiple of a biplane.

(4) Delete a block of a biplane (that is a symmetric design with $\lambda = 2$). What we get is a design in which every two blocks intersect in either 1 or 2 points. We thus get a quasi-symmetric design. Unfortunately, due to the fact that only a finite number of biplanes is known as of now, this will only produce a finite number of quasi-symmetric designs. The famous Hall-Connor embedding theorem shows that such designs can always be embedded into a unique biplane by supplying the missing block.

(5) Let D be a point-block incidence structure constructed from $PG(n, q)$ a projective geometry of dimension n over a field with q elements by taking as points the points of the geometry and as blocks the subspaces of codimension two (where $n \geq 3$). Parameters of D as a quasi-symmetric design are:

$$v = \frac{q^{n+1} - 1}{q - 1}, \quad k = \frac{q^{n-1} - 1}{q - 1}$$

$$\lambda = \frac{q^{n-1} - 1}{q - 1}, \quad x = \frac{q^{n-3} - 1}{q - 1}, \quad y = \frac{q^{n-2} - 1}{q - 1}.$$

There are other classes of examples particularly the affine geometries (where $x = 0$) There is also a classical object called the Witt design on 23 points which is associated with the Mathieu group M_{23} on 23 letters.

Define the block graph Γ of a quasi-symmetric design D by taking as vertices of Γ the blocks of D. Make two vertices adjacent iff the corresponding blocks intersect in x points. Let N denote the incidence matrix of D and A, the adjacency matrix of Γ. Recall that we have already established the following matrix equations:

$$NN^t = (r - \lambda)I \; \lambda J, \quad N^t J = kJ, \quad NJ = rJ$$

Here, D is quasi-symmetric and hence the following matrix equation connects $N^t N$ and A:

$$\begin{aligned} N^t N &= kI \; + xA + y(J - I - A) \\ &= (k - y)I + (x - y)A + yJ \end{aligned}$$

Since N^tN commutes with J (check this!) it follows that A also commutes with J which is just the same thing as saying that Γ is a regular graph. If we can now show that besides the degree of regularity, A has exactly two other eigenvalues, then Γ must be a strongly regular graph. But NN^t and N^tN have the set of non-zero eigenvalues. So, N^tN has eigenvalues $rk, r - \lambda$ and 0. The eigenvalue rk is a simple eigenvalue of N^tN that corresponds to the largest simple eigenvalue of A which is also the degree of regularity of Γ. Hence A has exactly two other eigenvalues proving that the block graph is strongly regular.

The connection between quasi-symmetric designs and strongly regular graphs has paved way for a lot of recent interesting research work in the area of quasi-symmetric designs. Some prominent results in this area include the classification of quasi-symmetric designs with $x = 0$ and those for which the strongly regular graph contains no triangles. This study was initially undertaken by Baartmans and M. S. Shrikhande. This problem also has connections with the design extension theorem of Cameron mentioned earlier in this paper.

Theorem 8.9. ([22]) *Let* $\mathbf{D}$ *be a proper quasi-symmetric design with block intersection numbers* $x = 0$ *and* $y \geq 2$. *Then* $\mathbf{D}$ *has only a finite number of possibilities under any one of the following assumptions.*

(1) k *is fixed.*
(2) y *is fixed.*
(3) λ *is fixed.*

This theme has been greatly explored in a number of recent publications of Pawale. It was proved by Pawale and the author ([17]) as well as Calderbank and Morton that:

Theorem 8.10. *Up to isomorphism, the only quasi-symmetric* 3*-design with* $(x, y) = (1, 3)$ *are the unique* $4 - (23, 7, 1)$*-design or its block residual.*

A paper of Sane and M. S. Shrikhande ([23]) began the task of classifying quasi-symmetric 3-designs. Various finiteness results were proved in that paper. It should be noted here that extremely few cases of the existence of a quasi-symmetric 3-designs are known, In fact, upto complementation, these are just the Witt designs mentioned in the above theorem. Classification of quasi-symmetric 3-designs with $x = 1$, and Pawale's work in his Ph.D. thesis has also completed this for all quasi-symmetric 3-designs with x upto

100. Pawale has also obtained a classification of all quasi-symmetric 3-designs whose block graph is triangle-free and essentially with the exception of the Witt designs these are the Hadamard 3-designs. The classification programme of quasi-symmetric 3-designs has an obvious connection with what are called tight designs. Specifically, a tight $2s$-design is one that has s blocks intersection numbers. It was proved by Ray-Chaudhuri and Wilson that any $2s$-design must have at least s block intersection numbers (generalization of Fisher inequality for higher designs).

Recall that an s-quasi-multiple design $\mathbf{D}'$ has parameters

$$v' = v, b' = sv, r' = sk, k' = k, \lambda' = \lambda \quad (*)$$

Call such design *proper* if it is *not obtained* by simply repeating each block of a symmetric (v, k, λ)-design s times. Jungnickel and Tonchev considered proper s-quasi-multiples which are also *quasi-symmetric*. Let us call such a design *a special design*. Perhaps the existence question of special designs will shed some light on the connection between symmetric and quasi-symmetric designs. However, only one example of a special design seems to be known, the unique design with 22 points and 176 blocks which is a proper 8-fold quasi-multiple of a non-existent $(22, 7, 2)$ symmetric design. In that paper of Jungnickel and Tonchev, various results on special designs were proved. In particular, the authors proved that no such special designs which are quasi-multiples of projective planes exist. the author proved a classification result for special designs corresponding to biplanes, extending the results in Jungnickel and Tonchev. After partial results by Jungnickel and Tonchev ([12]) and Ionin and M. S. Shrikhande, the following result was proved by the author ([21]).

Theorem 8.11. *Let D be a s-fold quasi-multiple design with parameters $v, sv, sk, k, s\lambda$. Let D be also a quasi-symmetric design. Further assume that the parameters satisfy $(k, (s-1)\lambda) = 1$ (here (m, n) denotes the g.c.d. of m and n). Then D cannot be proper. That is, D* **must be a multiple of a symmetric design**.

Other than purely combinatorial techniques employed in the study of symmetric and quasi-symmetric designs, we also have techniques derived from linear algebra, number theory and coding theory. This article has discussed linear algebra of designs and strongly regular graphs in some detail. The number theoretic techniques stem mainly from the Hasse-Minkowski theory of quadratic form as applied to a binary matrix. Coding theoretic techniques mainly rely on the weight enumerator identities and deeper results in the algebra of codes that use invariant theory. Though it is a fact that 'most of the strongly regular graphs' have trivial automorphism groups, looking for symmetries in symmetric and quasi-symmetric designs as well as strongly regular graphs, is one of the strong motivations for studying these combinatorial objects. At a meta level, it is to be expected that this study will also reflect on the structures of many interesting finite groups (and vice versa), the study of which has been carried out extensively in the last hundred years.

Acknowledgments

The author wishes to record his thanks to the referee for a very careful reading of the earlier version of this article and for many useful suggestions that helped improve the presentation.

References

[1] Assmuss, E. F. Jr. and Key, J. D. (1994). *Designs and their codes.* Cambridge Tracts in Mathematics. **103** Cambridge University Press.

[2] Baartmans, A. and Sane, S. (2006). A characterization of projective subspaces of codimension two as quasi-symmetric designs with good blocks. *Discrete Math.* **306** 1493-1501.

[3] Beth, Th., Jungnickel, D. and Lenz, H. (2000). *Design theory, Encyclopedia of Mathematics and it Applications.* **1** and **2** Cambridge University Press.

[4] Cameron, P. J. and van Lint, J. H. (1991). *Designs, graphs, codes and their links.* London Mathematical Society Student Texts 22, Cambridge University Press.

[5] Gharge, S. S. (2008). *Linear Codes of Combinatorial Configurations.* Ph.D. thesis, University of Mumbai.

[6] Gharge, S. and Sane, S. (2007). Quasi-affine symmetric designs. *Des. Codes Cryptogr.* **42** 145-166.

[7] Godsil, C. and Royle, G.(2001). *Algebraic graph theory.* Graduate Texts in Mathematics. **207** Springer-Verlag.

[8] Holzmann, W. H., Kharaghani, H. and Sane, S. (2006). On a class of quasi-symmetric designs. *J. Combin. Math. Combin. Comput.* **57** 103-106.

[9] Ionin, Y. J. (1998). A technique for constructing symmetric designs. *Des. Codes Cryptogr.* **14** 147-158.

[10] Ionin, Y. J. and Shrikhande, M. S. (1994). On a conjecture of Jungnickel and Tonchev for quasi-symmetric designs. *Journal of Combinatorial Designs.* **2/1** 49-59.

[11] Ionin, Y. J. and Shrikhande, M. S. (2006). *Combinatorics of Symmetric Designs.* (New Mathematical Monographs). Cambridge University Press.

[12] Jungnickel, D. and Tonchev, V. D. (1991). Intersection numbers of quasi-multiples of symmetric designs. In *Advances in Finite Geometries and Designs.* Proceedings of the Third Isle of Thorns Conference, 1990. (J. W. P. Hirschfeld, D. R. Hughes and J. A. Thas, Eds), Oxford University Press, Oxford, 227-236.

[13] Lam, C. W. H., Thiel, L. and Swiercz, S. (1989). The non-existence of finite projective planes of order 10. *Canadian Journal of Mathematics.* **XLI** 1117-1123.

[14] Pawale, R. M. (1991). Quasi-symmetric 3-designs with triangle-free graph. *Geom. Dedicata.* **37** 205-210.

[15] Pawale, R. M. (2004). Quasi-symmetric 3-designs with a fixed block intersection number. *Australas. J. Combin.* **30** 133-140.

[16] Pawale, R. M. (2007). Quasi-symmetric designs with fixed difference of block intersection numbers. *J. Combin. Des.* **15** 49-60.

[17] Pawale, R. M. and Sane, S. S. (1991). A short proof of a conjecture on quasi-symmetric 3-designs. *Discrete Math.* **96** 71-74.

[18] Rotman, J. J. (1999). *Introduction to the theory of groups.* Graduate Texts in Mathematics. Springer Verlag.

[19] Rudvalis, A. (1971). (v, k, λ)-graphs and polarities of (v, k, λ)-designs. *Math. Z.* **120** 224-230.

[20] Sane, S. (2000). Quasi-multiple quasi-symmetric designs corresponding to biplanes. Proceedings of the Thirty-first Southeastern International Conference on Combinatorics, Graph Theory and Computing (Boca Raton, FL,-2000). *Congr. Numer.* **146** 213-216.

[21] Sane, S. (2001). A proof of the Jungnickel-Tonchev conjecture on quasi-multiple quasi-symmetric designs. *Des. Codes Cryptogr.* **23** 291-296.

[22] Sane, S. S. and Shrikhande, M. S. (1986). Finiteness questions in quasi-symmetric designs. *J. Combin. Theory Ser. A.* **42** 252-258.

[23] Sane, S. S. and Shrikhande, M. S. (1987). Quasisymmetric $2, 3, 4$-designs. *Combinatorica.* **7** 291-301.

[24] Sane, S. S. and Shrikhande, M. S. (1993). Some characterizations of quasi-symmetric designs with a spread. *Des. Codes Cryptogr.* **3** 155-166.

[25] Sane, S. S., Shrikhande, S. S. and Singhi, N. M. (1985). Maximal arcs in designs. *Graphs Combin.* **1** 97-106.

[26] Shrikhande, M. S. and Sane, S. S. (1991). *Quasi-symmetric designs.* London Mathematical Society Lecture Notes No. 164, Cambridge University Press.

[27] Suzuki, M. (1982, 1986). *Group theory.* **I** and **II**. Springer Verlag.

[28] Van Lint, J. H. (2000). *An introduction to Coding theory.* Graduate Texts in Mathematics, Springer Verlag.

Chapter 9

Perturbation Determinant, Krein's Shift Function and Index Theorem

Kalyan B. Sinha

Jawaharlal Nehru Centre for Advanced Scientific Research, Bangalore-64, India
and
Indian Institute of Science, Bangalore-12, India
kbs_jaya@yahoo.co.in

The concept of a determinant in an infinite-dimensional space is introduced, leading to the perturbation determinant for a pair of self adjoint operators (H, H_0) as a meromorphic function in complex plane. The winding number of this function counts the difference of discrete spectra of H and H_0. As an application the generalized Levinson theorem connecting the number of bound states of H and the Krein's shift function at 0+ is proved.

9.1. Introduction

One can extend the notion of a determinant to that of an operator of the form $I+T$ where T is a finite rank or trace-class operator on an infinite dimensional Banach or Hilbert space respectively. This determinant satisfies many of the properties of the usual finite-dimensional one. This leads to a natural definition of a meromorphic function, called perturbation determinant, defined in an open subset of the complex plane for a pair $(T,\ T+A)$ where T is a closed operator and A is either finite-rank on a Banach space or trace class in a Hilbert space. This constitutes essentially the content of the next section which ends with a theorem, known as Weinstein-Aronsajn formula that connects the total number of eigenvalues of the pair $(T,\ T+A)$ (counting multiplicities) with the poles and zeros of the perturbation determinant of the pair.

The third section introduces the Witten index in terms of the trace of the difference of the resolvents of two self-adjoint operators on a Hilbert space and proves a theorem about the invariance of such an index under "small" perturbations. The fourth section deals with Krein's theorems where the trace of the difference of the resolvents mentioned above is given as an integral representation in terms of a real valued function defined on the union of the spectra of the two self adjoint operators. This function is called Krein's shift function and it is shown that under suitable conditions, the Witten index equals (up to a sign) the value of the Krein's shift function at $0+$. In the final section all these results are put together to show that in many situations of interest in the scattering theory in quantum mechanics, the shift function at $0+$ is an integer (the total number of bound states in physicist's language) so that the Witten index is a negative integer which is invariant under small perturbations.

9.2. Perturbation Determinant

Let X be a Banach space and let $T \in B(X)$, the Banach space of all linear bounded operators on X with Domain of $T = X$. Assume furthermore, that T is degenerate, i.e. range of $T \equiv Ran(T)$ is finite dimensional. In such a case, T can be represented as:

$$Tu = \sum_{j=1}^{m} < e_j,\ u > y_j \tag{9.1}$$

where $m = rank(T) \equiv \dim\ Ran(T)$; $\{y_j\}_{j=1}^m$ is a basis for $Ran(T)$ and $\{e_j\}_{j=1}^m$ are linearly independent elements of X^*. Clearly Ran(T) is invariant under the action of T and if we denote by T' the restriction of T to the $Ran(T)$, then we can define the determinant of $I+T$ by

$$\det(I+T) = \det(1'+T') \tag{9.2}$$

which implies that

$$\det(I+T) = \det[\delta_{jk} + \langle e_j,\ y_k\rangle]. \tag{9.3}$$

In the identity (9.3), $y_k \in Ran(T)$ is looked upon as a (finite - dimensional) subset of X and the right hand side is the usual determinant of matrices.

With these definitions, it is easy to derive the property of such determinants: $\det\{(I+S)(I+T)\} = \det(I+S)\det(I+T)$, where T and S are two degenerate operators in $\mathcal{B}(X)$. If one defines furthermore the trace of T as

$$trT \equiv trT' = \Sigma_{j=1}^m < e_j,\ y_j >, \tag{9.4}$$

then one has the trace property

$$trTA = trAT, \quad \forall A \in \mathcal{B}(X), \tag{9.5}$$

the proof of which reduces to the same for a pair of finite dimensional matrices. Furthermore the $m \times m$ dimensional matrix $((\delta_{jk} + < e_j,\ y_k >))$ does not have any zero eigenvalue. For its having zero eigenvalue is equivalent to the null space of $(I' + T')$ on $Ran(T)$ being non-trivial. Let $(1 + T)u = 0$ for $u \in Ran(T)$. Then by (9.1), $\Sigma_{j=1}^{m} < e_j,\ u > y_j = -u$ and if we set $u = y_k$, we get $y_k = -\Sigma_{j=1}^{m} < e_j,\ y_k > y_j$ to conclude that all the y_j's are not linearly independent contradicting the assumption of the m-dimensionality of $Ran(T)$. Thus we can define the operator $\ln(I + T)$ to be

$$\ln(((\delta_{jk} + < e_j,\ y_k >))) = \ln(I' + T') \in \mathcal{B}(X).$$

It is also clear from the definition that $\ln(I+T)$ is also a degenerate operator with $Ran \ln(I + T) = Ran(T)$. Therefore by the Jordan decomposition of T

$$\begin{aligned} \det(I + T) &\equiv \det(I' + T') = \Pi_{j=1}^{r}(1 + \lambda_j)^{m_j} \\ &= \exp(\Sigma_{j=1}^{r} m_j \ln(1 + \lambda_j)) = \exp(tr \ln(I' + T')) = \exp(tr \ln(I + T)), \end{aligned} \tag{9.6}$$

where $(\lambda_i,\ m_j)$ are the eigenvalues and the multiplicities respectively for T'. Some more details on these can be found in Kato's book ([3, 4]). The above observations, of course remain valid when X is a Hilbert space $\mathcal{H}$. Moreover, since the infinite product $\prod(1 + \lambda_j)^{m_j}$ converges to a non zero complex number if and only if $\Sigma_{j=1}^{\infty}|\lambda_j| < \infty$, it is easy to see that $\det(I+T)$ will be definable for all trace-class $T \in \mathcal{B}_1(\mathcal{H})$ i.e. T such that the singular values of T are summable. In such a case, $\det(I + T) = \exp(tr \ln(I + T))$.

Reverting back to the Banach space scenario, let T be a closed operator on the Banach space X and let A be relatively degenerate, i.e. for operator $A(T - z)^{-1} \in \mathcal{B}(X)$ and degenerate for some z (and hence for all z) in $\rho(T)$, the resolvent set of T. In such a case we define for $z \in \rho(T)$, the perturbation determinant.

$$\Delta(T,\ T + A; z) \equiv \Delta(z) = \det(I + A(T - z)^{-1}). \tag{9.7}$$

It is also clear that in a Hilbert space $\mathcal{H}$ for $A(T - Z)^{-1} \in \mathcal{B}_1(\mathcal{H})$ for some $z \in \rho(T)$ the definition (9.7) will also make sense and

$$\Delta(z) = \det(I + A(T - z)^{-1}) = \exp(tr \ln(I + A(T - z)^{-1})). \tag{9.8}$$

In either case, it is also easy to see that for $z \in \rho(T) \cap \rho(T+A)$,

$$\Delta(z) = \det((T+A-z)(T-z)^{-1}) = [\det((T-z)(T+A-z)^{-1})]^{-1}. \quad (9.9)$$

For proceeding further, we need the following two definitions. Let φ be complex-valued meromorphic function defined in a domain $\Omega \subseteq \mathbb{C}$ and we define the multiplicity function $\nu(z,\ \varphi)$ of φ by

$$\nu(z,\ \varphi) = \begin{cases} m, & \text{if z is a zero of } \varphi \text{ of order m,} \\ -m, & \text{if z is a pole of } \varphi \text{ of order m,} \\ 0 & \text{for all other } z \in \Omega. \end{cases} \quad (9.10)$$

Therefore $\nu(z,\ \varphi)$ takes values in $\mathbb{Z}$ except when $\varphi = 0$ identically, in which case we set $\nu(z,\ \varphi) = +\infty$. Also we define the multiplicity function $\mu(z,\ T)$ for a closed operator T on X by

$$\mu(z,\ t) = \begin{cases} 0, & \text{if } z \in \rho(T)\ , \\ dim\ P & \text{if z is an isolated point in } \sigma(T), \\ \infty & \text{otherwise,} \end{cases} \quad (9.11)$$

where P is the projection associated with the isolated point $z \in \sigma(T)$, given by the holomorphic functional calculus (see [4], p. 178). Then we have the theorem:

Theorem 9.1. *(Weinstein-Aronsajn formula): Let T be a closed operator on a Banach space X, and let A be a relatively degenerate operator on X. Furthermore, let $\Delta(z)$ be the perturbation determinant $\Delta(T,\ T+A; z)$, and let Ω be a domain consisting of $\rho(T)$ along with the isolated eigenvalues of T with finite multiplicities. Then Δ is meromorphic in Ω and we have*

$$\mu(z,\ T+A) = \mu(z,\ T) + \nu(z,\ \Delta), \quad \forall z \in \Omega. \quad (9.12)$$

Since by definition, $\mu(z,\ T)$ is finite for all $z \in \Omega$ we have the alternatives: either $\mu(z,\ T+A)$ is finite for all $z \in \Omega$ or it is equal to $+\infty$ identically. In the case of the first alternative, $\Omega \subseteq \rho(T+A)$ except for at most countably many discrete eigenvalues of $T+A$, while in the second $\Omega \subseteq \sigma(T+A)$.

Proof. In case $\Delta(z) = 0$ identically, the relation (9.9) implies that $(T+A-z)(T-z)^{-1}$ has a zero eigenvalue for all $z \in \Omega$ which leads to the conclusion that z is an eigenvalue for $T+A$ for all $z \in \rho(T) \cap \Omega$. But this means by definition that $\mu(z,\ T+A) = +\infty$ since either $\rho(T) \cap \Omega = \emptyset$ in which case z is not an eigenvalues of $T+A$ and $\Omega \subseteq \sigma(T)$ or $\rho(T) \cap \Omega \neq \emptyset$

when z is an eigenvalue, not isolated. Thus we can assume without loss of generality that Δ is not identically zero, we consider first $z \in \rho(T)$ such that $\Delta(z) \neq 0$ and set $B(z) = A(T-z)^{-1}$ so that by (9.8), $\Delta(z) = \det(I + B(z)) = \exp[tr \ln(I + B(z))]$. That $\Delta(z) \neq 0$ implies that $B(z)$ has no eigenvalues -1, and in such a case by holomorphic functional calculus ([4], p. 178)

$$\ln(I + B(z)) = \frac{1}{2\pi i}\int_C \ln(1+\varsigma)(\varsigma - B(z))^{-1} d\varsigma,$$

where C is a closed curve enclosing all the eigenvalues of $B(z)$ (they are finite in number since $B(z)$ is finite-rank) and excluding the point $\{-1\}$. We note that

$$\frac{d}{dz}\ln\Delta(z) = \frac{d}{dz} tr \ln(I + B(z)) = \frac{1}{2\pi i}\frac{d}{dz} tr \int_c \ln(1+\zeta)(\zeta - B(z))^{-1} d\zeta. \tag{9.13}$$

Interchanging the order of integration with taking trace and differentiation, which is permissible because the function $z \to (\zeta - B(z))^{-1}$ is differentiable in trace, we note that:

$$\begin{aligned}\frac{d}{dz}(\zeta - B(z))^{-1} &= (\zeta - B(z))^{-1}B'(z)(\zeta - B(z))^{-1}\\ &= (\zeta - B(z))^{-1}(A(T-z)^{-2})(\zeta - B(z))^{-1}\end{aligned}$$

and substituting this in (9.13) and using the trace property, we have

$$\begin{aligned}\frac{d}{dz}\ln\Delta(Z) &= \frac{1}{2\pi i}\int_C tr[\ln(1+\zeta)(\zeta - B(z))^{-2}B'(z)]d\zeta\\ &= \frac{1}{2\pi i}\int_C \frac{d}{d\zeta}\{-\ln(1+\zeta)tr((\zeta - B(z))^{-1}B'(z))\}d\zeta\\ &\quad + \frac{1}{2\pi i}\int_C (1+\zeta)^{-1} tr((\zeta - B(z))^{-1}B'(z))d\zeta\\ &= tr[(1+B(z))^{-1}B'(z)],\end{aligned} \tag{9.14}$$

where we have observed that the first integral in the second step vanishes since $[\ln(1+\zeta)\text{tr}\,\{(\zeta - B(z))^{-1}B'(z)\}]$ is an holomorphic function of ζ inside C. On the other hand,

$$\begin{aligned}&tr[(1+B(z))^{-1}B'(z)] = tr[(T-z)^{-1}(1 + A(T-z)^{-1})^{-1}(T-z)^{-1}]\\ &= tr(T + A - z)^{-1}A(T-z)^{-1} = -tr[(T + A - z)^{-1} - (T-z)^{-1}].\end{aligned}$$

Thus for $z \in \rho(T)$ such that $\Delta(z) \neq 0$,

$$tr[(T + A - z)^{-1} - (T - z)^{-1}] = -\frac{d}{dz} \ln \Delta(z). \tag{9.15}$$

In the above we have also noted that since $(T+A-z) = (I+B(z))(T-z)$ and since for such $z, I+B(z)$ is bounded invertible it follows that $z \in \rho(T+A)$. Finally, to obtain Weinstein-Aronsajn (W-A) formula, we need to integrate the relation (9.15) and realize that $\frac{1}{2\pi i} \int_C \frac{d}{dz} \ln \Delta(z) = \nu(\lambda,\ \Delta)$ for a small circle C enclosing $\lambda \in \Omega$, and also that $\frac{1}{2\pi i} \int_C (T - z)^{-1} dz = P$, the eigen-projection for the eigenvalue λ of T and $\mu(\lambda,\ T) = dimP = trP$. □

Corollary 9.1. *Let $\mathcal{H}$ be a separable Hilbert space, T a closed operator on $\mathcal{H}$ and $A \in B_1(\mathcal{H})$. Then also W-A formula (9.12) is true.*

The proof follows by observing that the passage from equation (9.13) to (9.14) remains valid by an application of dominated convergence theorem in trace-norm, to interchange the integral and trace ;and then the rest of the steps also remain valid.

Remark 9.1. Let $\mathcal{H}$ be a separable Hilbert space, T_1 and T_2 two closed operators on $\mathcal{H}$ such that $[(T_2 - z)^{-1} - (T_1 - z)^{-1}] \in \mathcal{B}_1(\mathcal{H})$ for some $z \in \rho(T_1) \cap \rho(T_2)$. Then $(T_2 - z)^{-1} - (T_1 - z)^{-1} \in \mathcal{B}_1(\mathcal{H})$ for all $z \in \rho(T_1) \cap \rho(T_2)$. This easily follows from the resolvent equation.

9.3. Witten Index and Its Invariance

For the rest of this article $\mathcal{H}$ is a separable Hilbert space, and let H and H_0 be two self joint operators on $\mathcal{H}$. We also assume that $[(H - z)^{-1} - (H_0 - z)^{-1}] \in B_1(\mathcal{H})$, the Banach space of trace-class operators on $\mathcal{H}$, for some z (and hence every z) in $\rho(H) \cap \rho(H_0)$. Then the Witten index $\delta(H,\ H_0)$ for the pair $(H,\ H_0)$ is defined as:

$$\delta(H,\ H_0) \equiv \lim_{z \to 0 : |Re\ z| \leq C|Im\ z|} [-ztr\{(H - z)^{-1} - (H_0 - z)^{-1}\}], \tag{9.16}$$

whenever this limit exists. This is called the non-tangential limit and using the resolvent equation, one can easily establish that the limit, if it exists for some non-tangential ray in the complex plane, then it exists for every other such ray and they are equal. Thus without loss of generality we shall assume that z is pure imaginary for the purpose of the limit.

For the rest of this section, $\mathcal{H} = L^2(R^3)$, $H_0 = -\Delta$, the self-adjoint operator on H associated with the Laplacian, and $V_j : j = 1, 2$ the operators of multiplication by real measurable functions $V_j \in L^2 \cap L^1(\mathbb{R}^3)$. It is well known ([6, 7]) that $H = H_0 + V_1$ and $H_\alpha = H + \alpha V_2$ for all $\alpha \in \mathbb{R}$ are self-adjoint operators with $D(H_\alpha) = D(H) = D(H_0)$, where we have written D(A) for the domain of definition of a linear (not necessarily bounded) operator A on $\mathcal{H}$. Next we note that $V_j(x) = sgn[V_j(x)]|V_j(x)|^{\frac{1}{2}}.\,|V_j(x)|^{\frac{1}{2}} \equiv V_j(x)^{\frac{1}{2}}|V_j(x)|^{\frac{1}{2}}$ so that $V_j^{\frac{1}{2}}, |V_j|^{\frac{1}{2}} \in L^2(\mathbb{R}^3)$.

Then we have the following sets of results which are either available in [6, 7] or easily provable.

Theorem 9.2. *Let H, H_0, V_j be as above. Then*
(i) $V_j^{\frac{1}{2}}(H_o - z)^{-1}|V_k|^{\frac{1}{2}} \equiv M_{jk}(z)$ are Hilbert -Schmidt and $\|M_{jk}\|_2 \leq \beta_{jk}$, independent of z for all $z \in \rho(H_0)$.

(ii) For all z with $Im\ z \neq 0$, the bounded operator $I + V_j^{\frac{1}{2}}(H_0 - Z)^{-1} \mid V_j \mid^{\frac{1}{2}}$ is bounded invertible. If furthermore, the point 0 is not an exceptional point for the self-adjoint operator $H_0 + V_j$ (see page 433 of [7]), then $\|[I + V_j^{\frac{1}{2}}(H_0 - z)^{-1}|V_j|^{\frac{1}{2}}]^{-1}\| \leq C_j$, independent of z with $Re\ z$ in a neighborhood of $\{0\}$.

(iii) For any z with $Im\ z \neq 0, [(H - z)^{-1} - (H_0 - z)^{-1}], [(H_\alpha - z)^{-1} - (H - z)^{-1}]$ and $[(H_\alpha - z)^{-1} - (H_0 + \alpha V_2 - z)^{-1}]$ are all trace - class.

(iv) Under the condition of part (ii), the Witten index $\delta(H_\alpha,\ H) = 0$ for sufficiently small $|\alpha|$ and therefore, $\delta(H_0 + \alpha V_2,\ H_0) = 0$ for sufficiently small $|\alpha|$ by setting $V_1 = 0$.

(v) The Witten index is invariant under "small perturbation", i.e. for sufficiently small $|\alpha|$

$$\delta(H_0 + V_1 + \alpha V_2,\ H_0 + \alpha V_2) = \delta(H_0 + V_1,\ H_0), \tag{9.17}$$

in the sense that if one exists so does the other and they are equal.

Proof. *Sketch:*
(i) By the hypothesis on the functions V_j, it is clear that $V_j \in L^{\frac{3}{2}}(\mathbb{R}^3)$. Also the operator $M_{jk}(z)$ is an integral operator with kernel

$$M_{jk}(z)(x, y) = V_j(x)^{\frac{1}{2}} \frac{e^{i\sqrt{z}|x-y|}}{4\pi|x - y|}|V_k(y)|^{\frac{1}{2}},$$

where the branch of the square root with $Im\sqrt{z} > 0$ is taken. Thus by Hardy-Littlewood-Sobolev inequality (see [6])

$$\|M_{jk}(z)\|_2^2 = \frac{1}{16\pi^2}\int |V_j(x)|\frac{e^{-2|Im\sqrt{z}||x-y|}}{|x-y|^2}|V_k(y)|dx\,dy$$
$$\leq C'\|V_j\|_{\frac{3}{2}}\|V_k\|_{\frac{3}{2}} \equiv \beta_{jk},$$

independent of z.

(ii) We have seen that $M_{jj}(z) \in \mathcal{B}_2(\mathcal{H})$, therefore compact. Also from the kernel of $M_{jj}(Z)$ and the choice of the branch of $\sqrt{z}$, it is clear that $\rho(H_0) \ni z \mapsto M_{jj}(z)$ is $\mathcal{B}_2$ analytic, since $V_j \in L^2$. Furthermore, from the resolvent equation $(H+V_j-z)^{-1}-(H_0-z)^{-1} = -(H_0-z)^{-1}V_j(H-z)^{-1}$ for $z \in \rho(H)\cap\rho(H_0)$, it follows that, for $Im\ z \neq 0$,

$$(1+V_j^{\frac{1}{2}}(H_0-Z)^{-1}|V_j|^{\frac{1}{2}})^{-1} = (1-V_j^{\frac{1}{2}}(H+V_j-z)^{-1}|V_j|^{\frac{1}{2}}) \in B(\mathcal{H}).$$

Therefore by the analytic Fredholm theorem ([6]),

$$[1+M_{jj}(\lambda+i0)]^{-1} \in \mathcal{B}(H)$$

for all $\lambda \in \mathbb{R}$ except for a closed set of Lebesgue measure 0. If for $\lambda = \lambda_0$, the above operator is not invertible then it follows that $g_{\lambda_0} \equiv (H_0-\lambda_0)^{-1}|V|^{\frac{1}{2}}f$ with f in the null space $N[1+M_{jj}(\lambda_0+i\,0)]$ is the eigenvector of the operator H_0+V_j. Since by hypothesis $\lambda_0 = 0$ is not an eigenvector of H_0+V_j, the part (ii) is proven.

(iii) Since if $V_j \in L^2(R^3)$, then $V_j(H_0-z)^{-1} \in \mathcal{B}_2$ and

$$\|V_j(H_0-z)^{-1}\|_2^2 \doteq \frac{1}{16\pi^2}\int\int |V_j(x)|^2\frac{e^{-2Im\sqrt{z}|x-y|}}{|x-y|^2}dx\,dy = C''\|V_j\|_2^2\frac{1}{Im\ \sqrt{z}}$$

for $Im\ z \neq 0$. This along with the resolvent equations verify (iii).

(iv) Since $D(H) = D(H_0) = D(H_0+V_2)$, for z with $Im\ z \neq 0$

$$(H-z)^{-1}-(H_0-z)^{-1} = -(H_0-z)^{-1}V_1(H-z)^{-1}$$

and by (i) and (ii) we have

$$\widetilde{M_{jk}}(z) \equiv V_j^{\frac{1}{2}}(H-z)^{-1}|V_k|^{\frac{1}{2}} = M_{jk}(z) - M_{j1}(z)\widetilde{M_{1k}}(z).$$

Thus for $Im\ z \neq 0$,

$$\widetilde{M_{11}}(z) = (1+M_{11}(z))^{-1}M_{11}(z)$$

and similarly $((V_1^{\frac{1}{2}}(H-z)^{-1}) = [1+M_{11}(z)]^{-1}(V_1^{\frac{1}{2}}(H_0-z)^{-1}) \in \mathcal{B}_2(\mathcal{H})$. One can show exactly in the same way that

$$V^{\frac{1}{2}}(H+\alpha V_2-z)^{-1} = [1-\alpha\widetilde{M_{22}}(z)]^{-1}[1+M_{11}(z)]^{-1}[V^{\frac{1}{2}}(H-z)^{-1}] \text{ and}$$

$$\widetilde{M_{12}}(z) = [1+M_{11}(z)]^{-1}]M_{12}(z). \tag{9.18}$$

By (ii) and equation (9.18), we get that, $\sup_{z\in[0,i]} \| \widetilde{M_{12}}(z) \|_2 \leq C'''\beta_{12}$ and then by choosing $|\alpha|$ sufficiently small,

$$||V^{\frac{1}{2}}(H+\alpha V_2-z)||_2 \leq (C''')^2(1-|\alpha|\beta_{12})^{-1}\|V^{\frac{1}{2}}(H_0-z)\|_2 \equiv D\,(Im\sqrt{z})^{-\frac{1}{2}}. \tag{9.19}$$

Finally putting all these together, we have that for $z = iy$,

$$\begin{aligned} &|tr[(H+\alpha V_2-iy)^{-1}-(H-iy)^{-1}]| \\ &\quad\leq |\alpha|\|(H-iy)^{-1}V_2(H+\alpha V_2-iy)^{-1}\|_1 \\ &\quad\leq |\alpha|\|V_2\|^{\frac{1}{2}}(H+iy)^{-1}\|_2\|V_2^{\frac{1}{2}}(H+\alpha V_2-iy)^{-1}\|_2 \leq D'\,y^{-\frac{1}{2}} \end{aligned}$$

and hence

$$\begin{aligned} &\lim_{y\to 0, z=iy} -z\,tr[(H+\alpha V_2-z)^{-1}-(H-z)^{-1}] \\ &\qquad = \lim_{y\to 0} -iy\,tr[(H+\alpha V_2-iy)^{-1}-(H-iy)^{-1}] = 0 \end{aligned}$$

for $|\alpha|$ sufficiently small. Setting $V_1 = 0$, for small enough $|\alpha|$ we get

$$\lim_{y\to 0-iy}[(H_0+\alpha V_2-i\,y)^{-1}-(H_0-iy)^{-1}] = 0.$$

(v) Finally, since

$$\begin{aligned} &tr[(H_0+V_1+\alpha V_2-iy)^{-1}-(H_0+\alpha V_2-iy)^{-1}] \\ &\qquad -tr[(H-iy)^{-1}-(H_0-iy)^{-1}] \\ &\qquad = tr[(H_0+V_1+\alpha V_2-iy)^{-1}-(H_0+V_1-iy)^{-1}] \\ &\qquad -tr[(H_0+\alpha V_2-iy)^{-1}-(H_0-iy)^{-1}] \end{aligned}$$

it follows by result of (iv) that under the assumption in (ii), for sufficiently small $|\alpha|$, $\delta(H+\alpha V_2,\ H_0+\alpha V_2) = \delta(H,\ H_0)$. □

In the special situation where H_1 and H_2 are positive self adjoint operators A^*A and AA^* respectively, with A a Fredholm operator on some Hilbert space, then the Witten index for the pair $\delta(H_1,\ H_2)$ is equal to the Fredholm index of A, which is an integer. But in a general case, the Witten index need not be an integer.

9.4. Krein's Shift Function

Given two self adjoint operators on a Hilbert space such that their difference is a trace-class operator, there is an L^1-function ξ (called Krein's Shift function) such that the integral of ξ equals the trace of the above difference. The value of ξ at zero, if it exists, has many properties of an index.

Proposition 9.1. *Let H_0 be a self-adjoint operator on a Hilbert space $\mathcal{H}(dim\mathcal{H} < \infty)$ and let V be another self adjoint operator. There exists a unique real-valued function $\xi \in L^\infty(\mathbb{R})$, supported in $\sigma(H_0) \cup \sigma(H_0 + V)$ such that for every polynomial p on $\mathbb{R}$.*

$$tr[p(H_0 + V) - p(H_0)] = \int p'(\lambda)\xi(\lambda)d\lambda.$$

In fact, by the spectral theorem and functional calculus, one has that

$$p(H_0 + V) - p(H_0) = \int p(\lambda)[E(d\lambda) - E_0(d\lambda)],$$

where E_0 and E are the spectral family of the self-adjoint operators H_0 and $H_0 + V$ respectively. Clearly these integrals are actually sums and the integral is over the union of the two spectra contained in some interval $[a, b]$. One can do integration by parts to get

$$p(H_0 + V) - p(H_0) = \int p'(\lambda)[E_0(\lambda) - E(\lambda)]d\lambda$$

and the expression for the trace follows with $\xi(\lambda) = tr[E_0(\lambda) - E(\lambda)]$.
In infinite dimensional $\mathcal{H}$, the story changes drastically, even for a pair of bounded self adjoint operators they can have only continuous spectra and moreover,they can be pair of unbounded self-adjoint operators. We state the next two theorems without proof, which can be found in [1, 3, 5].

Theorem 9.3. *Let H and H_0 be two self adjoint operators on a separable Hilbert space $\mathcal{H}$ such that $V \equiv H - H_0$ is trace-class (i.e $\in \mathcal{B}_1(\mathcal{H})$). Then there exists unique L^1-function ξ with support in $\sigma(H) \cup \sigma(H_0)$ such that*

(i) $trV = \int \xi(\lambda)d\lambda$,
(ii) $\int |\xi(\lambda)|d\lambda \leq \|V\|_1$,
(iii) for $\varphi : \mathbb{R} \to \mathbb{C}$ such that $\phi(\lambda) = \int \frac{e^{it\lambda}-1}{it} d\nu(t)+$ constant, with ν a complex measure in $\mathbb{R}$,

$$\varphi(H) - \varphi(H_0) \in \mathcal{B}_1(\mathcal{H}) \text{ and } tr[\varphi(H) - \varphi(H_0)] = \int \varphi'(\lambda)\xi(\lambda)d\lambda,$$

(iv) $\xi(\lambda) = \frac{1}{\pi}\lim_{\epsilon\downarrow 0} Im \ln \Delta(\lambda + i\epsilon)$ *for almost all* λ.

Theorem 9.4. *Let H and H_0 be two self adjoint operators such that $(H-z)^{-1} - (H_0 - z)^{-1} \in B_1(\mathcal{H})$ for some $z \in \rho(H) \cap \rho(H_0)$. Then there exists a real-valued measurable function ξ on $\mathbb{R}$, unique up to an additive constant such that*

(i) $\int \frac{|\xi(\lambda)|}{1+\lambda^2} < \infty$,

(ii) $tr\ [(H-z)^{-1} - (H_0 - z)^{-1}] = -\int \frac{\xi(\lambda)}{(\lambda - z)^2} dz$

(iii) For every function $\psi : \mathbb{R} \to \mathbb{C}$ with the property $\psi(\lambda) = (1+\lambda^2)^{-1}\varphi(\lambda)$ where φ satisfies the condition of (iii) of the previous theorem,

$$\varphi(H) - \varphi(H_0) \in \mathcal{B}_1(\mathcal{H}) \text{ and } tr[\psi(H) - \psi(H_0)] = \int \psi'(\lambda)\xi(\lambda)d\lambda.$$

The next theorem (cf. [2, 8]) connects the Witten index $\delta(H, H_0)$ and the Krein's shift function $\xi(H, H_0; \lambda)$ for the pair (H, H_0).

Theorem 9.5. *Let H and H_0 be self-adjoint operators bounded below on a Hilbert space $\mathcal{H}$ such that $[(H-z)^{-1} - (H_0 - z)^{-1}] \in \mathcal{B}_1(\mathcal{H})$ for $z \in \rho(H) \cap \rho(H_2)$. Assume furthermore that ξ is locally bounded and piecewise continuous. Then $\delta(H, H_0)$ exists and is equal to $\xi(H, H_0; 0_-) - \xi(H, H_0; 0_+)$.*

Proof. Let $f(z) = -z\, tr[(H-z)^{-1} - (H_0 - z)^{-1}] = -z\int_{-b}^{\infty} \frac{\xi(\lambda)}{(\lambda - z)^2} d\lambda$, where $b = -\min(\inf \sigma(H), \inf \sigma(H_0))$ and we have written $\xi(\lambda)$ for $\xi(H, H_0; \lambda)$. Thus by an integration by parts, we have that

$$z\int_{-b}^{0} \frac{\xi(\lambda)}{(\lambda - z)^2} d\lambda = -\frac{z}{b+z}\xi(-b) + \xi(0_-) + \int_{-b}^{0} \frac{z}{\lambda - z} d\xi(\lambda).$$

In the above we have used the fact that under the hypothesis of the theorem, ξ is locally a signed measure. The non-tangential condition $|Re\ z| \le C|Im\ z|$ leads to the bound $|\frac{z}{\lambda - z}| \le \sqrt{C^2 + 1}$ and therefore, by the dominated convergence theorem,

$$\lim_{z\to 0:|Re\ z|\le C|Im\ z|} -z\int_{-b}^{0} \frac{\xi(\lambda)}{(\lambda - z)^2} d\lambda = \xi(0_-).$$

An identical argument proves that

$$\lim_{z\to 0:|Re z|\le C|Im\ z|} \int_{0}^{a} \frac{\xi(\lambda)}{\lambda^2 + 1} d\lambda = -\xi(0_+).$$

On the other hand for $|z| < a$

$$|z\int_{a}^{\infty} \frac{\xi(\lambda)}{(\lambda - z)^2} d\lambda| = |z\int_{a}^{\infty} \frac{\lambda^2 + 1}{(\lambda - z)^2}\frac{\xi(\lambda)}{\lambda^2 + 1} d\lambda| \le \frac{|z|(a^2 + 1)}{(a - |z|)^2}\int_{a}^{\infty} \frac{|\xi(\lambda)|}{\lambda^2 + 1} d\lambda$$

and it goes to 0 as z tends to 0. Thus $\delta(H,\ H_0)$ exists in this case and

$$\delta(H,\ H_0) = \xi(0_-) - \xi(0_+).$$ □

9.5. Application to Quantum Mechanics and Generalized Levinson's Theorem

Next, we go back to the quantum mechanical case addressed in Theorem 9.2: $\mathcal{H} = L^2(\mathbb{R}^3)$, $H_0 = -\Delta$, the potentials $V_j \in L^2 \cap L^1(\mathbb{R}^3)$ for $j = 1, 2$. In such a case it is known ([6, 7]) that $H_0, H = H_0 + V_1, H_\alpha = H + \alpha V_2$ are all bounded below and that (real) negative spectra are respectively $\sigma_-(H_0) = \emptyset$ and $\sigma_-(H), \sigma_-(H_\alpha)$ consists of finite number of discrete eigenvalues with finite multiplicities. Therefore the open interval $(-\kappa,\ 0) \in \rho(H)$ and $\rho(H_\alpha)$ for $\kappa > 0$ small enough and therefore by the definition of ξ (see Theorem 4.2), $\xi(\lambda) = 0$ for $\lambda \in (-\kappa,\ 0)$ and thus $\xi(H,\ H_0; 0_-) = \xi(H_\alpha,\ H_0; 0_-) = 0$. In such a case the conclusion of the Theorem 4.4 would be that the Witten index $\delta(H,\ H_0)$ exist and is equal to $-\xi(H,\ H_0; 0_+)$, provided the other assumption on the shift function ξ, viz. that ξ is locally bounded and piecewise continuous, is valid.
If these extra hypothesis on $\xi(H, H_0; \lambda)$ and on $\xi(H_\alpha,\ H_0 + \alpha V_2; \lambda)$ are satisfied, then combining all the above results we conclude that the Witten indices $\delta(H,\ H_0)$ and $\delta(H_\alpha,\ H_0 + \alpha V_2)$ exist and are equal, and therefore,

$$\delta(H_\alpha,\ H_0 + \alpha V_2) = -\xi(H_\alpha,\ H_0 + \alpha V_2; 0+) = \delta(H,\ H_0) = -\xi(H,\ H_0; 0_+). \tag{9.20}$$

The above relation (9.20), which on the one hand describes the invariance of the Witten index under small perturbations, can be understood, on the other hand as a consequence of what may be called Levinson's theorem which we go on to describe now. Originally Levinson proved a similar theorem in $\mathcal{H}$=$L^2(\mathbb{R})$ i.e. where H_0 and H are ordinary differential operators. The starting point is W-A formula or rather the relation (9.15) with a pair of self adjoint operators H_0 and $H \equiv H_0 + V$ on a Hilbert space $\mathcal{H}$ replacing T and $T + A$ such that $[(H - z)^{-1} - (H_0 - z)^{-1}]$ is trace class for some (and hence for all) $z \in \rho(H) \cap \rho(H_0)$; this along with Krein's Theorem 9.4 leads to

$$tr[(H - z)^{-1} - (H_0 - z)^{-1} = -\frac{d}{dz} \ln(\Delta) = -\int \frac{\xi(\lambda)}{(\lambda - z)^2} d\lambda. \tag{9.21}$$

Theorem 9.6. *Let H and H_0 be two self adjoint operators on a Hilbert space such that $(H - z_0)^{-1} - (H_0 - z_0)^{-1} \in \mathcal{B}_1(\mathcal{H})$ for some $z_0 \in$*

$\rho(H)\bigcap\rho(H_0)$. Assume furthermore the following:

(i) H and H_0 bounded below, i.e. $H, H_0 \geq -\beta$, the discrete spectrum of H contained in $[-b,-a]$ with $0 < a < b < \beta < \infty$ and H_0 has no point spectrum,

(ii) the associated Krein's shift function $\xi(H, H_0; \lambda) \equiv \xi(\lambda)$ locally bounded and piecewise continuous,

(iii) $\xi(+\infty) \equiv \lim_{\lambda\to+\infty}\xi(\lambda)$ exists and is equal to 0.

Then $\xi(H, H_0; 0_+) = n$, where $n = \Sigma_{i:\lambda_i\in\sigma_{disc}(H)}\mu(\lambda_i, H)$.

Proof. The relation (9.15) is integrated along the contour given below.

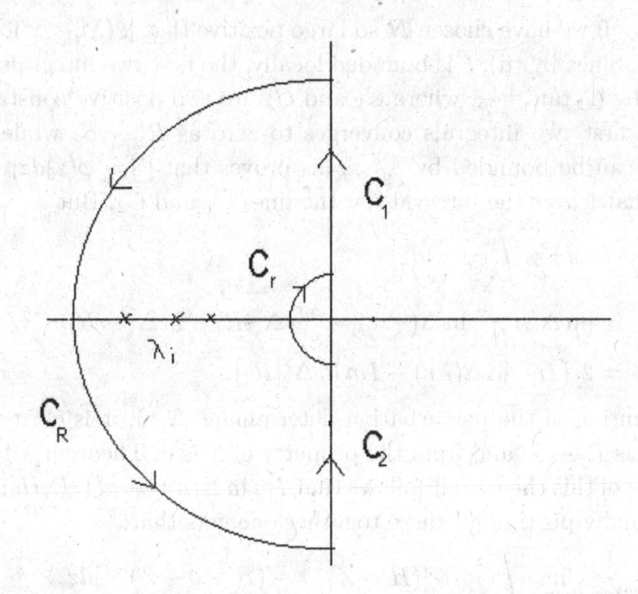

We note the following, the function $z \mapsto tr[(H-z)^{-1} - H_0 - z)^{-1}]$ is continuous on the contour $C_{r,R} = C_R \cup (C_1 \cup C_2) \cup C_r$, and is meromoprphic in the interior with poles at $\{\lambda_i\}$, the discrete spectrum of H with $\mu(\lambda_i,\ H) = dim\ P_i < \infty$ and

$$\frac{1}{2\pi i}\int_{C_{r,R}} tr[(H-z)^{-1} - (H_0-z)^{-1}]dz = \Sigma_i dim P_i = \Sigma_i\mu(\lambda_i,\ H) = n.$$

On the other hand, setting $\varphi(z) = tr[(H-z)^{-1} - (H_0-z)^{-1}]$, we note that $\int_{C_r}\varphi(z)dz = -\int_{C_r}\frac{d}{dz}\ln\Delta(z)dz$ and since $\frac{d}{dz}\ln\Delta(z)$ is holomorphic in compact region containing C_r, it is uniformly bounded in that region, say by c' so that $|\int_{C_r} \varphi(z)dz| \leq \pi\, r\, c' \to 0$ as $r \to 0$. For the integral on C_R, we use (9.21), to get that

$$|\int_{C_R} \varphi(z)dz| \leq \int_{C_R} |dz| \int \frac{|\xi(\lambda)|}{(\lambda^2 + R^2 - 2\lambda R\cos\theta)} d\lambda \leq \pi R \int \frac{|\xi(\lambda)|}{\lambda^2 + R^2} d\lambda.$$

Next we note that the support of the λ-integral is contained in $[-b, -a] \cup [0, \infty)$ by (i) and we write

$$R \int \frac{|\xi(\lambda)|}{\lambda^2 + R^2} d\lambda = R \int_{-b}^{-a} \frac{|\xi(\lambda)|}{\lambda^2 + R^2} d\lambda R \int_0^N \frac{|\xi(\lambda)|}{\lambda^2 + R^2} d\lambda$$

$$+ R \int_N^\infty \frac{|\xi(\lambda)|}{\lambda^2 + R^2} d\lambda$$

where given $\epsilon > 0$ we have chosen N so large positive that $|\xi(\lambda)| < \epsilon$ for all $\lambda > N$ by (iii). Since by (ii), ξ is bounded locally, the first two integrals can be estimated by $C_1 \tan^{-1} \frac{C_2}{R}$ where C_1 and C_2 are two positive constants. Therefore the first two integrals converges to zero as $R \to \infty$ while the third integral can be bounded by $\frac{\pi\ \epsilon}{2}$. This proves that $|\int_{C_R} \varphi(z)dz| \to 0$ as $R \to \infty$. That leaves the integrals on the lines C_1 and C_2. But

$$\begin{aligned}
&\int_{C_1} \varphi(z)dz + \int_{C_2} \varphi(z)dz \\
&\quad = [\ln\Delta(ir) - \ln\Delta(-ir)] - [\ln\Delta(iR) - \ln\Delta(-iR)] \\
&\quad = 2i\{Im\ \ln\Delta(i\,r) - Im\ln\Delta(iR)\}.
\end{aligned}$$

From the definition of the perturbation determinant $\Delta(z)$, it is clear that $\ln\Delta(iR) \to 0$ as $R \to \infty$ and from the property of ξ as in Theorem 9.4 and assumption (ii) of this theorem it follows that $Im\ln\Delta(ir) \to \pi\xi(H,\ H_0; 0+)$ as $r \to 0+$. Finally putting all these together one gets that:

$$\begin{aligned}
n &= \frac{1}{2\pi i} \lim_{r\to 0+} \int_{C_{r,\ R}} tr[(H - Z)^{-1} - (H - 0 - Z)^{-1}]dz \\
&= \frac{2i}{2\pi i} \pi\xi(H,\ H_0; 0_+) = \xi(H,\ H_0; 0_+),
\end{aligned}$$

which is the generalized Levinson's theorem. □

Combining this with the Theorem 9.2 and the discussions at the beginning of this section, we get the following corollary relevant to the quantum mechanical situation.

Corollary 9.2. *Let $H_0 = -\Delta,\ H = H_0 + V_1$ and $H_\alpha = H + \alpha V_2$ (with α real) be the three self adjoint operators on the Hilbert space $\mathcal{H} = L^2(\mathbb{R}^3)$ with V_j (j=1,2) real valued functions in $L^2 \cap L^1(\mathbb{R}^3)$ as in Theorem 3.1.*

Assume furthermore as in the discussion in the first paragraph of this section that all the relevant Krein's shift function satisfy the properties of (ii) and (iii) of Theorem 5.1, viz. that they are locally bounded and piecewise continuous and they converge to zero at infinity. Then

$$\delta(H_\alpha,\ H_0 + \alpha V_2) = \delta(H,\ H_0) = -\xi(H,\ H_0; 0_+) = -n$$

for sufficiently small $|\alpha|$ *and where* $n = \Sigma_i \mu(\lambda_i,\ H)$ *as in Theorem 9.6.*

The proof follows by combining the earlier discussions and Theorem 9.6. In the language of physics, the corollary states that the number of bound states (i.e. the discrete spectrum, counting multiplicities) of the Hamiltonian operator $H_\alpha = H_0 + \alpha V_1 + \alpha V_2$ for $|\alpha|$ sufficiently small is independent of α.

References

[1] Mohapatra, A. and Sinha, K. B. (1994). Spectral shift function and trace formula in Special issue on Spectral and Inverse Spectral Theory. *Proc. Ind. Acad. Sc. (Math. Sc.).* **104** 819-853.

[2] Gesztesy, F. and Simon, B. (1988). Topological invariance of the Witten index. *J. Funct. Anal.* **79** 91-102.

[3] Gesztesy, F., Latushkin, Y., Mitrea, M. and Zinchenko, M. (2005). Non-self-adjoint operators, infinite determinants, and some applications. *Russian J. Math. Phys.* **12** 443-471.

[4] Kato, T. (1966). *Perturbation theory for linear operators.* Die Grundlehren der mathematischen Wissenschaften, Band **132**, Springer-Verlag, New York.

[5] Kreĭn, M. G. (1953). On the trace formula in perturbation theory. (Russian) *Mat. Sbornik N. S.* 33 (**75**) 597-626.

[6] Reed, M. and Simon, Barry. (1972, 1975, 1979, 1978). *Methods of modern mathematical physics.* **I, II, III, IV**. Academic Press, New York-London.

[7] Amrein, W. O., Jauch, J. M. and Sinha, K. B. (1977). *Scattering Theory in Quantum Mechanics. Physical Principles and Mathematical Methods.* Lecture Notes and Supplements in Physics, No. 16. W. A. Benjamin, Inc., Reading, Mass.-London-Amsterdam.

[8] Sinha, K. B. (1995). Index of a pair of projections and applications. *Current trends in mathematics and physics.* 198-207, Narosa, New Delhi.

Chapter 10

Zero Cycles and Complete Intersection Points on Affine Varieties

V. Srinivas

School of Mathematics,
Tata Institute of Fundamental Research,
Mumbai, India
srinivas@math.tifr.res.in

Let us begin with some elementary observations. Let A be a *Dedekind domain* (see [8], Ch. V). Two basic examples are obtained as follows.

(i) Let K be a finitely generated extension field of transcendence degree 1 of an algebraically closed field k, $t \in K$ a transcendence basis, and

$$A = \text{integral closure of } k[t] \text{ in } K.$$

These rings arise as the coordinate rings of *nonsingular affine algebraic curves* over k. In this lecture, we focus on this example and some natural generalizations.

(ii) Let K be an algebraic number field (finite algebraic extension of $\mathbb{Q}$), and let

$$A = \text{integral closure of } \mathbb{Z} \text{ in } K.$$

This is called the ring of *algebraic integers* in K.

For a Dedekind domain A, let

$$\begin{aligned}\text{Div}\,(A) &= \text{Free abelian group on maximal ideals of } A\\ &= \text{Group of } \textit{divisors} \text{ of } A.\end{aligned}$$

If $a \in A$ is a non-zero element, it defines a *principal ideal* aA, which has a factorization of the form $aA = \mathfrak{M}_1^{n_1} \cdots \mathfrak{M}_r^{n_r}$, where $\mathfrak{M}_i$ are distinct maximal ideals, and n_i are positive integers (if a is a unit, then $aA = A$, and we take $r = 0$). Equivalently, aA has a primary decomposition $aA = \mathfrak{M}_1^{n_1} \cap \cdots \cap \mathfrak{M}_r^{n_r}$ (the equivalence is a property of Dedekind domains).

The assignment

$$a \mapsto \sum_{i=1}^{r} n_i[\mathfrak{M}_i] \in \mathrm{Div}\,(A)$$

gives a semigroup homomorphism

$$A \setminus \{0\} \to \mathrm{Div}\,(A).$$

The subgroup $\mathrm{P}\,(A) \subset \mathrm{Div}\,(A)$ generated by the image of $A \setminus \{0\}$ is called the group of *principal divisors*.

We may now define

$$\begin{aligned} \mathrm{Cl}\,(A) &= \textit{Divisor class group} \text{ of } A \\ &= \frac{\mathrm{Div}\,(A)}{\mathrm{P}\,(A)}. \end{aligned}$$

Theorem 10.1. *For a Dedekind domain A, the following properties are equivalent.*

(i) A has the unique factorization *property.*

(ii) Every maximal ideal of A is a principal ideal.

(iii) $\mathrm{Cl}\,(A) = 0$.

Thus the divisor class group measures the failure of the unique factorization property.

The divisor class group of a Dedekind domain has an interpretation in terms of *algebraic K-theory* (see [5]).

Recall that a *projective* A-module is a direct summand of a free A-module. Let $K_0(A)$ be the Grothendieck group of finitely generated projective A-modules, defined as a quotient

$$K_0(A) = \frac{F(A)}{R(A)},$$

with

$$F(A) = \begin{array}{l}\text{Free abelian group on isomorphism classes of} \\ \text{finitely generated projective } A\text{-modules}\end{array}$$

$$R(A) = \begin{array}{l}\text{subgroup of } F(A) \text{ generated by classes } [P_2] - [P_1] - [P_3] \\ \text{for all exact sequences } 0 \to P_1 \to P_2 \to P_3 \to 0.\end{array}$$

Since a Dedekind domain A has Krull dimenion 1 (i.e., every non-zero prime ideal is a maximal ideal), one can show that there is an isomorphism

$$K_0(A) \cong \mathbb{Z} \oplus \mathrm{Cl}\,(A).$$

The map $K_0(A) \to \mathbb{Z}$ is induced by

$$[P] \mapsto \operatorname{rank} P = \dim_K P \otimes_A K,$$

where K is the quotient field of A.

To define the map $K_0(A) \to \mathrm{Cl}(A)$, one first identifies $\mathrm{Cl}(A)$ with the *Picard group*

$$\mathrm{Pic}(A) = \text{isomorphism classes of projective } A\text{-modules of rank } 1,$$

which is a group with respect to the tensor product, with the class of the free module A as identity element.

If $\mathfrak{M}$ is any maximal ideal of A, then it is in fact a projective A-module of rank 1, and so has a class in $\mathrm{Pic}(A)$. This defines a homomorphism $\mathrm{Div}(A) \to \mathrm{Pic}(A)$, which vanishes on $\mathrm{P}(A)$ because a principal ideal aA is a free module, which represents the identity element of $\mathrm{Pic}(A)$; thus, there is an induced homomorphism $\mathrm{Cl}(A) \to \mathrm{Pic}(A)$. For certain "traditional" reasons, one usually multiplies the above map $\mathrm{Cl}(A) \to \mathrm{Pic}(A)$ by -1; it can be shown that it is an isomorphism.

Now if P is a projective module, say of rank s, we can associate to it the element

$$[\det(P)] = [\overset{s}{\wedge} P] \in \mathrm{Pic}(A),$$

the *determinant* of the projective module P, and (using the above identification), there is a corresponding element in $\mathrm{Cl}(A)$. Thus we have a composition

$$K_0(A) \xrightarrow{\det} \mathrm{Pic}(A) \xrightarrow{\cong} \mathrm{Cl}(A).$$

To make a connection with geometry, let A be a Dedekind domain which is the coordinate ring of an affine algebraic curve C, over an algebraically closed field k ((see [2]). The points of C are in bijection with the maximal ideals of A. We write $C = \operatorname{Spec} A$ to denote this relationship. Finitely generated projective A-modules correspond to *vector bundles* on the curve C, giving an isomorphism of $K_0(A)$ with the Grothendieck group $K_0(C)$ of vector bundles.

There is an associated projective non-singular curve X over k, such that $C \cong X \setminus S$ for some finite, nonempty set S of points of X. If K is the quotient field of A, then the *points of* X are in bijection with the *discrete valuation subrings of* K which contain k (see [2], Ch. I). This determines X as a set, and the projective algebraic structure of X may also be determined intrinsically, using the algebraic structure of the field K, which is then identified with the field of *rational functions* on the curve X.

There is a similarly defined Grothendieck group $K_0(X)$ of vector bundles (or locally free sheaves) on X. If $\mathrm{Pic}\,(X)$ is the group of isomorphism classes of algebraic line bundles on X, one can show that there is an isomorphism

$$K_0(X) \cong \mathbb{Z} \oplus \mathrm{Pic}\,(X),$$

again given by the rank and determinant for vector bundles.

There is also an isomorphism $\mathrm{Pic}\,(X) \cong \mathrm{Cl}\,(X)$ (which we may view as the *algebraic first Chern class*), where $\mathrm{Cl}\,(X) = \mathrm{Div}\,(X)/\mathrm{P}\,(X)$, with

$$\begin{aligned}\mathrm{Div}\,(X) &= \text{Group of } \textit{Divisors} \text{ on } X \\ &= \text{Free abelian group on points of } X,\end{aligned}$$

$$\begin{aligned}\mathrm{P}\,(X) &= \textit{Principal divisors} \\ &= \text{Divisors of non-zero rational functions on } X.\end{aligned}$$

If f is a nonzero rational function on X, its divisor is

$$\mathrm{div}\,(f) = (\text{zeroes of } f) - (\text{poles of } f)$$

where the zeroes and poles are counted with multiplicities.

Finally, there is a relationship between the class groups of X and of $C = \mathrm{Spec}\, A = X \setminus S$, given by an isomorphism

$$\begin{aligned}\mathrm{Cl}\,(A) &= \mathrm{Cl}\,(C) \\ &\cong \frac{\mathrm{Cl}\,(X)}{\text{Subgroup generated by points of } S}.\end{aligned}$$

The geometry comes in now, in terms of a structure theorem for $\mathrm{Cl}\,(X)$. There is a degree homomorphism

$$\deg : \mathrm{Cl}\,(X) \to \mathbb{Z}.$$

$$\deg \sum n_i[x_i] \mapsto \sum_i n_i.$$

This is well-defined, because any nonzero rational function on X is known to have the same number of poles and zeroes, when counted with multiplicities.

Now an important theorem in algebraic geometry is that the kernel $\mathrm{Cl}\,(X)_{\deg 0}$ has the structure of an *abelian variety* (a projective algebraic group variety), called the *Jacobian variety*, which is denoted $J(X)$, and whose dimension is the *genus* of the non-singular projective curve X.

If the ground field k is $\mathbb{C}$, the field of complex numbers, then the complex points of X naturally form a *compact Riemann surface*, and the "algebraic genus" of X, as considered above, equals the "topological genus" of the underlying compact connected oriented 2-manifold. The Jacobian $J(X)$ is identified, as a real Lie group, with the torus $H^1(X, \mathbb{R}/\mathbb{Z})$, which has a natural complex structure coming from the theory of harmonic forms.

The genus may be defined algebraically as the dimension of the vector space of algebraic regular 1-forms on the curve X, or equivalently (using Serre duality) as the dimension of the sheaf cohomology $H^1(X, \mathcal{O}_X)$ of the sheaf $\mathcal{O}_X$ of algebraic regular functions on X.

We now have the following remarkable result:

Theorem 10.2. *Let $C = \operatorname{Spec} A$ be an affine algebraic curve over an algebraically closed field k, where A is a Dedekind domain (i.e., C is a* nonsingular *algebraic curve). Let X be the corresponding nonsingular projective algebraic curve over k. Then the following are equivalent:*

every maximal ideal of A is principal $\Leftrightarrow \operatorname{Cl}(A) = 0 \Leftrightarrow$
the projective curve X has genus 0, i.e., $H^1(X, \mathcal{O}_X) = 0$.

This is because of a property of abelian varieties: if the Jacobian $J(X)$ is nonzero (equivalently the genus of X is nonzero), then $J(X)$ is not a finitely generated group. This means that for A as in the Theorem, we either have that $\operatorname{Cl}(A) = 0$, or that $\operatorname{Cl}(A)$ is not finitely generated.

A refinement of the above result applies to an arbitrary reduced, finitely generated k-algebra A of Krull dimension 1, where k is an algebraically closed field.

Once again, one can uniquely associate to $C = \operatorname{SpecA}$ a projective algebraic curve X, such that $C = X \setminus S$ for a finite set S of nonsingular points of X. The dimension of the sheaf cohomology group $H^1(X, \mathcal{O}_X)$ is called the *arithmetic genus* of the projective curve X, and equals the dimension of a certain algebraic group variety called the *generalized Jacobian* of the (possibly singular) curve X.

A point $x \in C$ is a nonsingular point if for the corresponding maximal ideal $\mathfrak{M}_x$ of A, the localization $A_{\mathfrak{M}_x}$ is a discrete valuation ring; C has only a finite number of singular points. The maximal ideal corresponding to a nonsingular (or *smooth*) point will be called a *smooth maximal ideal.*

Theorem 10.3. *Let $C = \operatorname{Spec} A$ and X be as above. Then:*

every smooth maximal ideal of A is principal $\Leftrightarrow$
the projective curve X has arithmetic genus 0.

We now turn to the higher dimensional case.

If A is a finitely generated k-algebra which is an integrally closed domain (i.e., a normal domain) of dimension d, one can associate to it a divisor class group, defined using Weil divisors (free abelian group on irreducible $d-1$-dimensional subvarieties), with a relationship to the theory of the *Picard variety*. One result obtained from this theory is the following (it can be refined in several ways, which we do not go into here).

Theorem 10.4. *Let A be a finitely generated k-algebra, which is an integral domain. Then the following properties hold.*

(i) The group of units of A is of the form

$$A^* = k^* \times (\text{free abelian group of finite rank}).$$

(ii) Assume that A is normal. Let X be a normal projective algebraic k-variety containing $V = \operatorname{Spec} A$ as a dense Zariski open set. Then the divisor class group of A is finitely generated $\Leftrightarrow$ the Picard variety of X (in the sense of Weil) is trivial.

Another generalization, which is our main interest here, is to consider the *complete intersection property* for maximal ideals.

Let A be a reduced, finitely generated algebra of Krull dimension d, over an algebraically closed field k. Let $V = \operatorname{Spec} A$ be the affine variety associated to A, so that maximal ideals of A correspond to (closed) points of V. A point $x \in V$, corresponding to a maximal ideal $\mathfrak{M}_x$ in A, is called a *complete intersection point* if the ideal $\mathfrak{M}_x$ is of height d, and is generated by d elements $f_1, \ldots, f_d$. Geometrically, this means that if $H_i \subset V$ is the hypersurface defined by

$$H_i = \{y \in V | f_i(y) = 0\} = \operatorname{Spec} A/f_iA,$$

then $H_1 \cap \cdots \cap H_d = \{x\}$, and $x \in V$ is a nonsingular point, such that the hypersurfaces H_i are also nonsingular at x, and intersect *transversally*.

Note that when $d = \dim A = 1$, $\mathfrak{M}$ is a complete intersection $\Leftrightarrow$ $\mathfrak{M}$ is a principal ideal.

Recall that a point $x \in V$ is a *smooth* (or nonsingular) point if the local ring $\mathcal{O}_{V,x} = A_{\mathfrak{M}_x}$ is a regular local ring of dimension d, in the sense of commutative algebra; this means that $\mathfrak{M}_x$ has height d, and the localized maximal ideal $\mathfrak{M}_x A_{\mathfrak{M}_x}$ is generated by d elements.

Equivalently, there is an affine Zariski open subset $W \subset V$ containing x such that $x \in W$ is a complete intersection point. Thus, we may view a smooth point $x \in V$ as a "local complete intersection point". A maximal ideal $\mathfrak{M}$ of A is called a *smooth maximal ideal* if it corresponds to a smooth point of $V = \operatorname{Spec} A$.

The main question we want to discuss is the following.

Question: *Which k-algebras A of dimension d > 1 have the property that all smooth maximal ideals are complete intersections? In other words, when are all local complete intersection points on* $V = \operatorname{Spec} A$ *the same as the complete intersection points?*

There are several conjectures and results related to this Question. We first state a general "positive" result.

Theorem 10.5. *Let* $k = \overline{\mathbb{F}}_p$ *be the algebraic closure of the finite field* $\mathbb{F}_p$. *Then for any reduced finitely generated k-algebra A of dimension* $d > 1$, *every smooth maximal ideal is a complete intersection.*

In the case when $\dim A \geq 3$, or A is smooth of dimension 2, this is a result essentially due to M. P. Murthy. The higher dimensional case is reduced to the 2-dimensional case by showing that any smooth point of $V = \operatorname{Spec} A$ lies on a smooth affine surface $W \subset V$ such that the ideal of W in A is generated by $d-2$ elements (i.e., W is a complete intersection surface in V). This argument depends on the fact that we are dealing here with *affine* algebraic varieties. (see [6]).

The case of an arbitrary 2-dimensional algebra is a corollary of results of Amalendu Krishna and mine (see [3], [4]).

Next, we state two conjectures, which are affine versions of famous conjectures on 0-cycles.

Conjecture 10.1. (Bloch Conjecture). *Let* $k = \mathbb{C}$, *the complex numbers. Let* $V = \operatorname{Spec} A$ *be a non-singular affine* $\mathbb{C}$*-variety of dimension* $d > 1$, *and let* $X \supset V$ *be a smooth proper (or projective)* $\mathbb{C}$*-variety containing* V *as a dense open subset. Then:*

all maximal ideals of A are complete intersections

$\Leftrightarrow$ *X does not support any global regular (or holomorphic) differential d-forms*

$\Leftrightarrow H^d(X, \mathcal{O}_X) = 0.$

Here, $\mathcal{O}_X$ is the sheaf of algebraic regular functions on X. The non-existence of d-forms is equivalent to the cohomology vanishing condition, by Serre duality; the open question is the equivalence of either of these properties with the complete intersection property for maximal ideals.

This conjecture has been verified in several "non-trivial" examples (for example, if $V = \operatorname{Spec} A$ is a "small enough" Zariski open subset of the

Kummer variety of an odd (> 1) dimensional abelian variety over $\mathbb{C}$, all smooth maximal ideals of A are complete intersections).

One consequence of the conjecture is that, for smooth affine $\mathbb{C}$-varieties, the property that all maximal ideals are complete intersections is a *birational* invariant (that is, it depends only on the quotient field of A, as a $\mathbb{C}$-algebra). This birational invariance can be proved to hold in dimension 2, using a result of Roitman; in dimensions ≥ 3, it is unknown in general.

Conjecture 10.2. (Bloch-Beilinson Conjecture) *Let $k = \overline{\mathbb{Q}}$ be the field of algebraic numbers (algebraic closure of the field of rational numbers). Then for any finitely generated* smooth *k-algebra of dimension $d > 1$, every maximal ideal is a complete intersection.*

This very deep conjecture has not yet been verified in any "nontrivial" example (i.e., one where there do exist smooth maximal ideals of $A \otimes_{\overline{\mathbb{Q}}} \mathbb{C}$ which are not complete intersections).

However, it is part of a more extensive set of interrelated conjectures (see [1]) relating *K-groups of motives over algebraic number fields* and *special values of L-functions*, and there are nontrivial examples where some other parts of this system of conjectures can be verified. This is viewed as indirect evidence for the above conjecture.

I will now relate these conjectures to algebraic cycles and K-theory.

The first step is a result of Murthy, giving a K-theoretic interpretation of the complete intersection property.

Recall that $K_0(A)$ denotes the Grothendieck group of finitely generated projective A-modules. If M is an arbitrary finitely generated A-module, recall that M has *finite projective dimension* if there exists a finite projective resolution of M, i.e., an exact sequence

$$0 \to P_r \to P_{r-1} \to \cdots \to P_0 \to M \to 0$$

where the P_i are finitely generated projective A-modules. Then M has a well-defined class $[M] \in K_0(A)$, obtained by choosing any such resolution, and defining

$$[M] = \sum_{i=0}^{r} (-1)^i [P_i] \in K_0(A).$$

Recall also that a maximal ideal $\mathfrak{M}$ has finite projective dimension precisely when the local ring $A_{\mathfrak{M}}$ is a regular local ring.

Theorem 10.6. ([7]) *Let A be a reduced finitely generated algebra over an algebraically closed field. A smooth maximal ideal $\mathfrak{M}$ of A is a complete intersection $\Leftrightarrow$ its class in $K_0(A)$ vanishes.*

Let A be a reduced, finitely generated algebra, of Krull dimension d, over an algebraically closed field k. We can associate to it the group $F^dK_0(A)$, the subgroup of $K_0(A)$ generated by the classes of smooth maximal ideals. If $V = \text{Spec } A$, then $F^dK_0(A)$ is a quotient of the free abelian group on smooth points of V, modulo a suitable equivalence relation. Murthy's theorem in fact is stronger: if V is non-singular, it identifies the above equivalence relation with *rational equivalence*, coming from algebraic geometry.

Recall that the *Chow group* of *0-cycles modulo rational equivalence* on a smooth d-dimensional variety X is

$$CH^d(X) = \frac{Z^d(X)}{R^d(X)},$$

where

$$Z^d(X) = \text{Free abelian group on points of } X,$$

and $R^d(X) \subset Z^d(X)$ is the subgroup generated by

$$\text{div}\,(f)_C = (\text{zeroes of } f) - (\text{poles of } f),$$

for all curves $C \subset X$, and nonzero rational functions f on C.

The Grothendieck group $K_0(X)$ of algebraic vector bundles on X equals the Grothendieck group of coherent sheaves on X, since X is smooth. Let $F^dK_0(X)$ be the subgroup of $K_0(X)$ generated by the classes of points on X. The induced surjective map $Z^d(X) \to F^dK_0(X)$ is easily seen to yield a surjection

$$\psi_d : CH^d(X) \to F^dK_0(X).$$

Grothendieck's algebraic theory of Chern classes, and the Riemann-Roch Theorem ("without denominators"), implies that the dth algebraic *Chern class* gives a homomorphism

$$c_d : F^dK_0(X) \to CH^d(X),$$

so that the compositions $\psi_d \circ c_d$ and $c_d \circ \psi_d$ both equal multplication by $(-1)^{d-1}(d-1)!$.

In particular, ψ_d and c_d are both isomorphisms modulo torsion.

Now assume $V = \operatorname{Spec} A$ is an affine open subset of a nonsingular projective k-variety X of dimension d. Clearly

$$CH^d(V) = \frac{CH^d(X)}{\text{subgroup generated by points of } X \setminus V}.$$

Roitman's Theorem on torsion 0-cycles, extended by Milne to arbitrary characteristic, gives a description of the torsion in $CH^d(X)$, using which it can be shown that $CH^d(V)$ is a torsion free, divisible abelian group (i.e., a vector space over $\mathbb{Q}$). In particular, we see that the map $\psi_d : CH^d(V) \to F^d K_0(V)$ is an isomorphism.

Thus, by Murthy's theorem, all maximal ideals of A are complete intersections $\Leftrightarrow CH^d(X)$ is generated by points of $X \setminus V$.

We now restate the Bloch and Bloch-Beilinson Conjectures in their "original" forms.

Conjecture 10.3. (Bloch Conjecture) *Let X be a projective smooth variety over $\mathbb{C}$. Suppose that, for some integer $r \geq 0$, X has no nonzero regular (or holomorphic) s-forms for any $s > r$. Then for any "sufficiently large" subvariety $Z \subset X$ of dimension r, we have $CH^d(X \setminus Z) = 0$.*

For a smooth projective complex surface X, this conjecture states that if X has no holomorphic 2-forms, then $CH^2(X \setminus C) = 0$ for some curve C in X. This has been verified in a few situations, for example, for surfaces of Kodaira dimension ≤ 1 (Bloch, Kas, Lieberman), for general Godeaux surfaces (Voisin), and in some other cases. In higher dimensions, Roitman proved it for complete intersections in projective space, and there are a few other isolated examples, like the Kummer variety associated to an odd dimensional abelian variety (Bloch and myself).

Conjecture 10.4. (Bloch-Beilinson Conjecture) *Let X be a smooth projective variety of dimension d over $\overline{\mathbb{Q}}$. Then $CH^d(X)$ is "finite dimensional"; in particular, there is a curve $C \subset X$ so that $CH^d(X \setminus C) = 0$.*

As remarked earlier, there is only indirect evidence for this conjecture: it has not been verified for any smooth projective surface over $\overline{\mathbb{Q}}$ which supports a non-zero 2-form (e.g., any hypersurface in projective 3-space of degree ≥ 4).

To exhibit one such nontrivial example is already an interesting open question.

From the algebraic viewpoint, it seems restrictive to work only with smooth varieties. In any case, it is unknown in characteristic $p > 0$ that a smooth affine variety V can be realized as an open subset of a smooth proper variety X (in characteristic 0, this follows from Hironaka's theorem on *resolution of singularities*).

Inspite of this, it is possible to make a systematic study of the singular case, and to try to extend the above conjectures.

For the purposes of this chapter, let me focus on one very special situation. Let

$$Z \subset \mathbb{P}^N_k$$

be a non-singular projective algebraic k-variety, and

$$\begin{aligned} A &= \oplus_{n\geq 0} A_n \\ &= \text{homogeneous coordinate ring of } Z. \end{aligned}$$

The affine variety $V = \operatorname{Spec} A$ is the "affine cone" over Z with "vertex" corresponding to the unique graded maximal ideal $\mathfrak{M} = \oplus_{n>0} A_n$, and the vertex is the unique singular point of V. The projective cone $C(Z)$ over Z with the same vertex naturally contains V as an open subset, whose complement is a divisor isomorphic to Z, and the vertex is again the only singular point of $C(Z)$.

The first part of the following theorem is obtained using results from my paper [3] with Amalendu Krishna, in the 2-dimensional case, and a preprint of Krishna's in the higher dimensional case. The result over $\overline{\mathbb{Q}}$ is in [4].

Theorem 10.7. *(i) Let $k = \mathbb{C}$. Assume that V is Cohen-Macaulay (for example, $d = 2$ and A is normal) of dimension $d \leq 3$. Then every smooth maximal ideal of A is a complete intersection $\Leftrightarrow H^{d-1}(Z, \mathcal{O}_Z(1)) = 0$ $\Leftrightarrow H^d(C(Z), \mathcal{O}_{C(Z)}) = 0$*
(ii) Let $k = \overline{\mathbb{Q}}$. Then every smooth maximal ideal of A is a complete intersection.

Here, (i) is analogous to the Bloch Conjecture, while (ii) is analogous to the Bloch-Beilinson Conjecture.

Let us close with two examples.

Example 1 ([3])

$$A = \frac{\overline{\mathbb{Q}}[x, y, z]}{(x^4 + y^4 + z^4)}.$$

Here, all smooth maximal ideals of A are complete intersections, while "most" smooth maximal ideals of $A \otimes_{\overline{\mathbb{Q}}} \mathbb{C}$ are *not* complete intersections. The complete intersection smooth maximal ideas are those determined by points on the rulings of the affine cone over points of the Fermat Quartic curve with $\overline{\mathbb{Q}}$ coordinates. This is a consequence of Theorem 10.7.

Example 2

$$A = \frac{\overline{\mathbb{Q}}[x, y, z]}{(xyz(1 - x - y - z))}.$$

Again, all smooth maximal ideals of A are complete intersections, while "most" smooth maximal ideals of $A \otimes_{\overline{\mathbb{Q}}} \mathbb{C}$ are *not* complete intersections. In fact, there is an identification

$$F^2K_0(A \otimes_{\overline{\mathbb{Q}}} k) = K_2(k),$$

where K_2 denotes the Milnor K_2. Now one has the result of Garland (vastly generalized by Borel) that $K_2(\overline{\mathbb{Q}}) = 0$, while $K_2(\mathbb{C})$ is "very large".

References

[1] Rapoport, M., Schappacher, N. and Schneider, P. (Eds.) (1988). *Beilinson's conjectures on special values of L-functions.* Perspectives in Math. **4** Academic Press, Boston.

[2] Hartshorne, R. *Algebraic Geometry.* Grad. Texts in Math. **52** Springer-Verlag.

[3] Krishna, A. and Srinivas, V. (2002). Zero cycles and K-theory on normal surfaces. *Annals of Math.* **156**.

[4] Krishna, A. and Srinivas, V. (2007). Zero cycles on singular varieties. In *Algebraic Cycles and Motives, Vol. 1*, London Math. Soc. Lect. Note Series **343** Cambridge, 264-277.

[5] Milnor, J. (1971). *Introduction to Algebraic K-Theory.* Ann. Math. Studies. **72** Princeton.

[6] Murthy, M. P., Mohan Kumar, N., and Roy, A. (1988). In *Algebraic geometry and commutative algebra.* **I** (in honour of Masayoshi Nagata), Kinokuniya, Tokyo, 281-287.

[7] Murthy, M. P. (1994). Zero cycles and projective modules. *Annals of Math.* 405-434.

[8] Zariski, O. and Samuel, P. *Commutative Algebra.* **I**, Grad. Texts in Math. **28** Springer-Verlag.

Chapter 11

Root Numbers and Rational Points on Elliptic Curves

R. Sujatha
School of Mathematics,
Tata Institute of Fundamental Research,
Homi Bhabha Road, Mumbai 400 005, India
sujatha@math.tifr.res.in

This article is broadly based on the talk that was given at the symposium 'Perspectives in the Mathematical Sciences', held at Indian Statistical Institute (ISI), Bangalore, to celebrate the Platinum Jubilee of the ISI. I would like to thank the organisers, especially Professor N.S.N. Sastry, for inviting me to lecture on this occasion. The article is intended for a general audience and does not contain any new results. We have aimed at highlighting several new results that have been proved on this topic by various authors, in the last few years. The proofs have been omitted and the interested reader is referred to the research papers, which we have tried to list rather extensively.

11.1. Elliptic Curves and the Birch and Swinnerton-Dyer Conjecture

Let F be a finite extension of $\mathbb{Q}$ and let E/F be an elliptic curve. Recall that E has an affine equation

$$E : y^2 = f(x),$$

where $f(x) \in F[x]$ is a cubic polynomial with distinct roots. A famous result of Mordell asserts that the group $E(F)$ of F-rational points of E is a finitely generated abelian group. Let $g_{E/F}$ denote the rank of $E(F)$. Associated with E is the complex L-function $L(E/F, s)$ of E, which is defined by an Euler product. This function converges only for $\mathrm{Re}(s) > 3/2$, but is conjectured to have an entire continuation (see [22]), and a functional equation relating its values at s and $2 - s$.

When $F = \mathbb{Q}$, thanks to deep results of Wiles ([24]) and [2], this conjecture is true, but it is only for elliptic curves with complex multiplication that it so far has been proven over all number fields F. Assuming the analytic continuation of $L(E/F, s)$, the analytic rank, which we denote by $r_{E/F}$, is defined to be the order of zero of $L(E/F, s)$ at $s = 1$, the centre of its critical strip. In the 1960's, based on rather compelling numerical evidence, Birch and Swinnerton-Dyer made the astonishing conjecture that

$$g_{E/F} = r_{E/F}. \tag{11.1}$$

A refined version of this conjecture even gives an exact formula for the leading Taylor coefficient of the L-function at $s = 1$. For more details, see [25]. An important part of this exact formula is the order of the Tate-Shafarevich group of E/F. For a finite extension K of F, the *Tate Shafarevich group of E over K*, denoted by $Ш(E/K)$, is defined by

$$Ш(E/K) = \text{Ker}\left(H^1(K, E) \to \prod_v H^1(K_v, E)\right). \tag{11.2}$$

Here, v varies over all the places of K and K_v denotes the completion of K at v, while $E := E(\bar{F})$ denotes the group of points of E over a fixed algebraic closure $\bar{F}$ of F, considered as a module over the Galois group of K. Finally, for any field K and a module M over the Galois group $G_K := \text{Gal}(\bar{K}/K)$, the first cohomology group is denoted by $H^1(K, M)$. The Tate-Shafarevich group is among the most mysterious groups occurring in the study of the arithmetic of elliptic curves and part of the full Birch and Swinnerton-Dyer conjecture is that it is always finite. However, it was only in the late 1980's that explicit examples of elliptic curves with finite Tate-Shafarevich group came to light. Kolyvagin, and independently Rubin, whose work was based on ideas of Thaine, gave these first examples. We remark here that we do not yet know the finiteness of the Tate-Shafarevich group for a single elliptic curve of rank at least 2.

11.2. Congruent Number Problem

Recall that a natural number $N \geq 1$ is said to be *congruent* if there exists a right angled triangle whose sides have rational length, and area N. In other words N is congruent if there exists rational numbers a, b, c in $\mathbb{Q}$ such that $a^2 + b^2 = c^2$ and $ab/2 = N$. One of the oldest problems in number theory is to explicitly give an algorithm which would determine whether a given number is congruent. While Arab manuscripts dating back to the tenth century A.D. give a long list of examples of congruent numbers, it is almost certain that the ancient Indians too grappled with congruent numbers and knew of many examples. For more on this subject, the reader is referred to the book by Koblitz ([13]) and the article by Coates ([3]). A folklore conjecture in this subject is the following, which remains open despite overwhelming numerical evidence:

Conjecture 11.1. *If N is a positive integer congruent to 5, 6, or 7 modulo 8, then N is congruent.*

The connection between congruent numbers and elliptic curves is the following. For any integer $N \geq 1$, consider the elliptic curve E over $\mathbb{Q}$ defined by

$$E_N : y^2 = x^3 - N^2 x.$$

Then N is congruent if and only if E has a rational point (x, y), x, $y \in \mathbb{Q}$ with $y \neq 0$. Indeed, if (a, b, c) are the lengths of the corresponding right angled triangle, with area N and $a^2 + b^2 = c^2$, then (x, y) with

$$x = \frac{N(a+c)}{b}, \quad y = \frac{2N^2(a+c)}{b^2}$$

is a point on E_N with $y \neq 0$. Such a point is well-known to give a point of infinite order on E_N and the theory of L-functions shows that for $N \equiv$ 5, 6, 7 mod 8, $L(E_N, s)$ has a zero of odd order (and therefore a zero) at $s = 1$. Thus, Conjecture 11.1 is seen to be a special case of the Birch and Swinnerton-Dyer conjecture.

Iwasawa theory is a p-adic theory that provides a systematic method to attack the Birch and Swinnerton-Dyer conjecture and has led to important results in the study of the arithmetic of elliptic curves. The main object of study here is the investigation of the Galois action on the dual Selmer group, viewed over certain infinite extensions of F. We refer the reader to [21], [9], and [3] for detailed accounts of the Iwasawa theory of elliptic curves. Hereafter, we fix an odd prime p. Recall that for an elliptic curve E/F and a finite extension K of F, the *p-Selmer group*, denoted $S_p(E/K)$, is defined as

$$S_p(E/K) = \text{Ker}\left(H^1(K, E_{p^\infty}) \longrightarrow \bigoplus_v H^1(K_v, E)\right)$$

where $E_{p^\infty} := \bigcup_n E_{p^n}(\bar{F})$ denotes the group of all p-power division points of $E(\bar{F})$ considered as a module over G_K. It is well-known that $S_p(E/K)$ is a cofinitely generated $\mathbb{Z}_p$-module and we define

$$s_{p,E/K} := \mathbb{Z}_p - \text{corank of } S_p(E/K). \tag{11.3}$$

For an infinite Galois extension K_∞ of F with Galois group $G := \mathrm{Gal}(K_\infty/F)$ a p-adic Lie group, the Selmer group $S_p(E/K_\infty)$ is defined as the direct limit over the Selmer groups $S_p(E/L)$, as L varies over finite Galois extensions of F contained in K_∞. It is clear that $S_p(E/K_\infty)$ is a discrete G-module, and its Pontryagin dual denoted $X_p(E/K_\infty)$ is a compact G-module, and is the *dual Selmer group*. It is even a finitely generated, compact module over the Iwasawa algebra $\Lambda(G)$ defined as the inverse limit

$$\Lambda(G) := \varprojlim \mathbb{Z}_p[G/G'],$$

the limit being taken over the group rings $\mathbb{Z}_p[G/G']$ as G' varies over open normal subgroups, with respect to the natural maps. The key idea in Iwasawa theory is to study the arithmetic of E over suitable infinite extensions via the dual Selmer groups considered as $\Lambda(G)$-modules. For a finite extension K of F, Kummer theory yields the well-known exact sequence of discrete G_K-modules

$$0 \to E(K) \otimes \mathbb{Q}_p/\mathbb{Z}_p \to S_p(E/K) \to Ш(E/K)(p) \to 0, \tag{11.4}$$

where $Ш(E/K)(p)$ denotes the p-primary torsion subgroup of $Ш(E/K)$. For an infinite p-adic Lie extension K_∞ of F, with Galois group G, a direct limit argument gives the exact sequence of G-modules

$$0 \to E(K_\infty) \otimes \mathbb{Q}_p/\mathbb{Z}_p \to S_p(E/K_\infty) \to Ш(E/K_\infty)(p) \to 0,$$

where $Ш(E/K_\infty)(p)$ denotes the p-primary part of the Tate-Shafarevich group of E over K_∞ and is defined as the direct limit of $Ш(E/K)(p)$ as K varies over finite Galois extensions of F in K_∞. Iwasawa theory can be used to prove the following result:

Theorem 11.1. *Let N be an integer ≥ 1 such that $L(E_N/\mathbb{Q}, 1) = 0$. If the p-primary part $Ш(E_N/\mathbb{Q})(p)$ is finite for some odd prime p, then N is congruent.*

The central idea here is to prove that $X_p(E_N/\mathbb{Q})$ has $\mathbb{Z}_p$-rank at least one. Then, using the hypothesis on $Ш(E_N)$, one deduces that the same is true of $E_N(\mathbb{Q})$. We shall return to this briefly in the next section.

11.3. Root Numbers and the Parity Conjecture

Let G_F denote the absolute Galois group as before. An *Artin representation* of G_F is a finite dimensional complex representation ρ of G_F which is trivial on an open subgroup and thus factors through a finite Galois extension of F. Unless necessary, the base field will not be specified and the associated representation space will be denoted by V_ρ. In a similar vein, as most of the results we need or state below are independent of the finite extension of the base field through which the Artin representation factors, we shall omit reference to the associated finite Galois extension, except in cases where it might be necessary to specify the extension.

Given an elliptic curve E/F and an Artin representation ρ of G_F, there is an associated twisted L-function defined by an Euler product and denoted by $L(E, \rho, s)$ (see [23] for details). Again, this twisted L-function is conjectured to be entire with a functional equation. More precisely, let

$$\tilde{L}(E,\rho,s) := \left(\frac{N(E,\rho)}{\pi^{2d_\rho}}\right)^{s/2} \Gamma\left(\frac{s}{2}\right)^{d_\rho} \Gamma\left(\frac{s+1}{2}\right)^{d_\rho} L(E,\rho,s);$$

here d_ρ is the dimension of V_ρ and $N(E,\rho)$ is the global conductor of the Galois representation associated to that of E twisted by V_ρ (cf. [6, 2.4]). The conjectured functional equation is

$$\tilde{L}(E,\rho,s) = w(E,\rho)\,\tilde{L}(E,\hat{\rho},2-s), \tag{11.5}$$

where $\hat{\rho}$ is the contragredient of ρ and $w(E,\rho)$ is the *root number*, which is an algebraic number of complex absolute value 1. We remark that even though the functional equation (11.5) is largely conjectural, the root number is nonetheless *well-defined*. Indeed, by the theorem of Deligne and Langlands, it can be written as a product of local root numbers, taken over all places v of F (see [23]). In particular, if ρ is self-dual (i.e. $\rho = \hat{\rho}$), then the root number is equal to ± 1 and is often referred to as the *sign in the functional equation*.

Definition 11.1. Assuming that $L(E, \rho, s)$ is entire, the ρ-analytic rank is defined as $r(E,\rho) = \mathrm{ord}_{s=1} L(E,\rho,s)$. Note that if ρ is self-dual and irreducible, then

$$w(E,\rho) = (-1)^{r(E,\rho)}. \tag{11.6}$$

Now let ρ be an irreducible Artin representation and K/F a finite Galois extension through which ρ factors, with $G_{K/F} = \mathrm{Gal}(K/F)$.

Definition 11.2. The multiplicity of the irreducible representation ρ occurring in the $G_{K/F}$-module $E(K) \otimes \bar{\mathbb{Q}}_p$ is the *ρ-algebraic rank* and is denoted by $g_{E,\rho}$.

The invariant $g_{E,\rho}$ is in fact independent of the prime p.

Definition 11.3. The multiplicity of the contragredient representation $\hat{\rho}$ in $X(E/K) \otimes_{\mathbb{Z}_p} \bar{\mathbb{Q}}_p$ is denoted by $s_{p,E,\rho}$.

The refined form of Birch and Swinnerton-Dyer conjecture in this context is the conjecture that

$$g_{E,\rho} = r_{E,\rho} \quad \text{and} \quad g_{E,\rho} = s_{p,E,\rho} \text{ for all } p. \tag{11.7}$$

Note that when ρ is the trivial representation these invariants coincide with the invariants g_E, r_E and $s_{p,E}$ defined earlier.

The *ρ-parity conjecture* is the assertion that for self-dual, irreducible Artin representations ρ, we have

$$w(E,\rho) = (-1)^{s_{p,E,\rho}}, \text{ or equivalently that } s_{p,E,\rho} \equiv r_{E,\rho} \mod 2. \tag{11.8}$$

We stress that it is completely unknown whether $s_{p,E,\rho}$ has the same parity as $g_{E,\rho}$, or even whether the parity of $s_{p.E,\rho}$ is independent of p. When ρ is the trivial representation, we clearly recover (11.1) as a special case of (11.7). Finally, note that if the root number is -1 for some self-dual Artin representation, then the parity conjecture implies that the dual Selmer group $X_p(E/K)$ has positive $\mathbb{Z}_p$-rank for a finite extension K of F. Further, the Birch and Swinnerton-Dyer conjecture predicts that $E(K)$ is infinite. However, to deduce that the elliptic curve has a point of infinite order over K, by (11.4), one needs to know the finiteness of the p-primary part of the Tate-Shafarevich group of E. We refer the reader to the beautiful paper of Rohrlich ([19]) which elaborates on the interplay between Artin representations, root numbers and elliptic curves.

Suppose that E/F is an elliptic curve and we are given a p-adic Lie extension K_∞ of F (with p and odd prime), with Galois group G. Let ρ be an irreducible, self-dual, Artin representation which factors through a finite quotient of G. Then there is the notion of the *twisted dual Selmer group*, which we denote by $X_p(E/K_\infty, \rho)$ (see [4, §3]). The basic idea is to study the invariants $g_{E,\rho}$, $r_{E,\rho}$ and $s_{p,E,\rho}$ for E as ρ ranges over all self-dual representations of G. We remark that the study of the dual Selmer group twisted by an Artin representation ρ of G, [4, §3], considered as a $\Lambda(G)$-module (see §3), is an important ingredient in the determination of the invariant $s_{p,E,\rho}$. We refer the reader to the papers of Greenberg, Guo [10], [11] and [4] for more details.

11.4. Recent Results

In this section, we list some of the comparatively recent results towards the ρ-parity conjecture. Earlier affirmative results in this direction (when the representation ρ is trivial), have been proved by many authors, notably [1, 11, 16, 17]. More recently, interesting work has been done by [6, 7, 10, 12, 14, 15, 18].

In the next section, we shall illustrate some of these results with numerical examples and also consider some interesting applications. Here is a striking result due to Tim and Vladimir Dokchitser ([8]):

Theorem 11.2. *Let E be any elliptic curve over $\mathbb{Q}$ and p any prime number. Then $s_{p,E/\mathbb{Q}} \equiv r_{E/\mathbb{Q}} \bmod 2$.*

A completely different perspective is adopted in [4], where the results proved give some fragmentary evidence that a close connection exists between root numbers and the dual Selmer group of an elliptic curve over certain non-commutative Galois extensions of the base field F. In fact, the Galois extensions being non-commutative, provide us with a rich source of examples of infinite families of irreducible self-dual Artin representations of the corresponding Galois groups. We consider two such extensions below where the base field is assumed to be $\mathbb{Q}$ for simplicity.

Let p be an odd prime. Fix an integer $m \geq 1$ which is p-power free, and not divisible by any prime of additive reduction for E.

Definition 11.4. The *False Tate extension* of $\mathbb{Q}$ corresponding to m is defined as

$$F_\infty = \bigcup_{n\geq 1} \mathbb{Q}(\mu_{p^n}, m^{1/p^n}). \tag{11.9}$$

For $n \geq 1$, put

$$F_n = \mathbb{Q}(\mu_{p^n}, m^{1/p^n}), \quad K_n = \mathbb{Q}(\mu_{p^n}), \quad L_n = \mathbb{Q}(m^{1/p^n}), \tag{11.10}$$

and let K^{cyc} denote the field obtained by adjoining all the p-power roots of unity to $K := \mathbb{Q}(\mu_p)$. Put

$$G = \text{Gal}(F_\infty/\mathbb{Q}), \qquad H = \text{Gal}(F_\infty/K^{\text{cyc}}) \simeq \mathbb{Z}_p. \tag{11.11}$$

Our hypotheses above on m imply that the degree $[L_n : \mathbb{Q}] = p^n$ for all $n \geq 0$. The extension F_∞ is a p-adic Lie extension of $\mathbb{Q}$ with Galois group isomorphic to the semi-direct product of H and $\mathbb{Z}_p^\times$. As the group H is isomorphic to $\mathbb{Z}_p$, the Iwasawa algebra $\Lambda(H)$ is isomorphic to the power series ring in one variable over $\mathbb{Z}_p$.

We fix an odd prime p such that $E/\mathbb{Q}$ has good ordinary reduction at p. Further, we shall also assume that the quotient of the dual Selmer group by its p-primary torsion subgroup,

$$Y_p(E/F_\infty) = X_p(E/F_\infty)/X_p(E/F_\infty)(p) \tag{11.12}$$

is finitely generated as a $\Lambda(H)$-module. Indeed, as we shall discuss in the next section, there are interesting numerical examples where these assumptions are satisfied, and it is conjectured in [5] that this latter hypothesis always holds. The self-dual irreducible Artin representations of G are well-known in this case. Further, the twisted L-functions are all known to be entire, and satisfy the standard functional equation, thanks to deep results in automorphic forms (Langlands-Tunnell, Arthur-Clozel, Wiles, Breuil-Conrad-Diamond-Taylor). Also, the root numbers exhibit a surprisingly uniform behaviour as was shown by T. Dokchitser. Further, it can be shown that the parity of the root numbers is equal to that of the $\Lambda(H)$-rank of $Y_p(E/F_\infty)$. We also have [4, §4]

Theorem 11.3. *Let $E/\mathbb{Q}$ be an elliptic curve such that E has good ordinary reduction at p. Assume that $Y_p(E/F_\infty)$ (see (11.12)), is a finitely generated $\Lambda(H)$-module. Then for all self-dual, irreducible Artin representations ρ of G with dimension > 1, the ρ-parity conjecture holds, i.e.*

$$w(E,\rho) = (-1)^{s_{p,E,\rho}}.$$

The second infinite extension is obtained as follows. Let $E/\mathbb{Q}$ be an elliptic curve with potential good ordinary reduction at p, where $p \geq 5$. Assume also that E does not have complex multiplication. Define

$$F_\infty := \mathbb{Q}(E_{p^\infty}) \tag{11.13}$$

where

$$E_{p^\infty} := \bigcup_{n\geq 0} E_{p^n}$$

is the Galois module of all p-power division points of E. Let $G := \mathrm{Gal}(F_\infty/\mathbb{Q})$, which by a theorem of Serre is an open subgroup of $\mathrm{GL}_2(\mathbb{Z}_p)$. By the Weil pairing, $K^{\mathrm{cyc}} = \mathbb{Q}(\mu_{p^\infty})$ is a subfield of F_∞, and we put $H := \mathrm{Gal}(F_\infty/K^{\mathrm{cyc}})$. The module $Y_p(E/F_\infty)$ is again defined as in (11.12).

Recall that an Artin representation is said to be *orthogonal* if the underlying vector space carries a G-invariant, non-degenerate, symmetric bilinear form. The following theorem is a particular case of a more general result proved in [4, §6].

Theorem 11.4. *Assume that E admits an isogeny of degree p over $\mathbb{Q}$, where p is a prime of potential good ordinary reduction. Assume further that $Y_p(E/F_\infty)$ is a finitely generated $\Lambda(H)$-module. Further, assume that the image of G in $PGL_2(\mathbb{F}_p)$ has even order. Then the ρ-parity conjecture holds for any self-dual, irreducible, orthogonal Artin representation ρ of G of dimension greater than 1.*

It is intriguing to note that the proof of this theorem given in [4] is uncannily parallel to Rohrlich's computation of the root numbers $w(E, \rho)$ in this case [20], perhaps suggesting an as yet undiscovered deeper connection between Iwasawa theory and local root numbers.

Finally T. and V. Dokchitser (see [8]) have proven the following more general result by completely different methods, which do not involve Iwasawa theory.

Theorem 11.5. *Let $E/\mathbb{Q}$ be an elliptic curve with semi-stable reduction at 2 and 3, and let p be any prime. Let K be a finite Galois extension of $\mathbb{Q}$ such that the p-Sylow subgroup of* $\mathrm{Gal}(K/\mathbb{Q})$ *is normal and has abelian quotient. Then the ρ-parity conjecture holds for the prime p, and all orthogonal representations of* $\mathrm{Gal}(K/\mathbb{Q})$.

11.5. Examples and Applications

In this final section, we give some applications of the ρ-parity conjecture to obtain lower bounds on the $\mathbb{Z}_p$-ranks of dual Selmer groups, and discuss several numerical examples. It is also worth noting that Iwasawa theory can be used to give upper bounds for the $\mathbb{Z}_p$-coranks of the ρ-components of the Selmer group, as ρ varies over the irreducible Artin characters of a p-adic Lie extension. In this spirit, we end by stating a joint conjecture with J. Coates that proposes strong upper bounds for the multiplicities of Artin representations which can occur in the dual Selmer group of an elliptic curve, considered over finite extensions within a p-adic Lie extension.

Let $E/\mathbb{Q}$ be an elliptic curve and write N_E for the conductor of E. Let p be an odd prime. For each integer $n \geq 1$, let F_n be the fixed field of the centre of $\mathrm{Gal}(\mathbb{Q}(E_{p^n})/\mathbb{Q}))$. The following result is a special case of a more general result of [15, Corollary 2.5].

Theorem 11.6. *Assume that E has good ordinary reduction at p and a rational prime of order p. Suppose further that every prime of bad reduction of E has odd order in $\mathbb{F}_p^\times$, and that $-N_E$ is not a square mod p. Then there exists a positive rational number c independent of n such that for every $n \geq 1$, we have*

$$s_{p,E/F_n} \geq cp^{2n}.$$

As stated above, Iwasawa theory provides lower bounds for the coranks of the Selmer groups. Particularly striking is the case of the False Tate extension (see Definition 11.4), when the module $Y_p(E/F_\infty)$ (see (11.12)) has $\Lambda(H)$-rank 1 where H is as in (11.11). We remark that there are many numerical examples where this is the case. We then have the following theorem [4, Theorem 4.8], where we recall that $K = \mathbb{Q}(\mu_p)$.

Theorem 11.7. *Assume $Y(E/F_\infty)$ is a finitely generated $\Lambda(H)$-module of $\Lambda(H)$-rank 1. Then for all $n \geq 1$, we have*

$$s_{p,E/L_n} = n + s_{p,E/\mathbb{Q}}, \quad s_{p,E/F_n} = p^n - 1 + s_{p,E/K},$$

where the fields F_n and L_n are as in (11.10).

A numerical example of this theorem is given by the elliptic curve

$$E : y^2 + y = x^3 - x^2,$$

and for the prime $p = 3$ with $m = 11$. We deduce from the theorem that

$$s_{3,E/L_n} = n, \qquad \text{and } s_{3,E/F_n} = 3^n - 1, \qquad \text{for } n \geq 1.$$

Similarly, the assertions of the theorem hold for $p = 7$ and $m = 2$, showing that

$$s_{7,E/L_n} = n, \qquad \text{and } s_{7,E/F_n} = 7^n, \qquad \text{for } n \geq 1.$$

Here we have used the fact that $s_{p,E/K} = 1$.

There is also a lower bound in the case of the GL_2 extension F_∞ as defined in (11.13).

Theorem 11.8. *Let E be an elliptic curve without complex multiplication and p a prime of potential good ordinary reduction. Let*

$$F_\infty = F(E_{p^\infty}), \qquad F_n = \mathbb{Q}(E_{p^n}).$$

In addition to the hypotheses of Theorem 11.4, assume that $p \equiv 3 \bmod 4$. Then there exists $c > 0$ independent of n, such that

$$s_{p,E/F_n} \geq c.p^{2n} \quad (n \geq 1).$$

As a numerical example of both Theorem 11.4 and Theorem 11.8, take E to be the elliptic curve

$$y^2 + xy = x^3 - x - 1,$$

of conductor $N_E = 2.3.7^2$, with $p = 7$, and $F_\infty = \mathbb{Q}(E_{7^\infty})$. Then E achieves good ordinary reduction at the unique prime of $\mathbb{Q}(\mu_7)$ above 7. Moreover, μ_7 is a Galois submodule of E_7, and so E has an isogeny of degree 7 defined over $\mathbb{Q}$. It can be shown that $X_7(E/F_\infty)$ is a finitely generated $\Lambda(H)$-module where $H = \mathrm{Gal}(F_\infty/\mathbb{Q}(\mu_{7^\infty}))$. Further, the image in this case of G in $PGL_2(\mathbb{F}_p)$ has order 42. Hence all the hypotheses of Theorems 11.4 and 11.8 hold in this example. We remark that Rohrlich has shown that the cases $w(E,\rho) = +1$ and $w(E,\rho) = -1$ both occur for infinitely many self-dual irreducible Artin representations ρ of G.

We end with a conjecture proposed jointly with John Coates, which was suggested by Theorem 4.12 of [4]. Let F_∞ be a Galois extension of $\mathbb{Q}$ which is unramified outside a finite set of primes, and whose Galois group G is p-adic Lie group. We assume that F_∞ contains the cyclotomic $\mathbb{Z}_p$-extension $\mathbb{Q}^{\mathrm{cyc}}$ of $\mathbb{Q}$. Let $\mathfrak{X}$ denote the set of all one dimensional characters of $\mathrm{Gal}(\mathbb{Q}^{\mathrm{cyc}}/\mathbb{Q})$, and E be any elliptic curve defined over $\mathbb{Q}$.

Conjecture 11.2. *Assume E has potential good ordinary reduction at p. Then there exists an integer C, depending only on E and F_∞, such that, for all irreducible Artin representations ρ of G, we have*

$$\sum_{\chi \in \mathfrak{X}} s_{p,E,\rho\chi} \leq C. \tag{11.14}$$

Naturally, there is another version of the above conjecture in which we replace $s_{p,E.\rho\chi}$ in (11.14) by $r_{E,\rho\chi}$. Of course, the generalised Birch and Swinnerton-Dyer conjecture (11.7), would imply the equivalence of the two versions. We note finally that Theorem 4.12 of [4] shows that the first version of the above conjecture holds for the False Tate extension. Both forms are true for the cyclotomic $\mathbb{Z}_p$ extension of $\mathbb{Q}$ by virtue of well-known theorems of Kato and Rohrlich. At present, it is completely unknown for the extension $F_\infty = F(E_{p^\infty})$ when E does not admit complex multiplication.

References

[1] Birch, B. and Stephens, N. (1966). The parity of the rank of the Mordell-Weil group. *Topology.* **5** 295-299.

[2] Breuil. C., Conrad, B., Diamond, F. and Taylor, R. (2001). On the modularity of elliptic curves over $\mathbb{Q}$: wild 3-adic exercises. *Jour. AMS.* **14** 843-931.

[3] Coates, J. (1999). Fragments of the GL_2 Iwasawa theory of elliptic curves without complex multiplication. In *Arithmetic theory of elliptic curves.* Lecture Notes in Math. **1716** Springer, 1-50.

[4] Coates, J., Fukaya, T., Kato, K. and Sujatha, R. (2008). Root numbers, Selmer groups and noncommutative Iwasawa theory. *Jour. Alg. Geometry.* (To appear).

[5] Coates, J., Fukaya, T., Kato, K. and Sujatha, R. and Venjakob, O. (2005). The GL_2 main conjecture for elliptic curves without complex multiplication. *Publ. Math. IHES.* **101** 163-208.

[6] Dokchitser, T. and Dokchitser, V. (2006). Numerical computations in noncommutative Iwasawa theory, with Appendix by Coates, J., and Sujatha, R. *Proc. London Math. Soc.* **94** 211-272.

[7] Dokchitser, T. and Dokchitser, V. (2008). Regulator constants and the parity conjecture. (To appear).

[8] Dokchitser, T. and Dokchitser, V. (2008). On the Birch-Swinnerton-Dyer quotients modulo squares. (To appear).

[9] Greenberg, R. (1999). Iwasawa theory for elliptic curves. In *Arithmetic theory of elliptic curves.* Lecture Notes in Math. **1716** Springer, 51-144.

[10] Greenberg, R.(2007). Iwasawa theory, projective modules, and modular representations. Preprint.

[11] Guo, L. (1993). General Selmer groups and critical values of Hecke L-functions. *Math. Ann.* **297** 221-233.

[12] Kim, B. D. (2005). *The parity conjecture and algebraic functional equations for elliptic curves at supersingular reduction primes.* Ph.D Thesis. Stanford University.

[13] Koblitz, N. (1984). *Introduction to elliptic curves and modular forms.* Graduate Texts in Math. Springer.

[14] Mazur, B. and Rubin, K. (2008). Finding large Selmer ranks via an arithmetic theory of local constants. *Ann. of Math.* (To appear).

[15] Mazur, B. and Rubin, K. (2008). Growth of Selmer ranks in nonabelian extensions of number fields. *Duke Math. Journal.* (To appear).

[16] Monsky, P. (1996). Generalizing the Birch-Stephens theorem. *Math. Z.* **221** 415-420.

[17] Nekovář, J. (2006). Selmer complexes. *Astérisque.* **310**.

[18] Nekovář, J. (2007). On the parity of ranks of Selmer groups III. *Documenta Math.* **12** 243-274.

[19] Rohrlich, D. E. (1996). Galois theory, elliptic curves, and root numbers. *Compositio Math.* **100** 311-349.

[20] Rohrlich, D. E. (2008). Scarcity and abundance of trivial zeros in division towers. *Jour. Alg. Geometry.* (To appear).

[21] Rubin, K. (1999). Elliptic curves with complex multiplication and the conjecture of Birch and Swinnerton-Dyer. In *Arithmetic theory of elliptic curves.* Lecture Notes in Math. **1716** Springer, 167-234.

[22] Silverman, J. (1986). *The Arithmetic of Elliptic Curves.* Springer-Verlag, GTM **106**.

[23] Tate, J. (1979). Number Theoretic Background. *Proc. of Symp. in Pure Math.* **33** 3-26.

[24] Wiles, A. (1995). Modular elliptic curves and Fermat's last theorem. *Ann. of Math.* **141** 443-551.

[25] Wiles, A. (2006). The Birch and Swinnerton-Dyer conjecture. In *The Millennium Prize Problems.* (J. Carlson, A. Jaffe and A. Wiles, Eds.) Clay Math. Inst. and AMS, 31-44.

Chapter 12

von Neumann Algebras and Ergodic Theory

V. S. Sunder
The Institute of Mathematical Sciences,
C. I. T. Campus, Chennai 600113, India
sunder@imsc.res.in

There has been a long-standing and strong link between ergodic theory and von Neumann algebras (in particular, factors) dating back to the seminal work (cf. [18]) of Murray and von Neumann, specifically their construction of the first examples of factors of type II and type III. The bridge is provided by the celebrated group-measure space construction (or the crossed-product construction in modern parlance). In this survey, we shall commence with a discussion of some aspects of the magnificent edifice created by Murray and von Neumann, Dye, Krieger, Connes, Ornstein, Weiss, Feldman, Moore, ..., and conclude with an attempt[a] to describe some "rigidity" results of Gaboriau and Popa.

We commence proceedings with brief introductions to each of the topics von Neumann algeras, ergodic theory, the group-measure space construction and II_1 factors.

von Neumann algebras

A von Neumann algebra is a self-adjoint (i.e., $x \in M \Rightarrow x^* \in M$) unital (i.e., $1 \in M$) subalgebra M of the *-algebra $B(H)$ of all continuous linear operators on a Hilbert space[b] H, which satisfies any of the following equivalent requirements: [c]

---

[a]It is only natural that the picture portrayed here is coloured/flawed by the author's own perceptions/limitations of exposure, and it is almost sure that there have been many grave omissions, for all of which only the author's limitations are to blame, and the author apologises for any such errors or omissions.

[b]All our Hilbert spaces will be assumed to be *separable*.

[c]The equivalence of these three conditions — two topological, one algebraic — is von Neumann's celebrated *double commutant theorem.*

(1) M is closed in the *strong operator topology* — i.e., $x_i \in M, x \in B(H), \|(x_i - x)\xi\| \to 0 \forall \xi \in H \Rightarrow x \in M$
(2) M is closed in the *weak operator topology* — i.e., $x_i \in M, x \in B(H), \langle (x_i - x)\xi, \eta \rangle \to 0 \forall \xi, \eta \in H \Rightarrow x \in M$
(3) $M''(= (M')') = M$, where $S' = \{x \in B(H) : xs = sx \forall s \in S\}$ denotes the *commutant* of S.

The prototypical example of an abelian von Neumann algebra is given by the algebra $A = L^\infty(X, \mathcal{B}, \mu)$ of essentially bounded measurable functions on a standard probability space $(X, \mathcal{B}, \mu)$, viewed as a subalgebra of $B(L^2(X))$ via $f \cdot \xi = f\xi \forall f \in A, \xi \in L^2(X)$. At the other extreme from an abelian von Neumann algebra is a *factor*, i.e., a von Neumann algebra whose center $M \cap M'$ reduces to the scalar operators $\mathbb{C}$.

It was recognised early that an important component to a von Neumann algebra is the set $\mathcal{P}(M) = \{p \in M : p = p^* = p^2\}$ of its projections. Just as all measurable functions can be approximated by simple functions, it is true that the linear subspace spanned by $\mathcal{P}(M)$ is norm-dense in M. Two projections p, q are said to be (Murray-von Neumann) equivalent "rel M" — denoted by $p \sim_M q$ — if there exists a $u \in M$ such that $u^*u = p, uu^* = q$. It turns out that M is a factor if and only if any two projections are "comparable" in the sense that one is equivalent to a sub-projection of the other. Murray and von Neumann initially classified factors into types I (there exists a minimal projection), II (there do not exist minimal projections, but there do exist non-zero projections which are finite meaning they are not equivalent to any strictly smaller sub-projection) and III (there do not exist non-zero finite projections).

(The material in this section first appeared in the papers of von Neumann, either singly authored or co-authored with Murray: see [18].)

Ergodic theory

Ergodic theory deals with the study of transformations T on a measure space $(X, \mathcal{B}, \mu)$ — which we will always assume is a complete standard probability space; the map T is usually assumed to be bijective mod μ, bimeasurable and non-singular — i.e., there are μ-null sets N_1, N_2 such that T maps $X \setminus N_1$ 1-1 onto $X \setminus N_2$, and $E \in \mathcal{B} \Leftrightarrow T(E) \in \mathcal{B}$ and $\mu(T^{-1}(E)) = 0 \Leftrightarrow \mu(E) = 0$. A countable group Γ of such transformations γ is said to act *ergodically* if it satisfies any of the following equivalent conditions:

(1) $\mu(\gamma^{-1}(E) \Delta E) = 0 \forall \gamma \in \Gamma \Rightarrow \mu(E) = 0$ or 1

(2) $f = f \circ \gamma \forall \gamma \in \Gamma \Rightarrow f$ is constant a.e.
(3) $E, F \in \mathcal{B}, \mu(E) > 0, \mu(F) > 0 \Rightarrow \exists \gamma \in \Gamma$ such that $\mu(F \cap \gamma(E)) > 0$.

Group-measure space construction

Suppose Γ is a countable group of non-singular transformations of a standard Borel space $(X, \mathcal{B})$, equipped with a σ-finite measure μ. Let $H = \ell^2(\Gamma, L^2(X, \mathcal{B}, \mu))$; the equations

$$(\pi(f)\widetilde{\xi})(\gamma) = (f \circ \gamma)\widetilde{\xi}(\gamma)$$
$$(\lambda(\gamma_0)\widetilde{\xi})(\gamma) = \widetilde{\xi}(\gamma_0^{-1}\gamma)$$

respectively define a *-algebra representation of $A = L^\infty(X, \mathcal{B}, \mu)$ into $B(H)$ and a unitary representation of Γ into $B(H)$, and these representations satisfy the commutation relation

$$\lambda(\gamma)\pi(f) = \pi(f \circ \gamma^{-1})\lambda(\gamma) \tag{12.1}$$

The von Neumann algebra $M = (\lambda(\Gamma) \cup \pi(A))''$ generated by these two representations is denoted by $A \rtimes \Gamma$ and called the *crossed product* of A with Γ. Suppose the group Γ acts *freely*: i.e., for each $\gamma \neq 1$ in Γ, the set of points fixed by γ is assumed to be a μ-null set. Then, we have the following beautiful result due to von Neumann (cf. [18] or [16]):

Theorem 12.1. *$A \rtimes \Gamma$ is a factor if and only if Γ acts ergodically. Further, in this case:*

(1) The following conditions are equivalent:
(i) μ is atomic;
(ii) $M = A \rtimes \Gamma$ has a minimal projection
In this case, M is a factor of type I. Further M is said to be a factor of type $I_n, n \leq \infty$ if μ admits precisely n mutually disjoint atoms.

(2) The following conditions are equivalent:
(i) μ has no atoms, but there exists a σ-finite measure ν which is mutually absolutely continuous with μ, which is invariant under Γ (i.e., $\nu \circ \gamma^{-1} = \nu \forall \gamma$);
(ii) M is type II
In this case, 1 is a finite projection in M precisely when ν is a finite measure.

(3) M is type III if and only if there is no σ-finite measure ν which is mutually absolutely continuous with μ, which is invariant under Γ.

Thus, we have our first examples of factors of type *II* — both type II_1 (which is type *II* with 1 being a finite projection) and type II_∞ (which is type II with 1 not being a finite projection) — and type III from the following examples of groups Γ acting ergodically on Lebesgue spaces:

- (II_1) $\Gamma = \mathbb{Z}$ acting on $(S^1, \mathcal{B}_{S^1}, \frac{1}{2\pi}d\theta)$ via $n.e^{2\pi i\theta} = e^{2\pi i(\theta+n\alpha)}$ with α being irrational.
- (II_∞) $\Gamma = \mathbb{Q}$ acting on $(\mathbb{R}, \mathcal{B}_{\mathbb{R}}, dx)$ via translation $(r.x = r + x)$
- (III) $\Gamma = \mathbb{Q} \rtimes \mathbb{Q}^\times$ acting on $(\mathbb{R}, \mathcal{B}_{\mathbb{R}}, m = dx)$ via $(b, a).x = ax + b$. (The point here is that Γ does not preserve the measure m, while the *proper* subgroup $\Gamma_0 = \{(b, 1)\} \subset \Gamma$ preserves m and itself acts ergodically, and such a group Γ cannot admit any σ-finite equivalent invariant measure.)

II_1 factors

Note that the only finite factors are the factors of type $I_n, n < \infty$ or of type II_1. It is a fact that a factor M is of finite type if and only if it admits a *trace*, i.e., a linear functional tr such that $tr(1) = 1$ and $tr(xy) = tr(yx)\ \forall x, y \in M$ and $tr(x^*x) \geq 0\ \forall x \in M$; further, such a trace is automatically faithful ($0 \neq x \in M \Rightarrow tr(x^*x) > 0$) and unique. A type I_n factor is isomorphic to the full matrix algebra $M_n(\mathbb{C})$, and the corresponding "tr" is nothing but the usual matrix trace normalised by a factor of $1/n$. On the other hand II_1 factors are infinite-dimensional, but their "finiteness" results in many pleasant features.

What is also true of a finite factor is that if $p, q \in \mathcal{P}(M)$, then $p \sim_M q \Leftrightarrow tr(p) = tr(q)$. While the set $\{tr(p) : p \in \mathcal{P}(M)\}$ is nothing but $\{k/n : 0 \leq k \leq n\}$ in the I_n case, it turns out to be $[0, 1]$ in the II_1 case. A Hilbert space equipped with a normal (= appropriately continuous) *-representation of a II_1 factor M is referred to as an M-module. It turns out (as a perfect parallel with the case of I_n factors) that a module $\mathcal{H}$ over a II_1 factor M is classified, up to M-linear isomorphism, by a number $dim_M \mathcal{H}$ (which can be any number in $[0, \infty]$), the so-called *von Neumann dimension as an M-module.*

If $\lambda : \Gamma \rightarrow \ell^2(\Gamma)$ denotes the *left-regular representation* of a countable group Γ, then the equation

$$tr(x) = \langle x1, 1 \rangle$$

defines a faithful trace on the von Neumann algebra $L\Gamma = \lambda(\Gamma)''$ where 1 denotes the standard basis vector indexed by the identity element of Γ; and $L\Gamma$ is a II_1 factor if and only if the conjugacy class of every $\gamma \neq 1$ is infinite (*the ICC condition*).

Almost all the material, so far, in this section, is from the seminal work of von Neumann ([18]). Some of the details, in slightly more modern terminology, may also be found in [16], for instance.

Two questions:

(1) *What pairs of algebras (M, A) arise in the above manner?*
(2) *When do two ergodic dynamical systems $(X_i, \mathcal{B}_i, \mu_i, \Gamma_i), i = 1, 2$ yield isomorphic pairs (M_i, A_i) as above?*

The first question, or rather, a near relative (where one considers more general *crossed-products twisted by a 2-cocycle*) has been answered very satisfactorily in [9], and the answer turns out to be: precisely when A is a *Cartan subalgebra* of M — meaning that it has the following properties:

- A is a maximal abelian von Neumann subalgebra of M;
- The *normaliser* $\mathcal{N}_M(A) = \{u \in \mathcal{U}(M) : uAu^* = A\}$ (where $\mathcal{U}(M) = \{u \in M : u^*u = uu^* = 1\}$ is the unitary group of M) generates M as a von Neumann algebra: i.e., $M = \mathcal{N}_M(A)''$; and
- there exists a *faithful conditional expectation* of M onto A.

We shall say no more about the first question, since our concern is primarily with the second, whose answer turns out to be:

if and only if the two actions are orbit equivalent

The notion of orbit (or weak-) equivalence (see definition below) was introduced (and the validity of the answer established) in the measure-preserving context by Dye (cf. [5], [6]) and studied (and the validity of the answer established) in the non-singular case by Krieger (cf. [12], [13]).

Before getting to the pertinent definitions, we first make two blanket assumptions for the remainder of this paper.

All our measure spaces $(X, \mathcal{B}, \mu)$ will henceforth be assumed to be complete standard probability spaces equipped with a non-atomic probability measure; "isomorphisms between such triples are bijective (mod null sets), bimeasurable measure preserving transformations".

Definition 12.1.

(1) An *isomorphism* between two spaces $(X_1, \mathcal{B}_1, \mu_1)$ and $(X_2, \mathcal{B}_2, \mu_2)$ is a bijective bimeasurable map $\phi : X_1 \setminus N_1 \to X_2 \setminus N_2$, for μ_i-null sets N_i, such that $\mu_1 \circ \phi^{-1} = \mu_2$.

(2) A dynamical system is a tuple $(X, \mathcal{B}, \mu, \alpha, \Gamma)$ where Γ is a countable group, and $\alpha : \Gamma \to Aut(X, \mathcal{B}, \mu)$ is a homomorphism of groups.
(3) Two dynamical systems $(X_i, \mathcal{B}_i, \mu_i, \alpha_i, \Gamma_i), i = 1, 2$ are *conjugate* if there exists an isomorphism $\phi : X_1 \to X_2$ such that $\alpha_2(\Gamma_2) = \phi\alpha_1(\Gamma_1)\phi^{-1}$.
(4) Two dynamical systems $(X_i, \mathcal{B}_i, \mu_i, \alpha_i, \Gamma_i), i = 1, 2$ are *orbit equivalent* if there exists an isomorphism $\phi : X_1 \to X_2$ such that $\phi(\alpha_1(\Gamma_1)x) = \alpha_2(\Gamma_2)\phi(x)$ for μ_1- a.a x.

Every dynamical system $(X, \mathcal{B}, \mu, \alpha, \Gamma)$ gives rise to an equivalence relation — which we shall denote by $\mathcal{R}_\Gamma$ or $\mathcal{R}_\alpha$ — which is the Borel subset of $X \times X$ given by $\{(x, \alpha(\gamma)(x)) : x \in X, \gamma \in \Gamma\}$. This equivalence relation has countable equivalence classes. In fact, a result of [8] shows that any such standard equivalence relation (with countable classes) arises as *orbit equivalence* defined by a countable group Γ acting as Borel isomorphisms of $(X, \mathcal{B})$ — *although not necessarily freely* according to a result of Furman.

Question 2 above may be viewed as asking when two dynamical systems are orbit equivalent — i.e., when is there a Borel isomorphism $f : X_1 \to X_2$ such that $(f \times f)(\mathcal{R}_{\alpha_1}) = \mathcal{R}_{\alpha_2}$. Dye showed ([5]) that any two ergodic actions of $\mathbb{Z}$ are so isomorphic. A volume of work by several people (notably Dye, Connes, Feldman, Krieger, Vershik, ...) culminated in the following beautiful result proved by Ornstein and Weiss (cf. [14], see also [4]).

Theorem 12.2. (Ornstein-Weiss) *If Γ_1 and Γ_2 are infinite amenable groups, every ergodic action of Γ_1 is orbit equivalent to every ergodic action of Γ_2.*

Equivalence relations obtained from such actions of such groups are characterised by the following property of **hyperfiniteness***:*

there exists a sequence of standard equivalence relations $\mathcal{R}_n$ on X with finite equivalence classes such that

$$\mathcal{R}_n \subset \mathcal{R}_{n+1} \forall n \text{ and } \mathcal{R} = \cup_n \mathcal{R}_n.$$

Thus $\mathcal{R}_\Gamma$ remembers neither Γ nor α if Γ is an infinte amenable group and α is an ergodic action. On the other hand, at the other end of the spectrum, many people (Zimmer, Furman, Gaboriau, and later Popa, Monod, Ozawa, ...) have obtained "rigidity results" which say something like this: if $\mathcal{R}_{\alpha_i}$ are orbit equivalent, then under some conditions on the Γ_i, these two dynamical systems must actually be conjugate! (For an example, see Popa's *strong rigidity theorems* (cf. [17]), which say something like this:

Certain kinds of free ergodic actions of certain kinds of groups G are such that if the resulting equivalence relation $\mathcal{R}$ has the property that the *'induced relation'* $\mathcal{R}_Y$ *(defined in the third paragraph below) is isomorphic to* $\mathcal{R}_\Gamma$ *for some Borel subset* Y *and some free ergodic action of some countable group* Γ*, then* Y *must have full measure, and the actions of* Γ *and* G *must be conjugate through a group isomorphism.*

It follows that for a relation $\mathcal{R}$ as in this strong rigidity theorem, the restriction $\mathcal{R}_Y$ to a Borel subset with $0 < \mu(Y) < 1$ can never be obtained from a free ergodic action of any countable group Γ, thus furnishing another proof of Furman's result mentioned earlier.

The key notions used in Gaboriau's work are *stable orbit equivalence*, *measurable equivalence* and ℓ^2*-Betti numbers*, upon which we now briefly dwell.

It is well known that if the action is ergodic, then the "space of orbits" (= the quotient of X by the relation of being in the same orbit) does not have a "good Borel structure", i.e., is not standard. The space $\mathcal{R}$ is a good substitute. Now, if A is a Borel subset of positive measure in X, then A meets almost every orbit, so by the philosophy expressed in the previous sentence, the *induced relation* $\mathcal{R}_A := \mathcal{R} \cap (A \times A)$ is an equally good description of the "space of orbits". Let us call ergodic equivalence relations $\mathcal{R}_i$ on standard probability spaces $(X_i, \mathcal{B}_i, \mu_i)$ (for $i = 1, 2$) *stably orbit equivalent* (or simply SOE) if there exist Borel subsets $A_i \in \mathcal{B}_i$ of positive measure, a positive constant c and a Borel isomorphism $f : A_1 \to A_2$ such that $\mu_2 \circ f = c\mu_1$ on A_1 and $(f \times f)(\mathcal{R}_{A_1}) = \mathcal{R}_{A_2}$; and c is called the compression constant of the SOE.

On the other hand, call two countable groups $\Gamma_i, i = 1, 2$ *measurably equivalent* (or simply ME) if they admit commuting free actions on a standard measure space $(X, \mathcal{B}, \mu)$[d] which admit a *fundamental domain* F_i of finite measure; call the ratio $\frac{\mu(F_2)}{\mu(F_1)}$ the compression constant of the ME.

The two notions of equivalence defined in the preceding paragraphs turn out to be closely related, and we have the following result, proved originally by Furman (cf. [7], [10]):

Theorem 12.3. Γ_1 *is ME to* Γ_2 *with compression constant* c *if and only if* Γ_1 *and* Γ_2 *admit free actions on standard probability space such that the associated equivalence relations are SOE with compression constant* c.

Now, we briefly discuss ℓ^2-Betti numbers. These were first introduced by Atiyah in the context of actions of countable groups on manifolds with compact quotients; he relied on the von Neumann dimension $dim_{L\Gamma}\mathcal{H}_n$ of the Hilbert space of harmonic L^2-forms of degree n, which has the structure

[d]Here the measure is allowed to be infinite (but should be σ-finite).

of a module over the von Neumann algebra $L\Gamma$ (generated by the regular representation of Γ). This was later considerably extended by Cheeger and Gromov, who studied actions of countable groups on general topological spaces, and succeeded in defining the sequence $\{\beta_n(\Gamma)\}$ of ℓ^2-Betti numbers of any countable group.

Next, Gaboriau defined the ℓ^2-Betti numbers $\beta_n(\mathcal{R})$ of any standard equivalence relation with invariant measure. He was helped in this by the work of Feldman and Moore, where a von Neumann algebra $L\mathcal{R}$ with a finite faithful normal trace had been naturally associated to a standard equivalence relation with invariant probability measure. (If $\mathcal{R} = \mathcal{R}_\Gamma$ for an ergodic action preserving a probability measure space $(X, \mathcal{B}, \mu)$, then $L\mathcal{R}$ is just the II_1 factor given by the crossed product construction.) Gaboriau considers a *universal* $\mathcal{R}$*-simplicial complex* $E\mathcal{R}$ and essentially observes that the space of ℓ^2-chains has a natural structure of an $L\mathcal{R}$-module, defines $\beta_n(\mathcal{R})$ as the $L\mathcal{R}$*-dimension* of the corresponding reduced ℓ^2-homology groups of $E\mathcal{R}$, and proves:

Theorem 12.4. (Gaboriau) *If an equivalence relation* $\mathcal{R}$ *is produced by a free action of* Γ *which preserves a probability measure, then*

$$\beta_n(\mathcal{R}) = \beta_n(\Gamma).$$

Gaboriau goes on to prove that the ratio of corresponding ℓ^2-Betti numbers of two ME groups agrees with the compression constant of the ME.

Thus we find that if free actions of countable groups Γ_j yield equivalence relations $\mathcal{R}_j, j = 1, 2$ which are orbit equivalent, and hence SOE with compression constant 1, then the groups Γ_j must be ME with compression constant 1.

Coming back to orbit equivalence, we deduce the following fact from the foregoing discussion:

The ℓ^2*-Betti numbers of orbit equivalent free actions are equal.*

The simplest example of groups in the same ME class is furnished by any two lattices, not necessarily co-compact, of a locally compact second countable group (as seen by their actions by left-, resp., right- multiplications on the ambient group). Gaboriau obtains many rigidity results, a sample being:

Corollary 12.1. (Gaboriau)

(1) No lattice in $SP(n, 1)$ *is ME to a lattice in* $SP(p, 1)$ *if* $n \neq p$.
(2) No lattice in $SU(n, 1)$ *is ME to a lattice in* $SU(p, 1)$ *if* $n \neq p$.

(3) No lattice in $SO(2n,1)$ is ME to a lattice in $SO(2p,1)$ if $n \neq p$.

Proof. It is known from the work of Borel (cf. [2]) that

$$\beta_i(\Gamma(SP(m,1)) \neq 0 \Leftrightarrow i = 2m$$
$$\beta_i(\Gamma(SU(m,1)) \neq 0 \Leftrightarrow i = m$$
$$\beta_i(\Gamma(SO(2m,1)) \neq 0 \Leftrightarrow i = m$$

where we write $\Gamma(G)$ to denote any lattice in G. □

Finally, we should mention that Gaboriau's results have been used ingeniously by Sorin Popa to settle a long-standing conjecture of Kadison's — regarding the existence of II_1 factors with trivial fundamental group.

If M is a II_1 factor, there is a natural definition of the so-called *amplification* $M_d(M)$ (or the $d \times d$ matrix algebra over M) where d is any positive real number. For instance, it may be identified with the (II_1 factor $End_M(\mathcal{H}_d)$ of) M-linear operators on the M-module $\mathcal{H}_d$ with $dim_M \mathcal{H}_d = d$. von Neumann already realised the importance of the object, called the *fundamental group*[e] $\mathcal{F}(M)$ of M, and defined by

$$\mathcal{F}(M) = \{d > 0 : M \cong M_d(M)\}.$$

Popa showed that there are many examples of II_1 factors of the form $L\mathcal{R}_\alpha$ (arising from free ergodic actions α of suitable ICC groups) which do indeed have trivial fundamental group. An example of such an action is the natural action of $SL(2,\mathbb{Z})$ on $\mathbb{T}^2$. In fact, Gaboriau and Popa have even shown (cf. [11]) that (each finitely generated non-abelian free group) $\mathbb{F}_n$ admits uncountably many free ergodic actions α_i preserving a probability measure, which are pairwise not SOE, such that $L\mathcal{R}_{\alpha_i}$ has trivial fundamental group for each i. Much more of the subsequent exciting developments, as well as pertinent literature, may be found in the article [17] by Vaes.

References

[1] Atiyah, M. (1976). Elliptic operators, discrete groups and von Neumann algebras. *Colloque 'Analyse et Topologie' en l'Honneur de Henri Cartan* (Orsay, 1974), 43-72, *Asterisque* 32-33. Paris: Soc. Math. France.

[2] Borel, A. (1985). The L^2-cohomology of negatively curved Riemannian symmetric spaces. *Ann. Acad. Sci. Fenn. Ser. A I Math.* **10** 95-105.

[e] $\mathcal{F}(M)$ is a multiplicative subgroup of $\mathbb{R}^\times$.

[3] Cheeger, J. and Gromov, M. (1986). L_2-cohomology and group cohomology. *Topology.* **25** 189-215.
[4] Connes, A., Feldman, J. and Weiss, B. (1981). An amenable equivalence relation is generated by a single transformation. *Ergodic Theory Dynamical Systems.* **1** 431-450.
[5] Dye, H. (1959). On groups of measure preserving transformations, I. *Amer. J. Math.* **81** 119-159.
[6] Dye, H. (1963). On groups of measure preserving transformations, II. *Amer. J. Math.* **85** 551-576.
[7] Furman, A. (1999). Orbit equivalence rigidity, *Ann. of Math.* **150** 1083-1108.
[8] Feldman, J. and Moore, C. (1977). Ergodic equivalence relations, cohomology, and von Neumann algebras, I. *Trans. Amer. Math. Soc.* **234** 289-324.
[9] Feldman, J. and Moore, C. (1977). Ergodic equivalence relations, cohomology, and von Neumann algebras, II. *Trans. Amer. Math. Soc.* **234** 325-359.
[10] Gaboriau, D. (2002). Invariants ℓ^2 de relations d'equivalence et de groupes. *Publ. Math. I.H.E.S.* **95** 93-150.
[11] Gaboriau, D. and Popa, S. (2005). An uncountable family of nonorbit equivalent actions of $\mathbb{F}_n$. *J. Amer. Math. Soc.* **18** 547–559 (electronic).
[12] Krieger, W. (1969). On non-singular transformations of a measure space, I. *Z. Wahrscheinlichkeitstheorie und Verw. Gebiete.* **11** 83-97.
[13] Krieger, W. (1969). On non-singular transformations of a measure space, I. *Z. Wahrscheinlichkeitstheorie und Verw. Gebiete.* **11** 98-119.
[14] Ornstein, D. and Weiss, B. (1980). Ergodic Theory of amenable group actions, I. The Rohlin lemma. *Bull. Amer. Math. Soc.* (New Ser.) **2** 161-164.
[15] Popa, S. (2006). On a class of type II_1 factors with Betti numbers invariants. *Ann. of Math.* **163** 809-899.
[16] Sunder, V. S. (1987). *An Invitation to von Neumann algebras.* Springer Verlag.
[17] Vaes, S. (2007). Rigidity results for Bernouilli actions and their von Neumann algebras, (after Sorin Popa). *Asterisque.* **311** 237-294, also arXiv: math. OA/0603434 v2.
[18] von Neumann, J. (1961). *Collected works vol. III. Rings of Operators.* Pergamon Press.

Chapter 13

Gutzmer's Formula and the Segal-Bargmann Transform

S. Thangavelu
Department of Mathematics,
Indian Institute of Science,
*Bangalore 560 012, India**
veluma@math.iisc.ernet.in

We discuss the Segal-Bargmann transform associated to the Laplacian on the Heisenberg group. Using analogues of Gutzmer's formula for Hermite and special Hermite expansions, we obtain several characterisations of the image of $L^2(\mathbb{H}^n)$ under the Segal-Bargmann transform.

13.1. Introduction

In 1888 August Gutzmer ([5]) published a formula which in modern notation amounts to the following: If F is a 2π-periodic holomorphic function on the complex plane, then for every $y \in \mathbb{R}$ we have

$$\int_0^{2\pi} |F(x+iy)|^2 dx = \sum_{k=-\infty}^{\infty} |\hat{F}(k)|^2 e^{-2ky}$$

under suitable assumptions on the Fourier coefficients $\hat{F}(k)$ of the restriction of F to the real line $\mathbb{R}$. The above is just Parseval's identity applied to $F_y(x) = F(x+iy)$. Integrating the above identity over $\mathbb{R}$ with respect to $e^{-\frac{1}{2t}y^2}dy$ we obtain

$$\int_{\mathbb{R}} \int_0^{2\pi} |F(x+iy)|^2 e^{-\frac{1}{2t}y^2} dxdy = \sum_{k=-\infty}^{\infty} |\hat{F}(k)|^2 e^{2tk^2}.$$

*Dedicated to 222 A.

If $F = f * p_t$ where $f \in L^2(S^1)$ and p_t is the heat kernel defined by the series

$$p_t(x) = \sum_{-\infty}^{\infty} e^{-k^2 t} e^{ikx}$$

then the above becomes

$$\int_{\mathbb{R}} \int_0^{2\pi} |f * p_t(x+iy)|^2 e^{-\frac{1}{2t}y^2} dx dy = c \int_0^{2\pi} |f(x)|^2 dx$$

which is a characterisation of the image of $L^2(S^1)$ under the Segal-Bargmann transform which takes f into the holomorphic function $F(x + iy) = f * p_t(x + iy)$.

More generally, let G be a Lie group, Δ a (non-negative) Laplacian on G and $p_t, t > 0$ the heat kernel associated to Δ. Given $f \in L^2(G)$, the function $f * p_t$, initially defined on G extends holomorphically to certain $G-$invariant domain Ω contained in the complexification $G_{\mathbb{C}}$ of G. This transform, taking f into the holomorphic extension of $f * p_t$, is known as the Segal-Bargmann transform and is of interest to mathematical physicists. It is also called the heat kernel transform for obvious reasons.

This transform was studied by Bargmann ([1]) (and by Segal independently) when $G = \mathbb{R}^n$. They showed that the image of $L^2(\mathbb{R}^n)$ under this transform is a weighted Bergman space- namely, the space of all entire functions on $\mathbb{C}^n$ which are square integrable with respect to a positive weight function w_t. To be precise, they showed that

$$\int_{\mathbb{R}^{2n}} |f * p_t(x+iy)|^2 p_{t/2}(y) dx dy = c_n \int_{\mathbb{R}^n} |f(x)|^2 dx.$$

In 1994 similar results were obtained for all compact Lie groups by Hall ([6]) and later in 1999 Stenzel ([11]) treated the case of Riemannian symmetric spaces of compact type.

In the Euclidean set up, with the standard heat kernel

$$p_t(x) = (4\pi t)^{-n/2} e^{-\frac{1}{4t}|x|^2}$$

the result of Segal and Bargmann can be described as follows. A function F on $\mathbb{R}^n$ can be factored as $F = f * p_t$ if and only if it can be extended to $\mathbb{C}^n$ as an entire function and

$$\int_{\mathbb{R}^{2n}} |F(x+iy)|^2 p_{t/2}(y) dx dy < \infty.$$

Another way of saying the same is that

$$\int_{\mathbb{R}^{2n}} |\hat{F}(\xi)|^2 e^{-2y\cdot\xi} p_{t/2}(y)dyd\xi < \infty.$$

The equivalence follows from the fact that

$$\int_{\mathbb{R}^n} |F(x+iy)|^2 dx = \int_{\mathbb{R}^n} |\hat{F}(\xi)|^2 e^{-2y\cdot\xi} d\xi$$

which is the analogue of Gutzmer's formula for the Fourier transform. Note that it is again a simple consequence of the Plancherel theorem.

In 1978 M. Lassalle ([9], [10]) established analogues of Gutzmer's formula for compact Lie groups and compact symmetric spaces. His formulas were used by Faraut in [4] to study Segal-Bargmann transform on compact symmetric spaces. The works of Hall ([6]) and Stenzel ([11]) did not use Lassalle's formula but as can be seen from Faraut ([4]) and the recent work [14] of the author it is clear that the use of Gutzmer's formula makes the proofs simple and transparent. In [2], [3] Faraut proved an analogue of Gutzmer's formula for non-compact symmetric spaces which was used by Krötz, Olafsson and Stanton ([8]) to study Segal-Bargmann transform.

The investigations of Krötz et al. ([8]) have revealed that the results which are true for compact symmetric spaces are no longer true in general. In the case of non-compact Riemannian symmetric spaces $X = G/K$, the solution of the heat equation $f * p_t$ associated to the Laplace-Beltrami operator does not extend to the complexification $X_{\mathbb{C}}$ but only to a domain Ξ called the complex crown. Even then, the image is not a weighted Bergman space; this was shown in the work of Krötz et al. [8]. Another surprising case is that of the Heisenberg group $\mathbb{H}^n$. For the full Laplacian on $\mathbb{H}^n$ it was shown in Krötz, Thangavelu and Xu ([7]) that the image of $L^2(\mathbb{H}^n)$ is not a weighted Bergman space in the usual sense. However, the image turned out to be a direct sum of two Bergman spaces defined in terms of oscillating weight functions.

In this article we restrict ourselves to the Segal-Bargmann transform on the Heisenberg group. We obtain several characterisations - some are new and some already known- using various analogues of Gutzmer's formula.

As a motivation, return to $\mathbb{R}^n$ and consider the two conditions characterising the image of $L^2(\mathbb{R}^n)$ under the Segal-Bargmann transform mentioned above. The first is in terms of holomorphic properties of F whereas the second one transfers the condition on the Fourier transform side. In view of Gutzmer's formula, both are equivalent. Let us look at the second condition more carefully. For each $\xi \in \mathbb{R}^n$ we have an irreducible unitary representation e_ξ realised on $\mathbb{C}$. In terms of this the second condition takes the following form: the function $e_\xi(x)\hat{F}(\xi) = \hat{F}(\xi)e^{ix\cdot\xi}$ extends to $\mathbb{C}^n$ as an entire function and

$$\int_{\mathbb{R}^n}\int_{\mathbb{R}^n} |e_\xi(x+iy)\hat{F}(\xi)|^2 p_{t/2}(y)dyd\xi < \infty.$$

This restatement of the characterisation motivates the following consideration.

Given a Lie group G let $\hat{G}$ stand for its unitary dual with Plancherel measure $d\mu(\pi)$. For $f \in L^2(G)$ and $\pi \in \hat{G}$ let $\pi(f)$ be its Fourier transform which is a Hilbert-Schmidt operator on $\mathcal{H}_\pi$, the Hilbert space on which π is realised. Then we have another representation of G, denoted by the same symbol π and realised on $\mathcal{S}_2$, the Hilbert space of Hilbert-Schmidt operators on $\mathcal{H}_\pi$. This is defined by $g \to \pi(g)T, T \in \mathcal{S}_2$. Unlike the abelian case, it is not true that this representation can be extended to $G_\mathbb{C}$ for all $T \in \mathcal{S}_2$. However, under further assumptions on T it might have a holomorphic extension. For example, when $T = \pi(F), F = f * p_t$ we hope that we get an operator valued holomorphic function on $G_\mathbb{C}$. Then the main problem for Segal-Bargmann transform reduces to characterise functions F that can be written as $f * p_t$ in terms of properties of the above function.

Taking the clue from the Euclidean case, instead of looking for a weight function w_t on $G_\mathbb{C}$ so that

$$\int_{G_\mathbb{C}} |f * p_t(z)|^2 w_t(z)dz = \int_G |f(g)|^2 dg$$

we can look for a weight function $w_t(z,\pi)$ on $G_\mathbb{C} \times \hat{G}$ such that

$$\int_{\hat{G}}\int_{G_\mathbb{C}} \|\pi(z)\pi(f * p_t)\|^2_{HS} w_t(z,\pi)dzd\mu(\pi) = \int_G |f(g)|^2 dg.$$

In other words, we ask if the image can be identified with a direct integral of (operator valued) weighted Bergman spaces. In the case of Heisenberg group, we already know from the work [7] that the image is a direct integral of twisted Bergman spaces. The new approach leads to another such result where the fibres are operator valued weighted Bergman spaces (see Theorem 2.5).

As we have noted above, both approaches lead to the same result for the Euclidean spaces. Using Gutzmer's formula, which is due to Lassalle ([9]), we can also prove the same (i.e., the equivalence of two approaches) for all compact symmetric spaces (and hence for all compact Lie groups). Our investigations with non-compact symmetric spaces does not lead to anything new. In this respect Heisenberg group seems to be special. We believe similar characterisations are true for all nilpotent Lie groups.

13.2. Segal-Bargmann Transform on the Heisenberg Group

The Heisenberg group $\mathbb{H}^n$ is just $\mathbb{R}^n \times \mathbb{R}^n \times \mathbb{R}$ with the group law given by

$$(x, u, \xi)(x', u', \xi') = (x + x', u + u', \frac{1}{2}(x \cdot u' - u \cdot x') + \xi + \xi') .$$

Here $x \cdot u = \sum_{j=1}^{n} x_j u_j$, as usual, denotes the standard pairing on $\mathbb{R}^n$. On $\mathbb{H}^n$ we consider the $(2n + 1)$ vector fileds

$$X_j = \partial_{x_j} + \frac{1}{2} u_j \partial_\xi, U_j = \partial_{u_j} - \frac{1}{2} x_j \partial_\xi, Z = \partial_\xi, j = 1, 2, ..., n$$

forming a basis for the Heisenberg Lie algebra. Recall that $[X_j, U_j] = Z$, all other brackets being zero. We consider the full Laplacian

$$\Delta = -\sum_{j=1}^{n} (X_j^2 + U_j^2) - Z^2$$

which is a non-negative elliptic differential operator which generates a diffusion semigroup. The heat kernel associated to this semigroup is explicitly given by

$$p_t(x, u, \xi) = c_n \int_{\mathbb{R}} e^{-i\lambda\xi} e^{-t\lambda^2} \left(\frac{\lambda}{\sinh \lambda t} \right)^n e^{-\frac{1}{4}\lambda(\coth t\lambda)(x \cdot x + u \cdot u)} \, d\lambda$$

with $c_n = (4\pi)^{-n}$ (see [7] for a proof).

In the sequal we use the notation

$$f^\lambda(x, u) = \int_{\mathbb{R}} e^{i\lambda\xi} f(x, u, \xi) d\xi$$

for a function f on $\mathbb{H}^n$. In this notation we have

$$p_t(x, u, \xi) = \int_{\mathbb{R}} e^{-i\lambda\xi} e^{-t\lambda^2} p_t^\lambda(x, u) \, d\lambda$$

where

$$p_t^\lambda(x, u) = c_n \left(\frac{\lambda}{\sinh \lambda t} \right)^n e^{-\frac{1}{4}\lambda(\coth t\lambda)(x \cdot x + u \cdot u)} .$$

From the above representation it is clear that $p_t(x, u, \xi)$ extends to the complexification $\mathbb{C}^n \times \mathbb{C}^n \times \mathbb{C}$ of $\mathbb{H}^n$ as an entire function $p_t(z, w, \zeta)$. Therefore, for any $f \in L^2(\mathbb{H}^n)$ the convolution $f * p_t$ also extends as an entire function. This map, which takes f into $f * p_t(z, w, \zeta)$ is the Segal-Bargmann transform or heat kernel transform for the Heisenberg group. Our problem is to characterise functions F in $L^2(\mathbb{H}^n)$ that can be factored as $f * p_t$ for some $f \in L^2(\mathbb{H}^n)$.

As $f * p_t$ extends to an entire function on $\mathbb{C}^n \times \mathbb{C}^n \times \mathbb{C}$ it is natural to ask if there is an analogue of Bargmann's result, namely, does there exist a non-negative weight function $W_t(z, w, \zeta)$ such that F on $\mathbb{H}^n$ can be factored as $F = f * p_t, f \in L^2(\mathbb{H}^n)$ if and only if F extends to $\mathbb{C}^n \times \mathbb{C}^n \times \mathbb{C}$ as an entire function which is square integrable with respect to W_t. In [7] the following result was proved, answering the above question in the negative.

Theorem 13.1. *The image of $L^2(\mathbb{H}^n)$ is not a weighted Bergman space.*

Our first characterisation is in terms of holomorphic properties of $F^\lambda(x, u)$. We consider the weight function

$$W_t^\lambda(z, w) = e^{\lambda(u\cdot y - v\cdot x)} p_{2t}^\lambda(2y, 2v)$$

where $z = x + iy, w = u + iv$. Let $\mathcal{B}_t^\lambda(\mathbb{C}^{2n})$ be the Hilbert space of entire functions G on $\mathbb{C}^{2n}$ which are square integrable with respect to W_t^λ. We equip $\mathcal{B}_t^\lambda(\mathbb{C}^{2n})$ with the norm

$$\|G\|_{\mathcal{B}_t^\lambda}^2 = \int_{\mathbb{C}^{2n}} |G(z, w)|^2 W_t^\lambda(z, w) dz dw.$$

These spaces are called twisted Bergman spaces. We then have the following result.

Theorem 13.2. *A function $F \in L^2(\mathbb{H}^n)$ can be factored as $F = f * p_t$ if and only if for every $\lambda \neq 0$, $F^\lambda(x, u)$ extends to an element of $\mathcal{B}_t^\lambda(\mathbb{C}^{2n})$ and*

$$\int_{\mathbb{R}} e^{2t\lambda^2} \|F^\lambda\|_{\mathcal{B}_t^\lambda}^2 d\lambda < \infty.$$

This result, proved in [7], shows that the image of $L^2(\mathbb{H}^n)$ under the Segal-Bargmann transform is a direct integral of twisted Bergman spaces. Moreover, it was proved in [7] that we have

$$\int_{\mathbb{R}} e^{2t\lambda^2} \|F^\lambda\|_{\mathcal{B}_t^\lambda}^2 d\lambda = \int_{\mathbb{H}^n} |f(g)|^2 dg$$

where f is the function related to F via $F = f * p_t$. The above result is in terms of the Fourier transform of F in the central variable. Our next result is in terms of the group Fourier transform of F.

In order to define the group Fourier transform of a function f on $\mathbb{H}^n$ we need to recall the relevant representations of $\mathbb{H}^n$. We use the Schrödinger picture. For every non-zero real λ we have an irreducible unitary representation π_λ of $\mathbb{H}^n$ which is realised on $L^2(\mathbb{R}^n)$. It is explicitly given by

$$\pi_\lambda(x,u,\xi)\varphi(v) = e^{i\lambda\xi}e^{i\lambda(x\cdot\xi+\frac{1}{2}x\cdot u)}\varphi(v+u)$$

for $\varphi \in L^2(\mathbb{R}^n)$. By a theorem of Stone and von Neumann any irreducible unitary representation of $\mathbb{H}^n$ which is nontrivial at the center is unitarily equivalent to some π_λ. Using these representations we define the Fourier transform of a function $f \in L^1(\mathbb{H}^n)$ as the operator valued function

$$\hat{f}(\lambda) = \int_{\mathbb{H}^n} f(g)\pi_\lambda(g)dg.$$

The Plancherel theorem says that the above definition can be extended to all $f \in L^2(\mathbb{H}^n)$, $\hat{f}(\lambda)$ is Hilbert-Schmidt for every λ and we have

$$\int_{\mathbb{H}^n} |f(g)|^2 dg = \int_{\mathbb{R}} \|\hat{f}(\lambda)\|_{HS}^2 d\mu(\lambda)$$

where $d\mu(\lambda) = (2\pi)^{-n-1}|\lambda|^n d\lambda$ is the Plancherel measure.

The Fourier transform of the heat kernel can be explicitly calculated in terms of the Hermite operators $H(\lambda) = -\Delta + \lambda^2|x|^2$. In fact $\hat{p_t}(\lambda) = e^{-t\lambda^2}e^{-tH(\lambda)}$ where $e^{-tH(\lambda)}$ is the Hermite semigroup generated by $H(\lambda)$. The Hilbert space $L^2(\mathbb{R}^n)$ has an orthonormal basis consisting of the Hermite functions $\Phi_\alpha^\lambda, \alpha \in \mathbb{N}^n$ which are eigenfunctions of $H(\lambda)$ with eigenvalues $(2|\alpha|+n)|\lambda|$. The image of $L^2(\mathbb{R}^n)$ under the Hermite semigroup $e^{-tH(\lambda)}$ has been characterised as a weighted Bergman space, see [16] for a proof. Let

$$U_t^\lambda(z) = 2^n(\sinh(4t\lambda))^{-n/2}e^{\lambda\tanh(2t\lambda)|x|^2-\lambda\coth(2t\lambda)|y|^2}$$

and define $\mathcal{H}_t^\lambda(\mathbb{C}^n)$ to be the Hilbert space of entire functions on $\mathbb{C}^n$ which are square integrable with respect to $U_t^\lambda(z)dz$. Then it is known that an entire function Φ on $\mathbb{C}^n$ belongs to $\mathcal{H}_t^\lambda(\mathbb{C}^n)$ if and only if $\Phi = e^{-tH(\lambda)}\varphi$ for some $\varphi \in L^2(\mathbb{R}^n)$. This space $\mathcal{H}_t^\lambda(\mathbb{C}^n)$ is called the Hermite Bergman space.

Theorem 13.3. *A function $F \in L^2(\mathbb{H}^n)$ can be factored as $F = f * p_t$ if and only if for every $\lambda \neq 0$, $\hat{F}(\lambda)^*$ maps $L^2(\mathbb{R}^n)$ into $\mathcal{H}_t^\lambda(\mathbb{C}^n)$ in such a way that*

$$\int_{\mathbb{R}} e^{2t\lambda^2}\left(\sum_{\alpha\in\mathbb{N}^n}\|\hat{F}(\lambda)^*\Phi_\alpha^\lambda\|_{\mathcal{H}_t^\lambda}^2\right)d\mu(\lambda) < \infty.$$

This result is not difficult to prove. The first assumption on $\hat{F}(\lambda)$ means, in view of the characterisation of $\mathcal{H}_t^\lambda(\mathbb{C}^n)$, that there is an operator valued function T_λ such that $\hat{F}(\lambda)^* = e^{-tH(\lambda)}T_\lambda$. The second assumption shows that T_λ is actually Hilbert-Schmidt and

$$\int_{\mathbb{R}} \|T_\lambda\|^2 e^{2t\lambda^2} d\mu(\lambda) < \infty.$$

Hence, we can find $f \in L^2(\mathbb{H}^n)$ such that $\hat{f}(\lambda) = e^{-t\lambda^2} T_\lambda^*$ and hence $F = f * p_t$ follows. The other implication is even simpler. The non trivial ingredient in the proof is the realisation of $\mathcal{H}_t^\lambda(\mathbb{C}^n)$ as the image of $L^2(\mathbb{R}^n)$ under $e^{-tH(\lambda)}$. This can be proved using an analogue of Gutzmer's formula for the Hermite expansions (see Section 3).

We can define another family of representations of $\mathbb{H}^n$ realised on $\mathcal{S}_2$, the Hilbert space of all Hilbert-Schmidt operators on $L^2(\mathbb{R}^n)$, by setting $\tilde{\pi}_\lambda(g)T = \pi_\lambda(g)T$ for $T \in \mathcal{S}_2$. This is clearly a unitary representation of $\mathbb{H}^n$. When T is an arbitrary element of $\mathcal{S}_2$ we may not be able to holomorphically continue the $\mathcal{S}_2$ valued function $(x, u, \xi) \to \tilde{\pi}_\lambda(x, u, \xi)T$. It turns out that such an extension is possible precisely when T can be factored as $T = Se^{-tH(\lambda)}$. This leads us to the following characterisation of the image of $L^2(\mathbb{H}^n)$ under the Segal-Bargmann transform. We let $\mathcal{C}_t^\lambda(\mathbb{C}^{2n}, \mathcal{S}_2)$ stand for the $\mathcal{S}_2$-valued weighted Bergman space consisting of entire functions $G(z, w)$ on $\mathbb{C}^{2n}$ for which

$$\|G\|_{\mathcal{C}_t^\lambda}^2 = \int_{\mathbb{R}^{2n}} \|G(iy, iv)\|_{HS}^2 \quad p_{2t}^\lambda(2y, 2v) dy dv < \infty.$$

Let $\mathcal{C}_{t,0}^\lambda(\mathbb{C}^{2n}, \mathcal{S}_2)$ be the subspace of $\mathcal{C}_t^\lambda(\mathbb{C}^{2n}, \mathcal{S}_2)$ consisting of functions $G(z, w)$ satisfying $G(z, w) = \tilde{\pi}_\lambda(z, w, 0)G(0, 0)$.

Theorem 13.4. *A function $F \in L^2(\mathbb{H}^n)$ can be factored as $F = f * p_t$ if and only if for every $\lambda \neq 0$, $\tilde{\pi}_\lambda(x, u, 0)\hat{F}(\lambda)$ can be extended to $\mathbb{C}^{2n}$ as an element of $\mathcal{C}_{t,0}^\lambda(\mathbb{C}^{2n}, \mathcal{S}_2)$ which satisfies*

$$\int_{\mathbb{R}} \|\tilde{\pi}_\lambda(\cdot, \cdot, 0)\hat{F}(\lambda)\|_{\mathcal{C}_{t,0}^\lambda}^2 e^{2t\lambda^2} d\mu(\lambda) < \infty.$$

By defining $w_t^\lambda(y, v, \eta) = p_{2t}^\lambda(2y, 2v)e^{-\frac{1}{2t}\eta^2}$ and $d\nu(y, v, \eta) = dy dv d\eta$ we can rewrite the above theorem as follows.

Theorem 13.5. *A function $F \in L^2(\mathbb{H}^n)$ can be factored as $F = f * p_t$ if and only if for every $\lambda \neq 0$, $\tilde{\pi}_\lambda(x, u, \xi)\hat{F}(\lambda)$ can be extended as entire* *function on* $\mathbb{C}^{2n+1}$ *which satisfies*

$$\int_{\mathbb{R}}\int_{\mathbb{R}^{2n+1}} \|\tilde{\pi}_\lambda(iy,iv,i\eta)\hat{F}(\lambda)^*\|_{HS}^2 w_t^\lambda(y,v,\eta)d\nu(y,v,\eta)d\mu(\lambda) < \infty.$$

When it happens, the above integral is a constant multiple of $\|f\|_2^2$.

A proof of this theorem requires an analogue of Gutzmer's formula for the Hermite expansions which will be stated in the next section. Similarly, Theorem 2.2 can be proved using Gutzmer's formula for special Hermite expansions. We can view the Heisenberg group as the homogeneous space G_n/K where G_n is the Heisenberg motion group and $K = U(n)$. We therefore have a Gutzmer's formula for entire functions on the complexification $\mathbb{C}^{2n+1}$ of $\mathbb{H}^n$ which has been used in proving a Paley-Wiener theorem for the inverse Fourier transform on $\mathbb{H}^n$, see [15]. We can also use it to prove the main result in [7], namely, the image of $L^2(\mathbb{H}^n)$ under the Segal-Bargmann transform is the direct sum of two weighted Bergman spaces defined via oscillating weight functions.

13.3. Gutzmer Formulas and Their Applications

The aim of this section is to state several analogues of Gutzmer's formula relevant to the Heisenberg group and use them to obtain various characterisations of the image of $L^2(\mathbb{H}^n)$ under the Segal-Bargmann transform. We begin with an analogue of Gutzmer's formula for the Hermite expansions.

Recall that the spectral decomposition of the Hermite operator $H(\lambda) = -\Delta + \lambda^2|x|^2$ is given by

$$H(\lambda) = \sum_{k=0}^{\infty}(2k+n)|\lambda|P_k(\lambda)$$

where $P_k(\lambda)$ is the orthogonal projection of $L^2(\mathbb{R}^n)$ onto the eigenspace spanned by $\{\Phi_\alpha^\lambda : |\alpha| = k\}$. Here Φ_α^λ are the normalised Hermite functions on $\mathbb{R}^n$ scaled by λ. For various facts about Hermite functions we refer to [12]. The following result, which is an analogue of Gutzmer's formula for the Hermite expansions, has been established in [16].

Theorem 13.6. *Let* $f \in L^2(\mathbb{R}^n)$ *has a holomorphic extension* F *to* $\mathbb{C}^n$. *Then for any* $z = x+iy, w = u+iv \in \mathbb{C}^n$ *we have*

$$\int_{\mathbb{R}^n}\int_K |\pi_\lambda(\sigma.(z,w))F(\xi)|^2 d\sigma d\xi$$

$$= e^{\lambda(u\cdot y - v\cdot x)} \sum_{k=0}^{\infty} \frac{k!(n-1)!}{(k+n-1)!} \varphi_k^\lambda(2iy, 2iv) \|P_k(\lambda)f\|_2^2$$

under some further assumptions on the function f.

In the above theorem $\varphi_k^\lambda(y, v)$ stands for the Laguerre function

$$L_k^{n-1}(\frac{1}{2}|\lambda|(y^2+v^2))e^{-\frac{1}{4}|\lambda|(y^2+v^2)}.$$

Note that we have used the notation y^2 in place of $|y|^2$ so that $\varphi_k^\lambda(y, v)$ can be holomorphically extended to $\varphi_k^\lambda(z, w)$. The group K appearing in the above formula is $Sp(n, \mathbb{R}) \cap O(2n, \mathbb{R})$, the intersection of the symplectic group and the orthogonal group. It has a natural action on $\mathbb{R}^n \times \mathbb{R}^n$ which extends to $\mathbb{C}^n \times \mathbb{C}^n$. The formula is valid if we assume, say, $f = e^{-tH(\lambda)}g$ for some function $g \in L^2(\mathbb{R}^n)$. As an immediate corollary of the above formula we obtain the following result mentioned earlier.

Theorem 13.7. *An entire function F belongs to $\mathcal{H}_t^\lambda(\mathbb{C}^n)$ if and only if $F = e^{-tH(\lambda)}g$ for some $g \in L^2(\mathbb{R}^n)$.*

For a proof of this theorem we refer to [16]. The only other ingredient needed is given in the following lemma proved in [13].

Lemma 13.1.

$$\int_{\mathbb{R}^{2n}} \varphi_k^\lambda(2iy, 2iv) p_{2t}^\lambda(2y, 2v)\, dy dv = c_n \frac{(k+n-1)!}{k!(n-1)!} e^{2(2k+n)|\lambda|t}.$$

We are now in a position give a sketch of the proof of Theorem 2.4. First of all we need to show that $\tilde{\pi}_\lambda(x, u, 0)\hat{F}(\lambda)^*$ extends to $\mathbb{C}^{2n}$ as an $\mathcal{S}_2$ valued entire function. If $F = f * p_t$ we are required to see if

$$tr\left(\tilde{\pi}_\lambda(x, u, 0)e^{-tH(\lambda)}\hat{f}(\lambda)^* T^*\right)$$

extends as an entire function for every $T \in \mathcal{S}_2$. This amounts to check, in view of inversion formula for the Weyl transform, if $p_t^\lambda *_\lambda g_\lambda$ is an entire function where $*_\lambda$ is the twisted convolution and g_λ is given by the relation $\pi_\lambda(g_\lambda) = \hat{f}(\lambda)^* T^*$. But this is the case since p_t^λ is a Gaussian.

We can calculate the $\mathcal{S}_2$ norm of $\tilde{\pi}_\lambda(z, w, 0)e^{-tH(\lambda)}\hat{f}(\lambda)^*$ using the Hermite basis $\{\Phi_\alpha^\lambda : \alpha \in \mathbb{N}^n\}$. Using Gutzmer's formula for the Hermite expansions and some properties of the special Hermite functions it is not hard to show that

$$\int_K \|\tilde{\pi}_\lambda(k.(z, w), 0)e^{-tH(\lambda)}\hat{f}(\lambda)^*\|_{HS}^2 dk$$

$$= e^{\lambda(u\cdot y - v\cdot x)} \sum_{k=0}^{\infty} e^{-2(2k+n)|\lambda|t} \frac{k!(n-1)!}{(k+n-1)!} \varphi_k^\lambda(2iy, 2iv) \|f^\lambda \times_\lambda \varphi_k^\lambda\|_2^2.$$

Setting $x = u = 0$ and integrating the above with respect to $p_{2t}^\lambda(2y, 2v)dydv$, we get

$$\int_{\mathbb{R}} e^{2t\lambda^2} \|\tilde{\pi}_\lambda(\cdot, \cdot, 0)\hat{F}(\lambda)^*\|_{\mathcal{C}_{t,0}^\lambda}^2 d\mu(\lambda) = c_n \int_{\mathbb{H}^n} |f(g)|^2 dg.$$

This proves one half of the theorem.

To prove the converse, suppose that $F \in L^2(\mathbb{H}^n)$ satisfies the hypothesis of the theorem. Then it is easy to see that it also satisfies the hypothesis of Theorem 2.3. Indeed, it follows from Gutzmer's formula that

$$\|g\|_{\mathcal{H}_t^\lambda}^2 = c \int_{\mathbb{R}^{2n}} \|\pi_\lambda(iy, iv)g\|_2^2 \; p_{2t}^\lambda(2y, 2v)dydv$$

for every $g \in \mathcal{H}_t^\lambda$. Hence we can appeal to Theorem 2.3 to conclude that $F = f * p_t$ for some $f \in L^2(\mathbb{H}^n)$.

Thus Gutzmer's formula for the Hermite expansions is the main ingredient in the proof of Theorem 2.4 and Theorem 2.5 follows immediately. We now show that Theorem 2.2 can be proved by using another Gutzmer formula, namely the one for special Hermite expansions.

Theorem 13.8. *Assume that the entire function f satisfies $\|f *_\lambda \varphi_k^\lambda\|_2 \leq Ce^{-((2k+n)|\lambda|)^{\frac{1}{2}}t}$ for all $t > 0$. Then we have*

$$\int_{\mathbb{R}^{2n}} \int_{U(n)} |f((x,u) + i\sigma(y,v))|^2 e^{\lambda[(x,u),\sigma(y,v)]} d\sigma dx du$$

$$= c_n \sum_{k=0}^{\infty} \|f *_\lambda \varphi_k^\lambda\|_2^2 \frac{k!(n-1)!}{(k+n-1)!} \varphi_k^\lambda(2iy, 2iv).$$

In the above theorem, $[(x,u),(y,v)]$ stands for the symplectic form $(u \cdot y - x \cdot v)$ on $\mathbb{R}^{2n}$. For a proof of this formula we refer to [13]. From this and Lemma 3.3 it follows that an entire function $F \in \mathcal{B}_t^\lambda$ if and only if $F = f *_\lambda p_t^\lambda$ for some $f \in L^2(\mathbb{C}^n)$. Theorem 2.2 is proved by observing that $(f * p_t)^\lambda = e^{-t\lambda^2} f^\lambda *_\lambda p_t^\lambda$.

We conclude this section with another Gutzmer formula for the full Heisenberg group. Let G_n be the Heisenberg motion group which is the semidirect product of $\mathbb{H}^n$ with the unitary group $U(n)$. This group has a natural action on $\mathbb{H}^n$ which extends to all entire functions on $\mathbb{C}^{2n+1}$.

Theorem 13.9. *Under suitable hypothesis on an entire function f on $\mathbb{C}^{2n+1}$ we have the following identity:*

$$\int_{G_n} |f(g.(z,w,\zeta))|^2 dg$$

$$= \int_{-\infty}^{\infty} e^{2\lambda\eta} e^{-\lambda(u\cdot y - v\cdot x)} \left(\sum_{k=0}^{\infty} \|f^\lambda *_\lambda \varphi_k^\lambda\|_2^2 \frac{k!(n-1)!}{(k+n-1)!} \varphi_k^\lambda(2iy, 2iv) \right) d\mu(\lambda)$$

where $\|f^\lambda *_\lambda \varphi_k^\lambda\|_2$ is the $L^2(\mathbb{C}^n)$ norm of $f^\lambda *_\lambda \varphi_k^\lambda$.

We remark that the above theorem is valid for all those functions for which the right hand side of the above identity is finite. For example, when the Fourier transform of the restriction of f to $\mathbb{H}^n$ is compactly supported (in a suitable sense) the theorem remains true. We refer to [13] for a proof of this theorem and also for the hypotheses under which it is true. The proof uses some calculations on special Hermite functions and a general theorem proved by Faraut ([2]). A similar result is valid for the semidirect product of the reduced Heisenberg group with $U(n)$ from which Gutzmer's formula for the special Hermite expansions can be deduced. Gutzmer's formula for the Heisenberg group has been used in [13] to prove a different characterisation of the image of $L^2(\mathbb{H}^n)$ under the Segal-Bargmann transform. A version of Paley-Wiener theorem for the inverse Fourier transform also follows from Gutzmer's formula.

References

[1] Bargmann, V. (1961). On Hilbert spaces of analytic functions and an associated integral transform, Part I. *Comm. Pure Appl. Math.* **14** 187-214.

[2] Faraut, J. (2002). Formule de Gutzmer pour la complexification d'un espace Riemannien symetrique. *Rend. Mat. Acc. Lincei s.9* **13** 233-241.

[3] Faraut, J. (2003). Analysis on the crown of a Riemannian symmetric space. *Amer. Math. Soc. Transl.* **210** 99-110.

[4] Faraut, J. (2003). Espaces Hilbertiens invariants de fonctions holomorphes. *Seminaires and Congres 7, Societe Math. de France.*

[5] Gutzmer, A. (1888). Ein satz uber potenzreihen. *Math. Ann.* **32** 596-600.

[6] Hall, B. (1994). The Segal-Bargmann coherent state transform for compact Lie groups. *J. Funct. Anal.* **122** 103-154.

[7] Krötz, B., S. Thangavelu and Xu, Y. (2005). The heat kernel transform for the Heisenberg group. *J. Funct. Anal.* **225**. 301-336

[8] Krötz, B., Olafsson, G. and Stanton, R. (2005). The image of the heat kernel transform on Riemannian symmetric spaces of noncompact type. *Int. Math. Res. Notes.* **22** 1307-1329.

[9] Lassalle, M. (1978). Series de Laurent des fonctions holomorphes dans la complexification d'un espace symetrique compact. *Ann. Sci. Ecole Norm. Sup.* **11** 167-210.

[10] Lassalle, M. (1985). L'espace de Hardy d'un domaine de Reinhardt generalise. *J. Funct. Anal.* **60** 309-340.

[11] Stenzel, M. (1999). The Segal-Bargmann transform on a symmetric space of compact type. *J. Funct. Anal.* **165** 44-58.

[12] Thangavelu, S. (1998). *Harmonic analysis on the Heisenberg group*, Prog. in Math. Vol. 159, Birkhäuser, Boston.

[13] Thangavelu, S. (2007). Gutzmer's formula and Poisson integrals on the Heisenberg group. *Pacific J. Math.* **231** 217-238

[14] Thangavelu, S. (2007). Holomorphic Sobolev spaces associated to compact symmetric spaces. *J. Funct. Anal.* **251** 438-462.

[15] Thangavelu, S. (2007). A Paley-Wiener theorem for the inverse Fourier transform on some homogeneous spaces. *Hiroshima Math. J.* **37** 145-159.

[16] Thangavelu, S. (2008). An analogue of Gutzmer's formula for the Hermite expansions.*Studia Math.* **185** 279-290.

Chapter 14

Finite Translation Generalized Quadrangles

J. A. Thas

Department of Pure Mathematics and Computer Algebra,
Ghent University, Krijgslaan 281 - S22, 9000 Gent, Belgium
jat@cage.ugent.be

My talk is a survey on finite translation generalized quadrangles. To each translation generalized quadrangle of order (s,t), with $s \neq 1 \neq t$, there corresponds a set $O(n,m,q)$ of q^m+1 $(n-1)$-dimensional subspaces of the projective space $\mathrm{PG}(2n+m-1,q)$ satisfying (i) every three subspaces generate a $\mathrm{PG}(3n-1,q)$ and (ii) for every such subspace π there is a subspace $\mathrm{PG}(n+m-1,q)$ containing π and having empty intersection with the other elements of $O(n,m,q)$. Conversely, every such $O(n,m,q)$ defines a finite translation generalized quadrangle. For each known example of $O(n,m,q)$ we have $m \in \{n,2n\}$, and for q even there are no other examples. Many papers were written on the case $m=2n$. Here emphasis is on the case $m=n$, and besides interesting and useful old results several new theorems are stated.

14.1. Finite Generalized Quadrangles

14.1.1. *Finite generalized quadrangles*

A (finite) *generalized quadrangle* (GQ) is an incidence structure $\mathcal{S} = (P,B,\mathrm{I})$ in which P and B are disjoint (nonempty) sets of objects called *points* and *lines* respectively, and for which I is a symmetric point-line *incidence relation* satisfying the following axioms.

(i) Each point is incident with $1+t$ lines $(t \geq 1)$ and two distinct points are incident with at most one line.
(ii) Each line is incident with $1+s$ points $(s \geq 1)$ and two distinct lines are incident with at most one point.
(iii) If x is a point and L is a line not incident with x, then there is a unique pair $(y,M) \in P \times B$ for which $x \mathrm{\ I\ } M \mathrm{\ I\ } y \mathrm{\ I\ } L$.

Generalized quadrangles were introduced by Tits ([42]) in the appendix of his celebrated work on triality.

The integers s and t are the *parameters* of the GQ and $\mathcal{S}$ is said to have *order* (s,t); if $s=t$, $\mathcal{S}$ is said to have *order* s. There is a *point-line duality* for GQs (of order (s,t)) for which in any definition or theorem the words "point" and "line" are interchanged and the parameters s and t are interchanged. Hence, we assume without further notice that the dual of a given theorem or definition has also been given.

If the point x is not incident with the line L, in a GQ $\mathcal{S}=(P,B,\mathrm{I})$, then we write $x \not\mathrm{I}\, L$.

Let $\mathcal{S}=(P,B,\mathrm{I})$ be a (finite) GQ of order (s,t). Then $\mathcal{S}$ has $v=|P|=(1+s)(1+st)$ points and $b=|B|=(1+t)(1+st)$ lines; see 1.2.1 of [24]. Also, $s+t$ divides $st(1+s)(1+t)$, and, for $s\neq 1\neq t$, we have $t\leq s^2$ and, dually, $s\leq t^2$ (inequalities of Higman, ([13])); see 1.2.2 and 1.2.3 of [24].

Given two (not necessarily distinct) points x,y of $\mathcal{S}$, we write $x\sim y$ and say that x and y are *collinear*, provided that there is some line L for which $x\ \mathrm{I}\ L\ \mathrm{I}\ y$. And $x\not\sim y$ means that x and y are not collinear. Dually, for $L,M\in B$, we write $L\sim M$ or $L\not\sim M$ according as L and M are *concurrent* or nonconcurrent, respectively. The line which is incident with distinct collinear points x,y is denoted by xy; the point which is incident with distinct concurrent lines L,M is denoted by either LM or $L\cap M$.

For $x\in P$, put $x^\perp=\{y\in P \,\|\, y\sim x\}$, and note that $x\in x^\perp$.

A *subquadrangle*, or also *sub*GQ, $\mathcal{S}'=(P',B',\mathrm{I}')$ of a GQ $\mathcal{S}=(P,B,\mathrm{I})$ of order (s,t), is a GQ for which $P'\subseteq P, B'\subseteq B$, and where I' is the restriction of I to $(P'\times B')\cup(B'\times P')$. If the GQ $\mathcal{S}'$ of order (s',t') is a subGQ of the GQ $\mathcal{S}$ of order (s,t), with $\mathcal{S}\neq\mathcal{S}'$, then either $s=s'$ or $s\geq s't'$, and, dually, either $t=t'$ or $t\geq s't'$; see 2.2.1 of [24].

Let $\mathcal{S}=(P,B,\mathrm{I})$ be a GQ. A *collineation* or *automorphism* of $\mathcal{S}$ is a permutation of $P\cup B$ that preserves P, B and incidence.

14.1.2. *Grids and dual grids*

A *grid* is an incidence structure $\mathcal{S}=(P,B,\mathrm{I})$ with $P=\{x_{ij} \,\|\, i=0,1,\ldots,s_1$ and $j=0,1,\ldots,s_2\}$, $s_1>0$ and $s_2>0$, $B=\{L_0,L_1,\ldots,L_{s_1},M_0,M_1,\ldots,M_{s_2}\}$, $x_{ij}\mathrm{I}L_k$ if and only if $i=k$, and $x_{ij}\mathrm{I}M_k$ if and only if $j=k$. We say that such a grid is an $(s_1+1)\times(s_2+1)$-*grid*. A grid is a GQ if and only if $s_1=s_2=s$; in such a case the GQ has order $(s,1)$. Any GQ of order $(s,1)$ is isomorphic to an $(s+1)\times(s+1)$-grid. A dual grid has parameters t_1,t_2, and it is a GQ if and only if $t_1=t_2=t$, in which case it is a GQ of order

$(1, t)$. Any GQ of order $(1, t)$ is isomorphic to a dual $(t+1) \times (t+1)$-grid. An ordinary quadrangle is a GQ of order (1,1) and is at the same time a grid and a dual grid. This is the motivation for the term "generalized quadrangle".

A GQ of order (s, t) is called *thin* if either $s = 1$ or $t = 1$; in the other case it is called *thick*.

14.1.3. *The classical generalized quadrangles*

We now give a brief description of three families of examples known as the *classical* GQs, all of which are associated with classical groups and were first recognized as GQs by Tits.

(i) Consider a nonsingular quadric Q of projective index 1, that is, of Witt index 2, of the projective space $\mathrm{PG}(d, q)$, with $d = 3, 4$ or 5. Then the points of Q together with the lines of Q (which are the subspaces of maximal dimension on Q) form a GQ $Q(d, q)$ with parameters

$$
\begin{aligned}
&s = q,\ t = 1,\ \ v = (q+1)^2, && b = 2(q+1), && \text{when } d = 3,\\
&s = q,\ t = q,\ \ v = (q+1)(q^2+1), && b = (q+1)(q^2+1), && \text{when } d = 4,\\
&s = q,\ t = q^2,\ v = (q+1)(q^3+1), && b = (q^2+1)(q^3+1), && \text{when } d = 5.
\end{aligned}
$$

Notice that $Q(3, q)$ is a grid.

(ii) Let H be a nonsingular Hermitian variety of the projective space $\mathrm{PG}(d, q^2)$, $d = 3$ or 4. Then the points of H together with the lines on H form a GQ $H(d, q^2)$ with parameters

$$
\begin{aligned}
&s = q^2, t = q,\ \ v = (q^2+1)(q^3+1), && b = (q+1)(q^3+1), && \text{when } d = 3,\\
&s = q^2, t = q^3, v = (q^2+1)(q^5+1), && b = (q^3+1)(q^5+1), && \text{when } d = 4.
\end{aligned}
$$

(iii) The points of $\mathrm{PG}(3, q)$, together with the totally isotropic lines with respect to a symplectic polarity, form a GQ $W(q)$ with parameters

$$s = q,\ t = q,,\ v = (q+1)(q^2+1),\ b = (q+1)(q^2+1).$$

Theorem 14.1.

(i) *The GQ $Q(4, q)$ is isomorphic to the dual of $W(q)$. Also, $Q(4, q)$ (or $W(q)$) is self-dual if and only if q is even.*

(ii) *The GQ $Q(5, q)$ is isomorphic to the dual of $H(3, q^2)$.*

For a proof of Theorem 14.1, see 3.2.1 and 3.2.3 of [24].

14.1.4. *Ovals, hyperovals and ovoids*

A *k-arc* of $\mathrm{PG}(2,q)$ is a set of k points of $\mathrm{PG}(2,q)$ no three of which are collinear. Then clearly $k \leq q+2$. By [5], for q odd, $k \leq q+1$. Further, any nonsingular conic of $\mathrm{PG}(2,q)$ is a $(q+1)$-arc. It can be shown that each $(q+1)$-arc K of $\mathrm{PG}(2,q)$, q even, extends to a $(q+2)$-arc $K \cup \{x\}$ (see, e.g., [15], p. 177); the point x, which is uniquely defined by K, is called the *kernel* or *nucleus* of K. The $(q+1)$-arcs of $\mathrm{PG}(2,q)$ are called *ovals*; the $(q+2)$-arcs of $\mathrm{PG}(2,q)$, q even, are called *hyperovals.*

For any k-arc K, with $3 \leq k \leq q+1$, choose three of its points as the triangle of reference $u_0u_1u_2$ of the coordinate system. The lines intersecting K in one point are called the *tangent lines* of K. A tangent line of K through one of u_0, u_1, u_2 has respective equation

$$X_1 - dX_2 = 0, X_2 - dX_0 = 0, X_0 - dX_1 = 0,$$

with $d \neq 0$. We call d the coordinate of such a line. Suppose the $t = q+2-k$ tangent lines at each of u_0, u_1, u_2 are

$$X_1 - a_iX_2 = 0, X_2 - b_iX_0 = 0, X_0 - c_iX_1 = 0,$$

$i = 1, 2, \ldots, t$. Then Segre ([27]) proved the following important lemma.

Lemma 14.1. (Lemma of Tangents) *The coordinates a_i, b_i, c_i of the tangent lines at u_0, u_1, u_2 of a k-arc K through these points satisfy*

$$\Pi_{i=1}^{t} a_ib_ic_i = -1.$$

For an oval K we have $t = 1$, and so the lemma becomes $abc = -1$. Geometrically this means that for q odd the triangles formed by three points of an oval and the tangent lines at these points, are in perspective, or, equivalently, that for any three distinct points u_0, u_1, u_2 on an oval there is a (unique) nonsingular conic containing these points u_0, u_1, u_2 and having as tangent lines at u_0, u_1, u_2 the tangent lines of the oval at u_0, u_1, u_2; for q even the condition means that the tangent lines at any three points of the oval are concurrent.

Relying on Lemma 14.1, Segre ([27]) proves the following celebrated result.

Theorem 14.2. *In* $\mathrm{PG}(2,q)$, *q odd, every oval is a nonsingular conic.*

For q even, Theorem 14.2 is valid if and only if $q \in \{2,4\}$; see, e.g., [35].

We now introduce the notion of "ovoid" as defined in [43]. An *ovoid* O of $\mathrm{PG}(3,q)$ is a set of points of $\mathrm{PG}(3,q)$, no three of which are collinear and such that for any point of O the union of the lines which meet O only in that point, that is, the *tangent lines* at that point, is a $\mathrm{PG}(2,q)$. If O is an ovoid, its number of points is q^2+1. A *k-cap* K of $\mathrm{PG}(3,q)$ is a set of k points no three of which are collinear. For any k-cap K of $\mathrm{PG}(3,q)$, with $q \neq 2, k \leq q^2+1$; for any k-cap K of $\mathrm{PG}(3,2), k \leq 8$ holds and the 8-caps of $\mathrm{PG}(3,2)$ are the complements of planes. For q odd this result is due to Bose ([5]), for q even to Qvist ([26]). A (q^2+1)-cap of $\mathrm{PG}(3,q)$, $q \neq 2$, is precisely an *ovoid* (cf. [3] or [14]); the *ovoids* of $\mathrm{PG}(3,2)$ are the sets of 5 points no 4 of which are coplanar. It is easy to show that each nonsingular elliptic quadric of $\mathrm{PG}(3,q)$ is an ovoid. By a celebrated theorem, due independently to Barlotti ([3]) and Panella ([19]), every ovoid in $\mathrm{PG}(3,q), q$ odd or $q=4$, is an elliptic quadric.

Theorem 14.3. *Each ovoid of* $\mathrm{PG}(3,q)$, *q odd, is an elliptic quadric.*

To the contrary, in the even case, Tits ([43]) showed that for any $q = 2^{2e+1}$, with $e \geq 1$, there exists an ovoid which is not an elliptic quadric; these ovoids are called *Tits ovoids*, or also *Suzuki-Tits ovoids*, and are related to the simple Suzuki groups $Sz(q)$. In fact, for $q=8$, Segre ([28]) discovered an ovoid which is not an elliptic quadric, and which was shown to be a Tits ovoid in [12]. For even q no other ovoids than the elliptic quadrics and the Tits ovoids are known.

If O is an ovoid in $\mathrm{PG}(3,q)$, then any plane π of $\mathrm{PG}(3,q)$ intersects O in either one point or in an oval. If $|\pi \cap O| = 1$, then we say that π is a *tangent plane* of O. At each of its points O has exactly one tangent plane. For more details, see [14]. Finally, a beautiful result due to Brown ([6]) tells us that any ovoid O of $\mathrm{PG}(3,q)$ containing at least one conic section, is an elliptic quadric.

14.1.5. *The generalized quadrangles $T_2(O)$ and $T_3(O)$ of Tits*

Let $d=2$ (respectively, $d=3$) and let O be an oval (respectively, an ovoid) of $\mathrm{PG}(d,q)$. Further, let $\mathrm{PG}(d,q) = H$ be embedded as a hyperplane in $\mathrm{PG}(d+1,q) = P$. Define

- *Points* as
  - (i) the points of $P \backslash H$,
  - (ii) the hyperplanes X of P for which $|X \cap O| = 1$, and
  - (iii) one new symbol (∞).

- *Lines* as

  (a) the lines of P which are not contained in H and meet O (necessarily in a unique point), and
  (b) the points of O.

- *Incidence* is defined as follows. A point of Type (i) is incident only with lines of Type (a); here the incidence is that of P. A point of Type (ii) is incident with all lines of Type (a) contained in it and with the unique element of O in it. The point (∞) is incident with no line of Type (a) and all lines of Type (b).

It is an easy exercise to show that the incidence structure $T_2(O)$ (respectively, $T_3(O)$) so defined is a GQ. The parameters are

$$s = q, t = q, \quad v = (q+1)(q^2+1), b = (q+1)(q^2+1), \quad \text{when } d = 2,$$
$$s = q, t = q^2, v = (q+1)(q^3+1), b = (q^2+1)(q^3+1), \quad \text{when } d = 3.$$

Theorem 14.4.

(i) *The* GQ $T_2(O)$ *is isomorphic to the* GQ $Q(4,q)$ *if and only if* O *is a nonsingular conic. The* GQ $T_2(O)$ *is isomorphic to the* GQ $W(q)$ *if and only if* q *is even and* O *is a conic.*
(ii) *The* GQ $T_3(O)$ *is isomorphic to the* GQ $Q(5,q)$ *if and only if* O *is a nonsingular elliptic quadric.*

For a proof of Theorem 14.4, see 3.2.2 and 3.2.4 of [24].

Remark 14.1.

(i) For q odd any oval is a nonsingular conic, hence for q odd we always have $T_2(O) \cong Q(4,q)$.
(ii) For q odd any ovoid is an elliptic quadric, hence for q odd we always have $T_3(O) \cong Q(5,q)$.

14.1.6. *The generalized quadrangles $T_2^*(O)$*

Let O be a hyperoval of $\mathrm{PG}(2,q)$, so q is even. Embed $\mathrm{PG}(2,q) = H$ as a hyperplane in $\mathrm{PG}(3,q) = P$.

- The points of the GQ $T_2^*(O)$ are the points of $P \backslash H$,
- lines of the GQ are the lines of P not in H which meet O, and
- the incidence is inherited from P.

Then $T_2^*(O)$ is a GQ with parameters

$$s = q - 1, t = q + 1, v = q^3, b = (q + 2)q^2.$$

14.1.7. *Orders of the known generalized quadrangles*

The orders of the known GQs are

$$\begin{array}{ll}
(s,1), & s \in \mathbb{N}\backslash\{0\}, \\
(1,t), & t \in \mathbb{N}\backslash\{0\}, \\
(q,q), & q \text{ any prime power}, \\
(q,q^2), & q \text{ any prime power}, \\
(q^2,q), & q \text{ any prime power}, \\
(q^2,q^3), & q \text{ any prime power}, \\
(q^3,q^2), & q \text{ any prime power}, \\
(q-1,q+1), & q \text{ any prime power}, \\
(q+1,q-1), & q \text{ any prime power}.
\end{array}$$

14.1.8. *Generalized quadrangles with small parameters*

The proofs of all the results in this section are contained in Chapter 6 of [24].

Let $\mathcal{S}$ be a GQ of order $(s,t), 1 < s \leq t$. If $s = 2$, then $t \in \{2,4\}$ and GQs of order $(2,2)$ and $(2,4)$ are unique. If $s = 3$, then $t \in \{3,5,6,9\}$. GQs of order $(3,5)$ and $(3,9)$ are unique; up to isomorphism there are two GQs of order 3. If $s = 4$, then $t \in \{4,6,8,11,12,16\}$. There is just one GQ of order 4, and for $t \in \{6,8,16\}$ a unique example is known. Nothing is known about $t = 11$ and $t = 12$.

14.2. Translation Generalized Quadrangles

14.2.1. *Translation generalized quadrangles*

For proofs of all results in this section we refer to Chapter 8 of [24] or Chapter 3 of [41].

Let $\mathcal{S} = (P, B, \mathrm{I})$ be a finite generalized quadrangle. A *whorl about a point* p of the GQ $\mathcal{S}$ is a collineation fixing p linewise. A *whorl about a line* L is a collineation fixing L pointwise. An *elation about the point* $p \in P$ is a whorl about p that fixes no point of $P \setminus p^\perp$. Dually, one defines *elations about lines*. If θ is an elation about p, then we will often say that p is the *center* of θ. By definition, the identical permutation is an elation (about every point and every line). If p is a point of the GQ $\mathcal{S}$, for which there exists a group of elations G about p which acts sharply transitively on the points of $P \setminus p^\perp$, then $\mathcal{S}$ is said to be an *elation generalized quadrangle* (EGQ) with *base-point* or *center* or *elation point* p and *elation group* (or *base-group*) G, and we sometimes write $(\mathcal{S}^{(p)}, G)$ or $\mathcal{S}^{(p)}$ for $\mathcal{S}$. Dually, we define the *base-line* of an EGQ. We only work with EGQs that are thick, and therefore we will not bother each time to mention this.

If a GQ $(\mathcal{S}^{(p)}, G)$ is an EGQ with elation point p, and if G is abelian, then we say that $\mathcal{S}$ is a *translation generalized quadrangle* (TGQ) with *base-point* or *translation point* or *center* p and *translation group* (or *base-group*) G. The elements of a translation group are called *translations*. It can be shown that the translation group G is uniquely defined by the translation point p.

Theorem 14.5. *For any* TGQ *of order* (s,t) *we have* $t \geq s$.

14.2.2. *The kernel of a translation generalized quadrangle*

Suppose $(\mathcal{S}^{(p)}, G)$ is a TGQ of order (s,t), $s \neq 1 \neq t$, with translation point p and translation group G, and let y be a point of $P \backslash p^\perp$. Let $L_0, L_1, \ldots, L_t$ be the lines incident with p, and define r_i and M_i by $L_i \mathrm{I} r_i \mathrm{I} M_i \mathrm{I} y$, $0 \leq i \leq t$. Put $H_i = \{\theta \in G \,\|\, M_i^\theta = M_i\}, H_i^* = \{\theta \in G \,\|\, r_i^\theta = r_i\}$, with $0 \leq i \leq t$; then $|H_i| = s$ and $|H_i^*| = st$. The *kernel* K of $\mathcal{S}^{(p)}$ (or of $(\mathcal{S}^{(p)}, G)$) is the set of all endomorphisms α of G for which $H_i^\alpha \subseteq H_i$, $0 \leq i \leq t$. With the usual addition and multiplication of endomorphisms K is a ring.

Theorem 14.6. *The ring* K *is a field, so that* $H_i^\alpha = H_i, (H_i^*)^\alpha = H_i^*$ *for all* $i = 0, 1, \ldots, t$ *and all* $\alpha \in K^0 = K \backslash \{0\}$.

For each subfield F of K there is a vector space (G, F) whose vectors are the elements of G, and whose scalars are the elements of F. Vector addition is the group operation in G, and scalar multiplication is defined by $g\alpha = g^\alpha, g \in G, \alpha \in F$. It is easy to verify that (G, F) is indeed a vector space. As H_i is a subspace of (G, F), we have $|H_i| \geq |F|$. It follows that $s \geq |K|$. There is an interesting corollary of (G, F) being a vector space.

Theorem 14.7. *The group* G *is elementary abelian, so* s *and* t *must be powers of the same prime.*

Finally we have the following result.

Theorem 14.8. *The multiplicative group K^0 induces the group of all whorls about p and y, with $p \not\sim y$.*

14.2.3. $T(n, m, q)$s and translation generalized quadrangles

In this section, we introduce the notion of $T(n, m, q)$, which is a natural generalization of the $T_d(O)$ constructions of Tits, $d \in \{2, 3\}$.

Suppose $H = \mathrm{PG}(2n+m-1, q)$ is the finite projective $(2n+m-1)$-space over $\mathrm{GF}(q)$. Now define a set $O = O(n, m, q)$ of subspaces as follows: O is a set of q^m+1 $(n-1)$-dimensional subspaces of H, denoted by $\mathrm{PG}^{(i)}(n-1, q)$, and often also by π_i, so that

(i) every three generate a $\mathrm{PG}(3n-1, q)$;
(ii) for every $i = 0, 1, \ldots, q^m$, there is a subspace $\mathrm{PG}^{(i)}(n+m-1, q)$, also denoted by τ_i, of H of dimension $n+m-1$, which contains $\mathrm{PG}^{(i)}(n-1, q)$ and which is disjoint from any $\mathrm{PG}^{(j)}(n-1, q)$ if $j \neq i$.

If O satisfies these conditions for $n = m$, then O is called a *pseudo-oval* or a *generalized oval* or an $[n-1]$-*oval* of $\mathrm{PG}(3n-1, q)$. A $[0]$-oval of $\mathrm{PG}(2, q)$ is just an oval of $\mathrm{PG}(2, q)$. For $n \neq m$, $O(n, m, q)$ is called a *pseudo-ovoid* or a *generalized ovoid* or an $[n-1]$-*ovoid* or an *egg* of $\mathrm{PG}(2n+m-1, q)$. A $[0]$-ovoid of $\mathrm{PG}(3, q)$ is just an ovoid of $\mathrm{PG}(3, q)$.

The space $\mathrm{PG}^{(i)}(n+m-1, q)$ is the *tangent space* of $O(n, m, q)$ at $\mathrm{PG}^{(i)}(n-1, q)$; it is uniquely determined by $O(n, m, q)$ and $\mathrm{PG}^{(i)}(n-1, q)$. Sometimes we will call an $O(n, n, q)$ also an "egg" or a "generalized ovoid" for the sake of convenience.

From any egg $O = O(n, m, q)$ arises a GQ $T(n, m, q) = T(O)$ which is a TGQ of order (q^n, q^m) for some base-point (∞). This goes as follows. Let H be embedded in a $\mathrm{PG}(2n+m, q) = H'$.

- The *points* are of three types.
  (i) The points of $H' \setminus H$.
  (ii) The subspaces $\mathrm{PG}(n+m, q)$ of H' which intersect H in a $\mathrm{PG}^{(i)}(n+m-1, q)$.
  (iii) A symbol (∞).
- The *lines* are of two types.
  (a) The subspaces $\mathrm{PG}(n, q)$ of $\mathrm{PG}(2n+m, q)$ which intersect H in an element of the egg.
  (b) The elements of the egg $O(n, m, q)$.

- *Incidence* is defined as follows. The point (∞) is incident with all the lines of Type (b) and with no other lines. A point of Type (ii) is incident with the unique line of Type (b) contained in it and with all the lines of Type (a) contained in it. Finally, a point of Type (i) is incident with the lines of Type (a) containing it.

Conversely, by interpreting any TGQ in terms of the vector space (G, F), defined in Section 14.2.2, it can be seen that any TGQ is a $T(n, m, q)$ associated to some $O(n, m, q)$.

Theorem 14.9. *The geometry $T(n, m, q)$ is a TGQ of order (q^n, q^m) with translation point (∞) and for which $\mathrm{GF}(q)$ is a subfield of the kernel. Moreover, the translations of $T(n, m, q)$ induce translations of the affine space $\mathrm{AG}(2n+m, q) = \mathrm{PG}(2n+m, q) \setminus \mathrm{PG}(2n+m-1, q)$. Conversely, every TGQ for which $\mathrm{GF}(q)$ is a subfield of the kernel is isomorphic to a $T(n, m, q)$. It follows that the theory of TGQs is "equivalent" to the theory of the sets $O(n, m, q)$.*

Remark 14.2.

(a) We emphasize that with the group of all homologies of $\mathrm{PG}(2n + m, q)$ having center y not in $\mathrm{PG}(2n + m - 1, q)$ and axis $\mathrm{PG}(2n + m - 1, q)$ corresponds in a natural way the multiplicative group of a subfield of the kernel.
(b) It is clear that $T(1, 1, q)$ is a $T_2(O)$ of Tits, and that $T(1, 2, q)$ is a $T_3(O)$ of Tits.
(c) If $\mathcal{S} \cong T(O)$, with $O = O(n, m, q)$, then for fixed n, m, q, we have that O is uniquely defined by $\mathcal{S}$.

Corollary 14.1. *For any $O(n, m, q)$ we have $n \leq m \leq 2n$.*

Corollary 14.2. *Let $\mathcal{S}^{(x)}$ be a TGQ of order (s, t), s a prime. Then either $\mathcal{S}^{(x)} \cong Q(4, s)$ or $\mathcal{S}^{(x)} \cong Q(5, s)$.*

Theorem 14.10. *Let $(\mathcal{S}^{(x)}, G)$ be a TGQ of order (s, t). Then $\mathcal{S}^{(x)} \cong T_d(O)$ for some $d \in \{2, 3\}$, if and only if for a fixed point y, $y \not\sim x$, the group of all whorls about x fixing y has order $s - 1$, that is, if and only if $|K| = s$.*

14.2.4. *Regular pseudo-ovals and regular pseudo-ovoids*

In the extension $\mathrm{PG}(2n+m-1,q^n)$ of $\mathrm{PG}(2n+m-1,q)$, with $m \in \{n,2n\}$, we consider n $(\frac{m}{n}+1)$-dimensional spaces $\mathrm{PG}^{(i)}(\frac{m}{n}+1,q^n) = \xi_i$, with $i=1,2,\ldots,n$, which are conjugate with respect to the extension $\mathrm{GF}(q^n)$ of $\mathrm{GF}(q)$, that is, which form an orbit of the Galois group corresponding to this extension, and which span $\mathrm{PG}(2n+m-1,q^n)$. In ξ_1 we consider an oval O_1 for $m=n$ and an ovoid O_1 for $m=2n$. Let $O_1 = \{x_0^{(1)}, x_1^{(1)}, \ldots, x_{q^m}^{(1)}\}$. Further, let $x_i^{(1)}, x_i^{(2)}, \ldots, x_i^{(n)}$, with $i=0,1,\ldots,q^m$, be conjugate with respect to the extension $\mathrm{GF}(q^n)$ of $\mathrm{GF}(q)$. The points $x_i^{(1)}, x_i^{(2)}, \ldots, x_i^{(n)}$ define an $(n-1)$-dimensional space $\mathrm{PG}^{(i)}(n-1,q) = \pi_i$ over $\mathrm{GF}(q)$, with $i = 0,1,\ldots,q^m$. Then $O = \{\pi_0,\pi_1,\ldots,\pi_{q^m}\}$ is a generalized oval of $\mathrm{PG}(3n-1,q)$ for $m=n$, and a generalized ovoid of $\mathrm{PG}(4n-1,q)$ for $m=2n$. Here, we speak of a *regular* or *elementary* pseudo-oval, respectively a *regular* or *elementary* pseudo-ovoid. If O_1 is a conic, then we speak of a *pseudo-conic* or a *classical* pseudo-oval; if O_1 is an elliptic quadric, then we speak of a *classical* pseudo-ovoid.

If $m=n$, then in the regular case $T(n,n,q) \cong T_2(O_1)$; if $m=2n$, then in the regular case $T(n,2n,q) \cong T_3(O_1)$.

For $m=n$ each known pseudo-oval is regular. Also for $m=2n$ with q even each known pseudo-ovoid is regular. For $m=2n$ with q odd nonregular pseudo-ovoids are known, see Section 14.4.

14.3. Important Properties of $O(n,m,q)$

For details and proofs we refer again to Chapter 8 of [24] and to Chapter 3 of [41]. We remark that pseudo-ovals were already introduced in 1971; see [32].

14.3.1. *Properties of $O(n,m,q)$*

In this section properties of $O(n,m,q)$, for any n and m, will be stated. Let $\mathrm{PG}(2n+m-1,q)$ be the projective space containing $O(n,m,q)$.

Theorem 14.11. *The following hold for any $O(n,m,q)$.*

(i) *Each hyperplane of* $\mathrm{PG}(2n+m-1,q)$ *which does not contain a tangent space of $O(n,m,q)$, contains either* 0 *or* $1+q^{m-n}$ *elements of $O(n,m,q)$. If $m=2n$, then each hyperplane of* $\mathrm{PG}(4n-1,q)$ *which does not contain a tangent space of $O(n,2n,q)$ contains exactly* $1+q^n$

elements of $O(n, 2n, q)$*. If* $m \neq 2n$*, then there are hyperplanes which contain no element of* $O(n, m, q)$*.*

(ii) *Either* $n = m$ *or* $n(a + 1) = ma$ *with* $a \in \mathbb{N}_0$ *and* a *odd.*

Corollary 14.3. *Let* $\widetilde{O}$ *be the union of all elements of any* $O(n, 2n, q)$ *and let* π *be any hyperplane of* $\mathrm{PG}(4n-1, q)$*. Then* $|\widetilde{O} \cap \pi| \in \{\gamma_1, \gamma_2\}$*, with* $\gamma_1 = (q^n - 1)(q^{2n-1} + 1)/(q-1)$ *and* $\gamma_1 - \gamma_2 = q^{2n-1}$*. Hence* $\widetilde{O}$ *defines a linear projective two-weight code and a strongly regular graph.*

Remark 14.3. For the construction of a linear projective two-weight code and a strongly regular graph from a pointset of $\mathrm{PG}(r, q)$ with two intersection numbers with respect to hyperplanes of $\mathrm{PG}(r, q)$, we refer, e.g., to [8].

14.3.2. *Properties of pseudo-ovals*

In particular, $O(n, n, q)$ has the following properties.

Theorem 14.12. *Let* $O = O(n, n, q)$ *be a pseudo-oval in* $\mathrm{PG}(3n-1, q)$*.*

(i) *If* q *is even, then all tangent spaces of* O *contain a common* $(n-1)$*-dimensional space, called the* **kernel** *or* **nucleus** *of* O*.*

(ii) *If* q *is odd, then each point of* $\mathrm{PG}(3n-1, q)$ *not in an element of* O *is contained in either 0 or 2 tangent spaces of* O*, and each hyperplane of* $\mathrm{PG}(3n-1, q)$ *not containing a tangent space of* O *contains either 0 or 2 elements of* O*.*

Theorem 14.13. *For* q *odd, the tangent spaces of a pseudo-oval* $O = O(n, n, q)$ *are the elements of a pseudo-oval* $O^* = O^*(n, n, q)$ *in the dual space of* $\mathrm{PG}(3n-1, q)$*.*

Definition 14.1. The pseudo-oval O^* is called the **translation dual** of the pseudo-oval O. The TGQ $T(O^*) = T^*(O)$ is also called the **translation dual** of the TGQ $T(O)$.

Remark 14.4. It is not known whether or not we have always $O \cong O^*$, respectively $T(O) \cong T^*(O)$. In any case, for each known $T(O)$, with $O = O(n, n, q)$ and q odd, we have $T(O) \cong Q(4, q^n)$.

Theorem 14.14. *The* TGQ $\mathcal{S}$ *of order* s*, with* s *odd, and its translation dual* $\mathcal{S}^*$ *have isomorphic kernels.*

14.3.3. *Properties of eggs*

In this section we mention fundamental results on eggs.

Theorem 14.15. *Let $O(n,m,q)$ be an egg, so $n \neq m$.*

(i) *Each point of* $\mathrm{PG}(2n+m-1,q)$ *which is not contained in an element of $O(n,m,q)$ belongs to either 0 or $q^{m-n}+1$ tangent spaces of the egg. If $m=2n$, then each point of* $\mathrm{PG}(4n-1,q)$ *not contained in an element of $O(n,2n,q)$ belongs to exactly q^n+1 tangent spaces of the egg. If $m \neq 2n$, then there are points contained in no tangent space of $O(n,m,q)$.*

(ii) *For q even we necessarily have $m=2n$. Hence for any $O(n,m,q)$ with q even, we have $m \in \{n,2n\}$, that is, for any* TGQ *$\mathcal{S}$ of order (s,t), with s and/or t even, we have $t \in \{s,s^2\}$.*

Theorem 14.16. *Every three distinct tangent spaces of $O(n,m,q)$, $m \neq n$, have as intersection a space of dimension $m-n-1$.*

By Theorem 14.16 the tangent spaces of $O(n,m,q)$, with $m \neq n$, are the elements of an egg $O^*(n,m,q)$ in the dual space of $\mathrm{PG}(2n+m-1,q)$.

Definition 14.2. The tangent spaces of an egg $O(n,m,q)$ in $\mathrm{PG}(2n+m-1,q)$ form an egg $O^*(n,m,q)$ in the dual space of $\mathrm{PG}(2n+m-1,q)$. So in addition to the TGQ $T(n,m,q)$, a TGQ $T^*(n,m,q)$ arises. The egg $O^*(n,m,q)=O^*$ will be called the **translation dual** of $O(n,m,q)=O$, and $T^*(n,m,q)=T^*(O)=T(O^*)$ will be called the **translation dual** of the GQ $T(n,m,q)=T(O)$.

Remark 14.5. For regular eggs O, we clearly have $O \cong O^*$, respectively $T(O) \cong T^*(O)$. For q odd there are examples with $O \ncong O^*$, respectively $T(O) \ncong T^*(O)$; see Section 14.4. For q even we have $O \cong O^*$ for all known examples (all known examples are regular).

Theorem 14.17. *The* TGQ *$\mathcal{S}$ of order (s,t), with $s \neq t$, and its translation dual $\mathcal{S}^*$ have isomorphic kernels.*

14.4. Eggs $O(n,2n,q)$: Fundamental Results and Characterizations

In this section we first mention two interesting characterizations of regular eggs; see [24]. For the other results we refer to [41].

14.4.1. *Characterizations of regular eggs $O(n, 2n, q)$*

Theorem 14.18. *Let $O(n, 2n, q) = O$ be an egg in* $\mathrm{PG}(4n-1, q)$.

(i) *The egg O is regular if and only if the $q^n + 1$ tangent spaces containing any given point z not contained in an element of O, have exactly $(q^n - 1)/(q-1)$ points in common.*
(ii) *The egg O is regular if and only if each* $\mathrm{PG}(3n-1, q)$ *containing at least three elements of O, contains exactly $q^n + 1$ elements of O.*

14.4.2. *Fundamental results on eggs*

This section contains fundamental results on eggs, all of which were discovered after the monograph by Payne and Thas was published.

An egg $O = O(n, 2n, q)$, respectively a TGQ $T(O)$, is called *good* at an element $\pi \in O$ if for every two distinct elements π' and π'' of $O \setminus \{\pi\}$ the $(3n-1)$-dimensional space $< \pi, \pi', \pi'' >$ generated by π, π', π'' contains exactly $q^n + 1$ elements of O; in such a case π is called a *good element* of O.

Theorem 14.19. ([34]) *The egg $O(n, 2n, q)$, with q even, is good if and only if its translation dual $O^*(n, 2n, q)$ is good.*

Theorem 14.20. ([34]) *Let O be an egg in* $\mathrm{PG}(4n-1, q), q$ *odd, which is good at its element π. Then the $q^{2n} + q^n$ pseudo-ovals on O containing π are classical.*

Remark 14.6. Relying on Theorem 14.20 a new infinite class of ovoids of $Q(4, s)$ and a new infinite class of translation planes was discovered (an ovoid of $Q(4, s)$ is a set of $s^2 + 1$ points of $Q(4, s)$ no two of which are on a common line of $Q(4, s)$); see [39].

For q even every known egg $O(n, 2n, q)$ is regular, so good at any of its elements; hence the corresponding TGQs are TGQs of Tits. For q odd, there are $O(n, 2n, q)$s which are not regular. In fact four infinite classes of $O(n, 2n, q)$s, q odd, are known, and two sporadic examples: the classical eggs, an infinite class due to Kantor ([18]) (here $O(n, 2n, q) \cong O^*(n, 2n, q), O(n, 2n, q)$ is good at some element, and for each odd q together with an automorphism $\sigma \neq 1$ of $\mathrm{GF}(q)$ there is a nonclassical example), an infinite class of $O(n, 2n, q)$s with $q = 3^h$ and $h > 1$, deduced from Ganley semifields (see [23]) (here $O(n, 2n, q) \not\cong O^*(n, 2n, q), O^*(n, 2n, q)$ is good at some element, while $O(n, 2n, q)$ is not good at any element), the translation duals $O^*(n, 2n, q)$ of the foregoing $O(n, 2n, q)$s (it was Payne who discovered in 1989 that $O(n, 2n, q) \not\cong O^*(n, 2n, q)$), and finally an egg $O(5, 10, 3)$ together with its non-isomorphic translation dual $O^*(5, 10, 3)$ (these two examples were deduced by Bader, Lunardon and Pinneri ([1]) from an ovoid of $Q(4, 3^5)$ discovered by Penttila and Williams ([25]) (one of these eggs is good).

Let F be a *flock* of the quadratic cone K in $\mathrm{PG}(3, q)$, that is, a partition of K minus its vertex into q nonsingular conics. In 1976 it was shown by Walker ([44]), and independently by Thas, that F defines a translation plane $\mathcal{P}(F)$ of order q^2. In 1987 Thas ([33]), relying on work of Kantor ([17,18]) and Payne ([21,22]), proved that with each flock F there corresponds a GQ $\mathcal{S}(F)$ of order (s^2, s), called a *flock* GQ. The paper [33] was the origin of an explosion of interest in the theory of GQs and led to many results, also because of the fact that now certain GQs of order (s^2, s) and certain translation planes of order s^2 were closely linked.

Theorem 14.21. ([16]) *A* TGQ $T(O)$ *of order* (s, s^2), *s even, is the point-line dual of a flock* GQ *if and only if O (respectively, $T(O)$) is classical.*

Theorem 14.22. ([34, 37]) *The* TGQ $T(O)$ *of order* (s, s^2), *s odd, is the point-line dual of a flock* GQ *if and only if the translation dual O^* of O is good at one of its elements.*

Remark 14.7. Theorem 14.22 is a particular case of a more general result of [37].

We also mention that there is a classification in [36] of good eggs of $\mathrm{PG}(4n-1, q)$, q odd, in terms of Veronese surfaces and their projections.

In Section 14.1.4 we mentioned the beautiful theorem of Brown on ovoids in $\mathrm{PG}(3, q)$, with q even. The next theorem generalizes this result to pseudo-ovoids.

Theorem 14.23. ([7]) *Let $O(n, 2n, q)$ be an egg in* $\mathrm{PG}(4n-1, q)$, *q even. Then $O(n, 2n, q)$ is classical if and only if it contains a pseudo-conic.*

Other interesting characterization theorems are the following.

Theorem 14.24. ([40]) *Let O be an egg in* $\mathrm{PG}(4n-1, q)$, *q even, which is good at $\pi \in O$. If the $q^{2n} + q^n$ pseudo-ovals on O containing π are regular, then O is regular.*

Theorem 14.25. ([4]) *Assume that $O(n, 2n, q)$, q odd, is good and that $q \geq 4n^2 - 8n + 2$. Then either $O(n, 2n, q)$ is classical or is an egg of Kantor type.*

Remark 14.8. Many interesting and useful characterizations of eggs involving subquadrangles and automorphism groups can be found in [24] and [41].

14.5. Pseudo-Ovals: Old and New Results

All known pseudo-ovals are regular; so for q odd all known pseudo-ovals are classical. Compared to eggs, not many results on pseudo-ovals are known. Here we collect some interesting results on these objects, including very recent ones.

14.5.1. *Old results*

Let $O(n,n,q) = \{\pi_0, \pi_1, \ldots, \pi_{q^n}\}$ be a pseudo-oval in $\mathrm{PG}(3n-1,q)$. The tangent space of $O(n,n,q)$ at π_i will be denoted by τ_i, with $i = 0, 1, \ldots, q^n$. Let $\mathrm{PG}(2n-1,q) \subset \mathrm{PG}(3n-1,q)$ be skew to π_i, $i \in \{0, 1, \ldots, q^n\}$. Further, let $\tau_i \cap \mathrm{PG}(2n-1,q) = \xi_i$, $< \pi_i, \pi_j > \cap \mathrm{PG}(2n-1,q) = \xi_j, j \neq i$. Then $\{\xi_0, \xi_1, \ldots, \xi_{q^n}\} = S_i$ is an $(n-1)$-spread of $\mathrm{PG}(2n-1,q)$, that is, a partition of $\mathrm{PG}(2n-1,q)$ consisting of q^n+1 $(n-1)$-dimensional subspaces.

Now let q be even and let η be the nucleus (cf. Section 14.3.2) of $O(n,n,q)$. Let $\mathrm{PG}(2n-1,q)$ be skew to η. If $\zeta_j = \mathrm{PG}(2n-1,q) \cap < \eta, \pi_j >$, then $\{\zeta_0, \zeta_1, \ldots, \zeta_{q^n}\} = S$ is an $(n-1)$-spread of $\mathrm{PG}(2n-1,q)$.

Let q be odd. Let $i \in \{0, 1, \ldots, q^n\}$. Put $\tau_i \cap \tau_j = \delta_j, j \neq i$. Then by Theorem 14.12, $\{\pi_i, \delta_0, \ldots, \delta_{i-1}, \delta_{i+1}, \ldots, \delta_{q^n}\} = S_i^*$ is an $(n-1)$-spread of τ_i.

Theorem 14.26. ([9]) *Consider a pseudo-oval $O(n,n,q)$, with q odd. Then at least one of the $(n-1)$-spreads $S_0, S_1, \ldots, S_{q^n}, S_0^*, S_1^*, \ldots, S_{q^n}^*$ is regular if and only if they are all regular if and only if $O(n,n,q)$ is classical if and only if the corresponding* GQ *$T(n,n,q)$ is isomorphic to the classical* GQ *$Q(4,q^n)$.*

Let $O(n,n,q) = O$ be a generalized oval in $\mathrm{PG}(3n-1,q)$, with q even. Then O is a *translation generalized oval*, with *axis* the tangent space $\mathrm{PG}^{(i)}(2n-1,q)$ of O at $\mathrm{PG}^{(i)}(n-1,q) \in O$, if there is a group of involutions of $\mathrm{PGL}(3n,q)$ with axis $\mathrm{PG}^{(i)}(2n-1,q)$, fixing O and acting regularly on $O \backslash \{\mathrm{PG}^{(i)}(n-1,q)\}$. It can be shown that it is sufficient that there is a group of involutions of $\mathrm{PGL}(3n,q)$ fixing $\mathrm{PG}^{(i)}(n-1,q)$ and acting regularly on the remaining elements of O. If $n=1$, then a translation generalized oval is just called a *translation oval*. All translation ovals of $\mathrm{PG}(2,q), q = 2^h$, were determined by Payne ([20]); choosing suitable coordinates, they are always of the form

$$\{(1,t,t^{2^i}) \parallel t \in \mathrm{GF}(q)\} \cup \{(0,0,1)\},$$

where i is fixed in $\{1,2,\dots,h-1\}$ and $(h,i)=1$.

Theorem 14.27. ([40]) *Let O be a generalized oval in* $\mathrm{PG}(3n-1,q)$, *with q even. Then O is a translation generalized oval if and only if the point-line dual of $T(O)$ is a* TGQ.

Remark 14.9. For $n=1$ the theorem can be found in the monograph [24].

Let $O=\{\pi,\pi_1,\dots,\pi_{q^n}\}$ be a generalized oval in $\mathrm{PG}(3n-1,q)$, with q even, and let η be the nucleus of O. Further, let τ be the tangent space of O at π. For each $i\in\{1,2,\dots,q^n\}$, the set

$$\{\pi,\eta\}\cup\{<\pi_i,\pi_j>\cap\tau \parallel i\neq j\}$$

is an $(n-1)$-spread of τ, denoted $\bar{S}_i$. If O is a translation generalized oval with axis τ, then all the spreads $\bar{S}_i$ coincide.

Suppose that all the spreads $\bar{S}_i$ coincide. We say that O is *projective* at τ if the following property holds :

Let γ be an element of $\bar{S}=\bar{S}_i$ (for all i), where $\pi\neq\gamma\neq\eta$, and let j,k be in $\{1,2,\dots,q^n\}$, $j\neq k$, such that $<\gamma,\pi_j>\neq<\gamma,\pi_k>$. As the spreads $\bar{S}_i$ coincide, there are elements $\pi_{j'}$ and $\pi_{k'}$ so that $\pi_{j'}\subset<\gamma,\pi_j>$, with $j\neq j'$, and $\pi_{k'}\subset<\gamma,\pi_k>$, with $k\neq k'$. Then $<\pi_j,\pi_k>\cap<\pi_{j'},\pi_{k'}>$ is an element of $\bar{S}$.

Theorem 14.28. ([40]) *Let O be a generalized oval in* $\mathrm{PG}(3n-1,q)$, *q even, and use the above notation. Then O is a translation generalized oval with axis τ if and only if all the spreads $\bar{S}_i$ coincide and O is projective at τ.*

Let $O=\{\pi,\pi_1,\dots,\pi_{q^n}\}$ be a generalized oval in $\mathrm{PG}(3n-1,q)$, with $q=2^h$, and let τ be the tangent space of O at π. Now we define a point-line incidence structure $A(O)$ as follows

- Points are the elements of $O\setminus\{\pi\}$;
- Lines are the pairs $\{\pi_i,\pi_j\}$ with $i,j\in\{1,2,\dots,q^n\}$ and $i\neq j$;
- Incidence is containment.

Hence $A(O)$ is the complete graph with vertex set $O \setminus \{\pi\}$. Further, two lines $\{\pi_i, \pi_j\}$ and $\{\pi_k, \pi_l\}$ are called *parallel* if $< \pi_i, \pi_j > \cap \tau = < \pi_k, \pi_l > \cap \tau$.

Theorem 14.29. ([40]) *The incidence structure $A(O)$ provided with parallelism is isomorphic to the hn-dimensional affine space* $\mathrm{AG}(hn, 2)$ *over* $\mathrm{GF}(2)$ *if and only if O is a translation generalized oval with axis τ.*

We also have the following characterization of classical generalized ovals in the even case.

Theorem 14.30. ([40]) *The pseudo-oval $O(n, n, q)$, q even, is classical if and only if $O(n, n, q)$ is 2-transitive.*

Relying on the previous theorem, Bamberg and Penttila ([2]) obtained the following stronger characterization.

Theorem 14.31. ([2]) *The pseudo-oval $O(n, n, q)$, q even, is classical if and only if $O(n, n, q)$ is transitive.*

14.5.2. *New results*

In this section yet unpublished results on pseudo-ovals will be discussed; see [38].

Let π_1, π_2, π_3 be mutually skew $(n-1)$-dimensional subspaces of $\mathrm{PG}(3n-1, q)$, let τ_i be a $(2n-1)$-dimensional space containing π_i but skew to π_j and π_k, with $\{i, j, k\} = \{1, 2, 3\}$, and let $\tau_i \cap \tau_j = \eta_k$ with $\{i, j, k\} = \{1, 2, 3\}$. The space generated by η_i and π_i will be denoted by ζ_i, with $i = 1, 2, 3$. If the $(2n-1)$-dimensional spaces $\zeta_1, \zeta_2, \zeta_3$ have a $(n-1)$-dimensional space in common, then we say that $\{\pi_1, \pi_2, \pi_3\}$ and $\{\tau_1, \tau_2, \tau_3\}$ are in *perspective*; if $\zeta_1, \zeta_2, \zeta_3$ have a nonempty intersection, then we say that $\{\pi_1, \pi_2, \pi_3\}$ and $\{\tau_1, \tau_2, \tau_3\}$ are in *semi-perspective*.

Let O be a regular generalized oval in $\mathrm{PG}(3n-1, q)$, with q odd. Then for any three distinct elements π_i, π_j, π_k of O the sets $\{\pi_i, \pi_j, \pi_k\}$ and $\{\tau_i, \tau_j, \tau_k\}$ are in perspective, where τ_l is the tangent space of O at π_l, $l \in \{i, j, k\}$. This follows immediately from the fact that in the odd case, by the Lemma of Tangents (that is, Lemma 14.1), this property holds for every oval. In the even case, by Theorem 14.12(i), the triples $\{\pi_i, \pi_j, \pi_k\}$ and $\{\tau_i, \tau_j, \tau_k\}$ are in perspective for any three distinct elements π_i, π_j, π_k of any generalized oval O (here $\zeta_1 \cap \zeta_2 \cap \zeta_3$ is the nucleus of O).

Now we state an interesting property on quadrics in $\mathrm{PG}(3n-1, q)$, $n \geq 1$.

Theorem 14.32. *Let Q be a nonsingular quadric in* PG$(3n-1,q)$, *q odd, and let π_1, π_2, π_3 be distinct $(n-1)$-dimensional spaces on Q which generate* PG$(3n-1,q)$. *The tangent space of Q at π_i is denoted by τ_i, with $i = 1,2,3$. Further, assume that $\pi_i \cap \tau_j = \emptyset$, for all $i \neq j$. Let $\eta_k = \tau_i \cap \tau_j$ with $\{i,j,k\} = \{1,2,3\}$ and let $\zeta_i = < \eta_i, \pi_i >$, with $i = 1,2,3$. Then the dimension of $\zeta_1 \cap \zeta_2 \cap \zeta_3$ has the same parity as $n-1$. In particular, if n is odd, then $\{\pi_1, \pi_2, \pi_3\}$ and $\{\tau_1, \tau_2, \tau_3\}$ are always in semi-perspective.*

Remark 14.10. Since we are mainly interested in generalized ovals we assume in Theorem 14.32 that q is odd. In the even case the statement is quite different, so for example if q is even and $n = 2$ we always have $\zeta_1 \cap \zeta_2 \cap \zeta_3 = \emptyset$.

Let q be odd. Then for each n even any classical generalized oval of PG$(3n-1,q)$ belongs to a nonsingular elliptic quadric $Q^-(3n-1,q)$ and to a nonsingular hyperbolic quadric $Q^+(3n-1,q)$, and for each n odd, any classical generalized oval of PG$(3n-1,q)$ belongs to a nonsingular parabolic quadric, see [31]. In each of these cases the tangent space at any element π of the generalized oval coincides with the tangent space at π of any of the corresponding quadrics. For $n = 2$ any classical generalized oval O is the intersection of some $Q^-(5,q)$ and some $Q^+(5,q)$; it is contained in $(q+1)/2$ nonsingular elliptic quadrics and $(q+1)/2$ nonsingular hyperbolic quadrics, and at each element L of O these quadrics have a common tangent space which coincides with the tangent space of O at L. Shult and Thas ([30]) prove that if O is a generalized oval of lines contained in a nonsingular hyperbolic quadric $Q^+(5,q)$ of PG$(5,q)$, q odd, then O is classical.

As for q even all tangent spaces of a generalized oval contain a common $(n-1)$-dimensional space, it follows that for $n > 1$ and q even, a generalized oval is never contained in a nonsingular quadric.

In recent years there was great interest in generalized ovals consisting of q^2+1 lines of $Q^-(5,q)$, with q odd. Such a generalized oval is equivalent to a set of q^2+1 points of the nonsingular Hermitian variety $H(3,q^2)$ in PG$(3,q^2)$, such that the plane defined by any three distinct points of this set is nontangent to $H(3,q^2)$. This object was studied in different contexts in [29], [11] and [10].

Theorem 14.33. *Let O be a generalized oval of* PG$(3n-1,q)$, *with q odd, contained in a nonsingular quadric Q of* PG$(3n-1,q)$. *If $\pi \in O$, then the tangent spaces at π of O and Q coincide.*

Corollary 14.4. *Let O be a generalized oval of* PG$(3n-1,q)$, *with q odd,* *contained in a nonsingular quadric Q of* PG$(3n-1,q)$. *Then any three distinct elements π_1,π_2,π_3 of O satisfy the requirements in the statement of Theorem 14.32.*

Next we state an elegant characterization of pseudo-conics, generalizing a characterization of conics from Section 14.1.4 (Theorem 14.2, relying on the geometric interpretation of Lemma 14.1).

Theorem 14.34. *Assume that $O=\{\pi_0,\pi_1,\ldots,\pi_{q^n}\}$ is a pseudo-oval of* PG$(3n-1,q)$, *q odd, and let τ_i be the tangent space of O at π_i, with $i=0,1,\ldots,q^n$. If for any three distinct i,j,k with $i,j,k\in\{0,1,\ldots,q^n\}$ the triples $\{\pi_i,\pi_j,\pi_k\}$ and $\{\tau_i,\tau_j,\tau_k\}$ are in perspective, then O is a pseudo-conic; clearly the converse also holds.*

We will now give the formulation of Theorem 14.34 in terms of generalized quadrangles.

Theorem 14.35. *Let $\mathcal{S}$ be a* TGQ *of order s, s odd and $s\neq 1$, with base-point x. Further, let L_1,L_2,L_3 be distinct lines incident with x and let y_i I L_i, $x\neq y_i$, with $i=1,2,3$. Then $\mathcal{S}$ is isomorphic to the* GQ *$Q(4,s)$ if and only if there are s points $z_1,z_2,\ldots,z_s$ not collinear with x such that the line M_{ij} incident with z_i and concurrent with xy_j contains a point collinear with y_k and y_l, with $\{j,k,l\}=\{1,2,3\}$ and $i=1,2,\ldots,s$.*

There is also an interesting interpretation of Theorem 14.34 in terms of Laguerre planes; see [38].

Now we will state another property on quadrics in PG$(3n-1,q)$.

Theorem 14.36. *Let π_0,π_1,π_2 be mutually skew $(n-1)$-dimensional subspaces of* PG$(3n-1,q)$, *and let τ_i be a $(2n-1)$-dimensional space containing π_i but skew to π_j and π_k, with $\{i,j,k\}=\{0,1,2\}$. Coordinates are chosen in such a way that $\pi_0=<e_0,e_1,\ldots,e_{n-1}>$, $\pi_1=<e_n,e_{n+1},\ldots,e_{2n-1}>$, $\pi_2=<e_{2n},e_{2n+1},\ldots,e_{3n-1}>$, where e_i has coordinate 1 in position $i+1$, and zero elsewhere. Further, let*

$$\tau_0:\begin{bmatrix}X_n\\ \vdots\\ X_{2n-1}\end{bmatrix}+\alpha\begin{bmatrix}X_{2n}\\ \vdots\\ X_{3n-1}\end{bmatrix}=0,$$

$$\tau_1 : \beta \begin{bmatrix} X_0 \\ \vdots \\ X_{n-1} \end{bmatrix} + \begin{bmatrix} X_{2n} \\ \vdots \\ X_{3n-1} \end{bmatrix} = 0,$$

$$\tau_2 : \begin{bmatrix} X_0 \\ \vdots \\ X_{n-1} \end{bmatrix} + \gamma \begin{bmatrix} X_n \\ \vdots \\ X_{2n-1} \end{bmatrix} = 0,$$

for certain $n \times n$ matrices α, β, γ (then $\det(\alpha\beta\gamma) \neq 0$). Then there is a quadric containing π_0, π_1, π_2 and having τ_i as tangent space at π_i, with $i = 0, 1, 2$, if and only if the matrix equation

$$Z\theta = Z^T, \text{with } \theta = \alpha\beta\gamma,$$

has a nonsingular solution for the $n \times n$ matrix Z. Moreover the quadric is nonsingular if and only if $\theta + \varepsilon$ is nonsingular, that is, if and only if $\tau_0 \cap \tau_1 \cap \tau_2 = \emptyset$ (here ε is the identity matrix).

Corollary 14.5. *For $n = 2$ such a quadric exists if and only if $\det(\theta) = 1$ with either $\mathrm{Tr}(\theta) \neq 2$ or $\theta = \varepsilon$, that is, with either $\det(\theta - \varepsilon) \neq 0$ or $\theta = \varepsilon$.*

Remark 14.11. The condition $\theta = \varepsilon$ is equivalent for $\{\pi_0, \pi_1, \pi_2\}$ and $\{\tau_0, \tau_1, \tau_2\}$ to be in perspective; $\det(\theta - \varepsilon)\det(\theta + \varepsilon) = 0$ is equivalent for $\{\pi_0, \pi_1, \pi_2\}$ and $\{\tau_0, \tau_1, \tau_2\}$ to be in semi-perspective.

Another interesting result towards the classification of all pseudo-ovals of PG$(5, q)$, with q odd, is the following result.

Theorem 14.37. *Let $O = \{L_0, L_1, \ldots, L_{q^2}\}$ be a generalized oval in PG$(5, q)$, with q odd, and let τ_i be the tangent space of O at L_i, with $i = 0, 1, \ldots, q^2$. Then for any three distinct i, j, k in $\{0, 1, \ldots, q^2\}$ there is either a quadric containing L_i, L_j, L_k and having $\{\tau_i, \tau_j, \tau_k\}$ as tangent spaces at respectively L_i, L_j, L_k, or $\{L_i, L_j, L_k\}$ and $\{\tau_i, \tau_j, \tau_k\}$ are in semi-perspective but not in perspective.*

Remark 14.12. In order to prove Theorem 14.37 a new type of "Lemma of Tangents" was developed.

References

[1] Bader, L., Lunardon, G. and Pinneri, I. (1999). A new semifield flock. *J. Combin. Theory Ser. A.* **86** 49-62.

[2] Bamberg, J. and Penttila, T. (2006). Transitive eggs. *Innov. Incid. Geom.* **4** 1-12.

[3] Barlotti, A. (1955). Un 'estensione del teorema di Segre-Kustaanheimo. *Boll. Un. Mat. Ital.* **10** 498-506.

[4] Blokhuis, A., Lavrauw, M. and Ball, S. (2004). On the classification of semifield flocks. *Adv. Math.* **180** 104-111.

[5] Bose, R. C. (1947). Mathematical theory of the symmetric factorial design. *Sankhyā.* **8** 107-166.

[6] Brown, M. R. (2000). Ovoids of $\text{PG}(3,q), q$ even, with a conic section. *J. London Math. Soc.* **62** 569-582.

[7] Brown, M. R. and Lavrauw, M. (2004). Eggs in $\text{PG}(4n-1,q)$, q even, containing a pseudo-conic. *Bull. London Math. Soc.* **36** 633-639.

[8] Calderbank, R. and Kantor, W. M. (1986). The geometry of two-weight codes. *Bull. London Math. Soc.* **18** 97-122.

[9] Casse, L. R. A., Thas, J. A. and Wild, P. R. (1985). (q^n+1)-sets of $\text{PG}(3n-1,q)$, generalized quadrangles and Laguerre planes. *Simon Stevin.* **59** 21-42.

[10] Cossidente, A., King, O. H. and Marino, G. (2006). Special sets of the Hermitian surface and Segre invariants. *European J. Combin.* **27** 629-634.

[11] Cossidente, A., Ebert, G. L., Marino, G. and Siciliano, A. (2006). Shult sets and translation ovoids of the Hermitian surface. *Adv. Geom.* (4) **6** 523-542.

[12] Fellegara, G. (1962). Gli ovaloidi in uno spazio tridimensionale di Galois di ordine 8. *Atti Accad. Naz. Lincei Rend. Cl. Sci. Fis. Mat. Natur.* **32** 170-176.

[13] Higman, D. G. (1971). Partial geometries, generalized quadrangles and strongly regular graphs. In *Atti Convegno di Geometriae Combinatorica e Sue Applicazioni* (Univ. Perugia, Perugia, 1970) Ist. Mat., Univ. Perugia, Perugia, 263-293.

[14] Hirschfeld, J. W. P. (1985). *Finite Projective Spaces of Three Dimensions.* Oxford Mathematical Monographs. The Clarendon Press, Oxford University Press, New York.

[15] Hirschfeld, J. W. P. (1998). *Projective Geometries over Finite Fields, Second Edition.* Oxford Mathematical Monographs, The Clarendon Press, Oxford University Press, New York.

[16] Johnson, N. L. (1987). Semifield flocks of quadratic cones. *Simon Stevin.* **61** 313-326.

[17] Kantor, W. M. (1980). Generalized quadrangles associated with $G_2(q)$. *J. Combin. Theory Ser. A.* **29** 212-219.

[18] Kantor, W. M. (1986). Some generalized quadrangles with parameters q^2, q. *Math. Z.* **192** 45-50.

[19] Panella, G. (1955). Caratterizzazione delle quadriche di uno spazio (tridimensionale) lineare sopra un corpo finito. *Boll. Un. Mat. Ital.* **10** 507-513.

[20] Payne, S. E. (1971). A complete determination of translation ovoids in finite Desarguian planes. *Atti Accad. Naz. Lincei Rend. Cl. Sci. Fis. Mat. Natur.* **51** 328-331.
[21] Payne, S. E. (1980). Generalized quadrangles as group coset geometries. *Congr. Numer.* **29** 717-734.
[22] Payne, S. E. (1985). A new infinite family of generalized quadrangles. *Congr. Numer.* **49** 115-128.
[23] Payne, S. E. (1989). An essay on skew translation generalized quadrangles. *Geom. Dedicata.* **32** 93-118.
[24] Payne, S. E. and Thas, J. A. (1984). *Finite Generalized Quadrangles.* Research Notes in Mathematics **110**, Pitman Advanced Publishing Program, Boston/London/Melbourne; Second edition (2009), European Mathematical Society, Series of Lectures in Mathematics.
[25] Penttila, T. and Williams, B. (2000). Ovoids of parabolic spaces. *Geom. Dedicata.* **82** 1-19.
[26] Qvist, B. (1952). Some remarks concerning curves of the second degree in a finite plane. *Ann. Acad. Sci. Fennicae Ser. A I. Math.-Phys.* **134** 1-27.
[27] Segre, B. (1954). Sulle ovali nei piani lineari finiti. *Atti Accad. Naz. Lincei Rend. Cl. Sci. Fis. Mat. Natur.* **17** 141-142.
[28] Segre, B. (1959). On complete caps and ovaloids in three-dimensional Galois spaces of characteristic two. *Acta Arith.* **5** 315-332.
[29] Shult, E. E. (2005). Problems by the wayside. *Discrete Math.* **294** 175-201.
[30] Shult, E. E. and Thas, J. A. (1994). m-systems of polar spaces. *J. Combin. Theory Ser. A.* **68** 184-204.
[31] Shult, E. E. and Thas, J. A. (1995). Constructions of polygons from buildings. *Proc. London Math. Soc.* **71** 397-440.
[32] Thas, J. A. (1971). The m-dimensional projective space $S_m(M_n(GF(q)))$ over the total matrix algebra $M_n(GF(q))$ of the $n \times n$-matrices with elements in the Galois field $GF(q)$. *Rend. Mat.* **4** 459-532.
[33] Thas, J. A. (1987). Generalized quadrangles and flocks of cones. *European J. Combin.* **8** 441-452.
[34] Thas, J. A. (1994). Generalized quadrangles of order (s, s^2), I. *J. Combin. Theory Ser. A.* **67** 140-160.
[35] Thas, J. A. (1995). Projective Geometry over a Finite Field. Chapter 7 of *Handbook of Incidence Geometry.* Edited by F. Buekenhout, North-Holland, Amsterdam, 383-431.
[36] Thas, J. A. (1997). Generalized quadrangles of order (s, s^2), II. *J. Combin. Theory Ser. A.* **79** 223-254.
[37] Thas, J. A. (1999). Generalized quadrangles of order (s, s^2), III. *J. Combin. Theory Ser. A.* **87** 247-272.
[38] Thas, J. A. (2008). Generalized ovals in PG$(3n - 1, q)$, with q odd. *Pure Appl. Math. Quart.*, to appear.
[39] Thas, J. A. and Payne, S. E. (1994). Spreads and ovoids in finite generalized quadrangles. *Geom. Dedicata.* **52** 227-253.
[40] Thas, J. A. and Thas, K. (2006). Translation generalized quadrangles in even characteristic. *Combinatorica.* **26** 709-732.

[41] Thas, J. A., Thas, K. and Van Maldeghem, H. (2006). *Translation Generalized Quadrangles.* Volume 26 of Series in Pure Mathematics. World Scientific, New Jersey/Singapore.

[42] Tits, J. (1959). Sur la trialité et certains groupes qui s'en déduisent. *Inst. Hautes Etudes Sci. Publ. Math.* **2** 13-60.

[43] Tits, J. (1962). Ovoïdes et groupes de Suzuki. *Arch. Math.* **13** 187-198.

[44] Walker, M. (1976). A class of translation planes. *Geom. Dedicata.* **5** 135-146.

Chapter 15

Super Geometry as the Basis for Super Symmetry

V. S. Varadarajan

University of California,
Los Angeles, USA
vsv@math.ucla.edu

15.1. Introduction

Mathematicians who have an interest in physics know that a new and extremely powerful collider is being readied at CERN, the European Center for Nuclear Research, in Geneva, Switzerland. It is an order of magnitude more powerful than Fermilab, the host of the currently most powerful such machine. It is expected that the new collider will make fundamental discoveries about the nature of physics in ultra-small distances and times, thereby revealing a lot about the structure of elementary particles. Among the questions that *may* be answered by the experiments with the new machine are the existence of the super particles which have been predicted by what the theoretical physicists call *super symmetry.* I say may in the above because it is not known at what energy level in the past super symmetry was broken, and so one cannot be sure that the collider would reveal the presence of super symmetry. Nevertheless, because of its beauty, it would be surprising if Nature has not made use of super symmetry (to modify a famous remark of Dirac about monopoles).

For the mathematician there are two basic questions.

(i) What is super symmetry and what has it to do with symmetry, one of the oldest parts of mathematics?
(ii) How does the presence of super symmetry lead to the existence of super partners of the known elementary particles?

In this short article I shall attempt to examine these two questions briefly.

There is nothing that is new in what I am going to say, and all of it is known to experts. But the general mathematician may not be aware of some of the things I shall talk about, and this article is intended for such a person. Almost all of what I shall say is based on joint work with Gianni Cassinelli, Alessandro Toigo and Claudio Carmeli of the theoretical physics group at the University of Genoa and INFN, Genoa, Italy.

It is very well known [1, 2] that the notion and uses of symmetry go back to very ancient times and are tied up with concepts of group theory. In the most primitive setting we have a group G acting on a set X, and one views the actions of the elements of the group as "symmetries" of X or symmetries of the system whose states are described by the points of X. If G is a Lie group one can often replace the action of G by the infinitesimal action of $\mathfrak{g}$, the Lie algebra of G. Sometimes we only have the action of $\mathfrak{g}$; this is the case when the Lie algebra is infinite dimensional and the corresponding group is more ambiguous to define because of topological difficulties, as for example, when $\mathfrak{g}$ is an affine Kac-moody Lie algebra or a Virasoro algebra.

In classical physics, and even in quantum mechanics, there is no necessity to question the use of flat Minkowskian geometry for spacetime, since gravitational forces are negligible in that scale. It is only when experiments began to probe extremely small distances that theories trying to understand and predict the experiments began to encounter serious conceptual difficulties. Physicists then began to look more closely into the structure of spacetime at ultrashort scales of distances and times. The **Planck scale** refers to distances of the order of $10^{-33}cm$ and times of the order of $10^{-43}sec$. At the Planck scale, no measurements are possible and so conventional models can no longer be relied upon to furnish a true description of phenomena. String theory attempts to work in a framework where the smallest objects are not point-like but extended, i.e., strings or (more recently) membranes. Spacetime geometry at the Planck scale is thus almost surely non-commutative because there are no points. No one has so far constructed a convincing geometrical theory which is noncommutative at the Planck scale but has the Riemann-Einstein geometry as a limit when the scale becomes the usual one.

Even at energies very much lower than the Planck scale, a better understanding of phenomena is obtained if we assume that *the geometry of spacetime is described locally by a set of coordinates consisting of the usual ones supplemented by a set of anticommuting (Grassmann) coordinates.* Why Grassmann coordinates? The Grassman coordinates model the Pauli exclusion principle for the Fermions in an embryonic form. The theory of classical fields on such a manifold would then provide a basis for quantization that will yield the exterior algebras characteristic of quantum descriptions of Fermionic states. Such a manifold is nowadays called a super manifold. The symmetries of super manifolds are super Lie groups.

It must be noted that the original super symmetries discovered by the physicists were *infinitesimal* and acted on matter fields which, because of the Dirac equation, were complex Fermiominc fields. If A, B are two such symmetries, the *anticommutator* $[A, B]_+ := AB + BA$ is also one. This is very different from the usual infinitesimal symmetries governed by Lie algebras, where, for given A, B, the usual *commutator* $[A, B] = AB - BA$ is also an infinitesimal symmetry. The physicists' examples are actually *super Lie algebras* and the whole theory acquired a beautiful shape when Kac [3] introduced the notion of a *super Lie algebra* and classified the simple ones. However the theory of these symmetries still lacked a geometric underpinning till Salam, Strathdee, and others came up with the beautiful idea that it is essential to generalize the notion of a classical manifold to include Grassmann coordinates in order to have the proper geometric foundation for the super symmetries [4, 5].

As I shall explain below, the physicists' treatment of super symmetry and its consequences can be fitted into a rigorous framework of super manifolds, super Lie groups, and their actions on super manifolds. The result is a beautiful extension of classical geometry with a possibility of applications to the most fundamental parts of the physical world. For some physicists' views see [6]. For a survey see [7].

15.2. Super Geometry

Normally, manifolds, groups, and their actions are described by first describing them at the set-theoretical level and then adding the structural features (smoothness etc). In super geometry also this is the way things are done; but to achieve this geometrically intuitive picture it is necessary to take into account the presence of the odd coordinates (Grassmann variables) in an essential manner. The basic language to describe super manifolds is that of *ringed spaces*, namely, a topological space with a sheaf of rings on it. To encode the Fermionic structure of matter the rings of the sheaf are required to be non commutative, more specifically, *super commutative*. This means that they are $\mathbf{Z}_2$-graded and satisfy the super commutativity axiom:

$$ab = (-1)^{p(a)p(b)} ba$$

where p is the parity function: $p(x)$ is 0 for x even and 1 for x odd. The standard examples are $\mathbf{R}^{p|q} = (\mathbf{R}^p, \mathcal{O}^{p|q})$ where the space is $\mathbf{R}^p$ and for any open set U, the ring of the sheaf $\mathcal{O}^{p|q}$ is $C^\infty(U) \otimes \Lambda(\xi_1, \ldots, \xi_q)$, $\Lambda(\xi_1, \ldots, \xi_q)$ being the Grassman algebra in the indeterminates $\xi_1, \ldots, \xi_q$. A general super manifold is a ringed space $(X, \mathcal{O})$ which is locally isomorphic to $\mathbf{R}^{p|q}$. The odd local coordinates ξ_j cannot be observed but their presence in the fundamental sheaf alters the consequences of the theory. This is like the Grothendieck notion of a scheme in algebraic geometry where the structure sheaf contains nilpotents which are not observable but which have fundamental geometrical consequences. By equating all odd coordinates to 0, we obtain from a super manifold M a classical smooth manifold M_0, which can be realized as a *submanifold* of M. M may thus be interpreted as a classical manifold M_0 together with a *Grassmann cloud* over it.

The super manifolds form a category in a natural manner. Of course I have only spoken about real C^∞ super manifolds but one can define in a similar fashion the real or complex analytic super manifolds also. A *super Lie group* is a *group object* in the category of super manifolds. More precisely, in order to endow a super manifold G with the structure of a super Lie group, we must specify two morphisms

$$\mu : G \times G \longrightarrow G, \qquad \iota : G \longrightarrow G$$

which are the multiplication and inverse maps, satisfying the appropriate axioms for associativity. In order to speak of symmetries of super manifolds we must introduce super group actions on super manifolds. In principle this is easy: a super Lie group G acts on a super manifold M if there is an action morphism

$$\alpha : G \times M \longrightarrow M$$

satisfying the appropriate associativity constraints expressed in terms of the maps α, μ, ι.

The super manifolds are not easy to work with because of the involved nature of their definition which makes geometric motivation very hard to come by. In order to overcome this difficulty one follows the path laid out by Grothendieck in algebraic geometry, namely, the language of the *functor of points*. If we have a category $\mathcal{C}$ and M is an object in it, one has associated to it the contravariant functor $T \longrightarrow M(T)$ where $M(T) = \mathrm{Hom}(T, M)$. Elements of $M(T)$ are called the *T-points of M*. If M_1, M_2 are two objects in $\mathcal{C}$, any morphism $M_1 \longrightarrow M_2$ gives rise to a natural map $M_1(T) \longrightarrow M_2(T)$ between their respective functors of points. It is a simple but fundamental result (*Yonedà's lemma*) that any natural map $M_1(T) \longrightarrow M_2(T)$ arises in this way from a unique morphism $M_1 \longrightarrow M_2$. In this manner M can be entirely replaced by the functor $T \longrightarrow M(T)$. The natural question is of course to ask which contravariant functors from $\mathcal{C}$ to the category of sets arise as the functors of points of objects of $\mathcal{C}$. Such a functor is called *representable*. In practice the basic questions always come down to showing that certain functors constructed naturally from given representable functors, are representable.

This language restores the geometric intuition completely. For instance, if $M_i (i = 1, 2, \ldots, n)$ are super manifolds, the functor

$$T \longrightarrow M_1(T) \times M_2(T) \times \cdots \times M_n(T)$$

is representable and the object representing it is the product super manifold $M_1 \times M_2 \times \cdots \times M_n$. Super Lie groups correspond to representable functors from the category of super manifolds to the category of groups. Thus, if G is a super Lie group, $G(T)$ is a group for every T and $T \longrightarrow G(T)$ is a functor into the category of groups. It is important to realize that for any fixed $T, G(T)$ has no structure other than that of a group, and that the super Lie group structure of G is to be recovered from the entire assignment $T \longrightarrow G(T)$ and its functorial nature. The classical part G_0 of a super Lie group G is a classical Lie group. It is a general principle that all topological issues regarding a super Lie group G are always resolvable in terms of its classical part G_0. For a super Lie action of G on M the description in terms of functor of points is equally intuitive: the action is equivalent to an action $G(T) \times M(T) \longrightarrow M(T)$ in the usual sense for each T, functorial in T.

Just as in the classical case one can associate *super Lie algebras* to super Lie groups. The assignment $G \longrightarrow \mathrm{Lie}(G)$ that takes a super Lie group to its (super) Lie algebra is functorial and as in the classical case captures the group structure in essence. Because of its linearity, the super Lie algebra is a much easier object to handle than the corresponding group, and physicists invariably work with the super Lie algebra. However what captures the complete information is the so-called *super Harish-Chandra pair* $(G_0, \mathfrak{g})$ associated to the super Lie group G: here G_0 is the classical Lie group underlying $G, \mathfrak{g} = \mathrm{Lie}(G)$, and the two objects are interlocked by the action of G_0 on $\mathfrak{g}$. It is a fundamental theorem that the assignment

$$G \longrightarrow (G_0, \mathfrak{g})$$

is functorial and is an *equivalence of categories* from the category of super Lie groups to the category of super Harish-Chandra pairs.

15.3. Super Symmetric Extensions of Relativistic Theories and Super Poincaré Groups

The relativistic invariance of the description of a physical system is usually guaranteed by giving an action of the Poincaré group on the space of states of the physical system. In order to obtain a super symmetric extension of such a theory it is first of all necessary to enlarge the Poincaré group to a super Lie group whose classical part is the Poincaré group. Such enlargements are called *super Poincaré groups* and I shall discuss their construction now.

Let me first of all describe how Lie algebras can be enlarged to super Lie algebras. If $\mathfrak{g}$ is a super Lie algebra with even and odd parts $\mathfrak{g}_0, \mathfrak{g}_1$, then we have the following (to be pedantic, the ground field should have characteristic different from 2 and 3; here we may take it to be $\mathbf{R}$ or $\mathbf{C}$).

(i) $\mathfrak{g}_0$ is a Lie algebra and $\mathfrak{g}_1$ is a $\mathfrak{g}_0$-module
(ii) The super bracket $[\cdot,\cdot]$ is a *symmetric* bilinear map of $\mathfrak{g}_1 \times \mathfrak{g}_1$ into $\mathfrak{g}_0$ which is $\mathfrak{g}_0$-equivariant
(iii) For all $X \in \mathfrak{g}_1$, we have $[X, [X, X]] = 0$.

Only the condition (iii) is non-linear and so requires care in handling. Let $\mathfrak{t}$ be an ideal of $\mathfrak{g}_0$ and let us assume that $\mathfrak{t}$ acts trivially on $\mathfrak{g}_1$ and $[\cdot,\cdot]$ in (ii) maps into $\mathfrak{t}$. Then (iii) is automatic. Although this is the preferred mode of construction of super Lie algebras in the physics literature, there are other methods and examples.

In the physical situations the Lie algebra $\mathfrak{g}_0$ is a Poincaré Lie algebra. By this we mean that $\mathfrak{g}_0 = \mathrm{Lie}(\mathrm{ISO}(V)^0))$ where $\mathrm{ISO}(V)^0$ is the inhomogeneous Lorentz group attached to V, namely, the semidirect product $V \times' \mathrm{SO}(V)^0$ where the superindex 0 denotes the connected component of the identity. Clearly $\mathfrak{g}_0$ is the semi direct sum $V \times' \mathrm{Lie}(\mathrm{SO}(V))$, and any module for $\mathrm{Lie}(\mathrm{SO}(V))$ can be viewed as a module for $\mathfrak{g}_0$ in which V acts trivially. The physical requirements need the following.

(iv) If the ground field is $\mathbf{R}$ and $\mathfrak{g}_0 = V \times' \mathrm{Lie}(\mathrm{SO}(V))$ where V is a real vector space with a non-degenerate quadratic form (the metric) of Minkowski signature, then $\mathfrak{g}_1$, as a $\mathfrak{g}_0$-module, is *spinorial.*

The meaning of spinoriality for a $\mathfrak{g}_0$-module is as follows. Briefly, a module for $\mathfrak{g}_0 = V \times' \mathrm{Lie}(\mathrm{SO}(V))$ is spinorial if V acts trivially and as a $\mathrm{Lie}(\mathrm{SO}(V))$-module it splits over $\mathbf{C}$ as a direct sum of spin modules. Let me recall what the spin modules are. The orthogonal Lie algebras are simple (except when $\dim(V) = 4$ when it is the direct sum of 2 orthogonal Lie algebras in dimension 3) and their irreducible modules are generated in a natural sense by the fundamental modules whose highest weights are delta functions on the Dynkin diagram. Thus there is one to each node of the diagram and the ones corresponding to the right extreme nodes are the spin modules. There are 2 of them when V is even dimensional and 1 when V is odd dimensional. If $\mathrm{Spin}(V)$ is the two-fold cover of $\mathrm{SO}(V)^0$, then these modules can also be described as the irreducible modules of $\mathfrak{g}_0$ of minimal dimension which do not come down to modules for $\mathrm{SO}(V)^0$ (they always lift to $\mathrm{Spin}(V)$).

The super Poincaré algebras and the associated super Poincaré groups are defined as follows. We start with

$$\mathfrak{g}_0 = V \times' \mathrm{Lie}(\mathrm{SO}(V)^0), \qquad G_0 = V \times' \mathrm{Spin}(V)$$

and select a module S for $\mathrm{Lie}(\mathrm{SO}(V)^0)$ which is viewed as a module for $\mathfrak{g}_0$ in which V acts trivially. Then $\mathfrak{g} := \mathfrak{g}_0 \oplus S$ is a super Poincaré algebra, and $(G_0, \mathfrak{g})$ is the corresponding *super Poincaré group* if:

(i) G_0 acts on $\mathfrak{g}_1 = S$ spinorially.
(ii) The odd bracket $\mathfrak{g}_1 \times \mathfrak{g}_1 \longrightarrow \mathfrak{t}_0$ maps into the closure of the open forward light come (minus the origin) (*positivity of energy*).

Theorem 15.1. *Given any spinorial module S there is a super Poincaré group $(G_0, \mathfrak{g})$ with $\mathfrak{g}_0 = \mathrm{Lie}(G_0), \mathfrak{g}_1 = S$. The bracket on $\mathfrak{g}_1$ is projectively unique if S is irreducible.*

Given V it is thus clear that there are many super Poincaré algebras with classical part equal to $V \times' \mathrm{Lie}(\mathrm{SO}(V))$. If S is irreducible this super Poincaré algebra is essentially unique. If r is the number of irreducible constituents of S, physicists speak of $N = r$ super symmetry. The case when V has the Minkowski signature is most important for applications but one can classify the modules S for all signatures. The simplest special case arises when $\dim(V) = 4$ and V has Minkowski signature $(+---)$. The group $\mathrm{Spin}(V)$ is then $\mathrm{SL}(2, \mathbf{C})_{\mathbf{R}}$ where the suffix $\mathbf{R}$ means that the group is viewed as a group over $\mathbf{R}$. There are now 2 spin modules, namely the two dimensional representation $\mathbf{2}$ and its complex conjugate $\overline{\mathbf{2}}$. The representation $\mathbf{2} \oplus \overline{\mathbf{2}}$ is then real and we write $\mathbf{M}$ for it, so that $\mathbf{M_C} := \mathbf{C} \otimes_{\mathbf{R}} \mathbf{M} = \mathbf{2} \oplus \overline{\mathbf{2}}$. The physicists call $\mathbf{M}$ the *Majorana spinor*. Any real spinorial module for $\mathrm{SL}(2, \mathbf{C})_{\mathbf{R}}$ is a direct sum of copies of $\mathbf{M}$ and so to get the smallest super Lie algebra extending $\mathfrak{g}_0$ we must take $\mathfrak{g}_1$ to be $\mathbf{M}$.

To see that we have a projectively unique *symmetric map* $\mathbf{M} \otimes \mathbf{M} \longrightarrow V$ which is equivariant with respect to the spin group, it is easy to reduce it to the complex case; there it follows from the decomposition

$$(\mathbf{M_C} \otimes \mathbf{M_C})^{\mathrm{symm}} = \mathbf{3} \oplus \overline{\mathbf{3}} \oplus \mathbf{4}$$

where $\mathbf{3}$ is the complex 3-dimensional irreducible representation, and $\mathbf{4}$ is the unique 4-dimensional irreducible which descends to $\mathrm{SO}(V)^0$.

This super Lie algebra was the first to be constructed historically. In order to determine all the super Lie algebras extending the Poincaré algebra in arbitrary dimension and Minkowski signature one must have complete information on the reality of spin modules and the circumstances when they admit symmetric invariant forms. It turns out that this can be done and so one has a description of all super Poincaré algebras [8] (hence the corresponding super Lie groups). For a general discussion of spinor calculus see [8, 9]. The reference [9] is a profound *tour de force* of everything mathematical that one wants to know about super symmetry (and more!). See also $6c, 6d, 6e, 6f, 6g, 6h$.

15.4. Classification of Super Particles

Let P be a Poincaré group or better, its two-fold cover $V \times' \mathrm{Spin}(V)$, and let G be a super Lie group whose classical part is P: $G_0 = P$. The first question is to determine the *free* elementary particles which admit super symmetry and to describe the states of such particles. Without super symmetry the classification of elementary particles is through their *mass* and *spin*. The Hilbert space of states of the particles must carry a representation of P and elementarity requires the representation to be irreducible. Thus the free elementary particles are in bijection with UIR's (unitary irreducible representations) of P. However not all UIR's define physically realizable particles: the mass must be real or the energy must be ≥ 0. Mathematically it is a question of determining the UIR's of a *semi direct product* $P = V \times' L$ where V is a real vector space (spacetime) and L is a closed subgroup of $\mathrm{GL}(V)$. For determining the UIR's of such groups we have available the Mackey-Wigner [16, 17] method of *little groups*. I shall now explain briefly this method. The group L acts on V and so, by duality, on the dual V^* (the momentum space). We can then speak of the *orbits* in V^* for this action of L. Fix a point $p \in V^*$. The stabilizer L_p of p in L is called the *little group at* p. Then to each UIR σ of L_p one can associate a UIR of P in a canonical manner which we denote by $U^{p,\sigma}$. The unitary equivalence class of $U^{p,\sigma}$

does not change if we change p to some other point in the orbit of p. The Hilbert space of σ is the space of *spin states* of the particle (represented by the UIR $U^{p,\sigma}$) for the given momentum p. The orbit of p and the unitary equivalence class of σ are the parameters of the UIR of P thus defined. In the case when V is a quadratic vector space with Minkowski signature and L is $\mathrm{Spin}(V)$, the orbits are given essentially as the level sets of the dual quadratic form on V^* and so represent sets of constant energy. In pathological cases the group P may have additional UIR's than the $U^{p,\sigma}$. However when the action of L on V^* is *regular* in a certain sense, there are no other UIR's of P except for the $U^{p,\sigma}$. Regularity means that we can find a Borel set in V^* meeting each orbit exactly once. It is a theorem of Effros [18] that this happens if and only if each orbit is locally closed, i.e., is open in its closure. The Minkowskian vector spaces are regular in this sense and thus we obtain the classification of elementary particles in terms of mass and spin. Notice that we have to rule out the negative energy particles by *fiat*.

The super Poincaré groups are semi direct products also. The subspace $\mathfrak{a} := V \oplus \mathfrak{g}_1$ is a sub super Lie algebra of $\mathfrak{g}$ since the bracket $\mathfrak{g}_1 \times \mathfrak{g}_1$ goes into V. It is often called a *super translation algebra.* It is easy to construct explicitly the corresponding super Lie group which we call A so that $A_0 = V$. The action of the spin group $\mathrm{Spin}(V)$ on $\mathfrak{a}$ lifts to an action on A and allows us to form the semi direct product $A \times' \mathrm{Spin}(V)$ which we denote by G. The question now is the classification of the UIR's of G.

This was essentially done by the physicists [19, 20] in the 1970's. They assumed that the Mackey-Wigner method of little groups would extend to the super symmetric context and determined the little groups and the space of states for the little groups for a given momentum vector (on which the little groups act). There were several novel features in the situation that were not present in the theory when no super symmetry is assumed.

(i) Only the orbits with energy ≥ 0 give rise to UIR's. Thus *super symmetry makes energy positive.*
(ii) For a given momentum p let $\mathcal{H}_p$ be the space of states of fixed spin; the new space of states at p turns out to be $\mathcal{H}_p \otimes W_p$ where W_p is an irreducible representation of a certain *Clifford algebra* canonically associated to p.
(iii) A super particle, when viewed as a quantum system for the usual Poincaré group, splits as a direct sum of particles in the usual sense, forming a *multiplet*; the multiplet contains pairs of particles with same

mass but opposite parity of spin (*super partners*), leading to one of the most striking consequences of super symmetry, namely, *the existence of super partners*. (In practice, because of super symmetry breaking, the partners acquire mass.)

How are we to understand these results on a rigorous basis? Clearly what one wants is an extension of Mackey's classical theory of UIR's for regular semi direct products to the super symmetric case. I shall devote the next section to a discussion of this approach [21].

15.5. UIR's of Regular Super Semi Direct Products

We begin with some remarks on super Hilbert spaces. All sesquilinear forms are linear in the first argument and conjugate linear in the second. A super Hilbert space is a super vector space $\mathcal{H} = \mathcal{H}_0 \oplus \mathcal{H}_1$ over $\mathbf{C}$ with a scalar product $(\cdot\,,\ \cdot)$ such that $\mathcal{H}$ is a Hilbert space under $(\cdot\,,\ \cdot)$, and $\mathcal{H}_i (i = 0, 1)$ are mutually orthogonal closed linear subspaces. If we define

$$\langle x, y\rangle = \begin{cases} 0 & \text{if } x \text{ and } y \text{ are of opposite parity} \\ (x, y) & \text{if } x \text{ and } y \text{ are even} \\ i(x, y) & \text{if } x \text{ and } y \text{ are odd,} \end{cases}$$

then $\langle x, y\rangle$ is an even super Hermitian form with

$$\langle y, x\rangle = (-1)^{p(x)p(y)}\overline{\langle x, y\rangle},\ \langle x, x\rangle > 0 (x \neq 0 \text{ even})$$

and

$$i^{-1}\langle x, x\rangle > 0 (x \neq 0 \text{ odd}).$$

If $T(\mathcal{H} \to \mathcal{H})$ is a bounded linear operator, we denote by T^* its Hilbert space adjoint and by $T^\dagger$ its super adjoint given by

$$\langle Tx, y\rangle = (-1)^{p(T)p(x)}\langle x, T^\dagger y\rangle.$$

Clearly $T^\dagger$ is bounded, $p(T) = p(T^\dagger)$, and $T^\dagger = T^*$ or $-iT^*$ according as T is even or odd. For unbounded T we define $T^\dagger$ in terms of T^* by the above formula. These definitions are equally consistent if we use $-i$ in place of i. But our convention is as above.

The first problem to understand is how to define the concept of a unitary representation (UR) of a super Lie group. The notion of a representation ρ of a super Lie algebra $\mathfrak{g}$ is quite clear. However if X in an odd element of $\mathfrak{g}$, then $[X, X]$ must go over under ρ to $\rho(X)^2$; but as $[X, X] \in \mathfrak{g}_0$, it will in general be represented by an *unbounded* operator in the Hilbert space of the representation, and so it becomes clear that X itself must be represented by an unbounded operator. This means that domain considerations enter the picture and so we must exercise great care in defining what we mean by a UR of a super Lie group. Initially we shall suppose we are given a UR π_0 of the classical group G_0 in a super Hilbert space $\mathcal{H}$ and that the operators $\rho(X)(X \in \mathfrak{g}_1)$ are all defined on $C^\infty(\pi_0)$, the space of differentiable vectors for π_0 in $\mathcal{H}$. We require the following properties.

(i) π_0 is an even unitary representation of G_0.
(ii) ρ is a linear map of $\mathfrak{g}_1$ into the subspace of odd endomorphisms of $C^\infty(\pi_0)$. Here $C^\infty(\pi_0)$ is the space of differentiable vectors for π_0 (it is super linear, i.e., it contains the odd and even components of each of its elements).
(iii) ρ satisfies the requirements below:
   (a) $\rho(g_0 X) = \pi_0(g_0)\rho(X)\pi_0(g_0)^{-1}$ $(X \in \mathfrak{g}_1, g_0 \in G_0)$ (compatibility of ρ with π_0).
   (b) $\rho(X)$ with domain $C^\infty(\pi_0)$ is symmetric for all $X \in \mathfrak{g}_1$. This means that the adjoint $\rho(X)^*$ is an extension of $\rho(X)$ for $X \in \mathfrak{g}_1$.
   (c) $-id\pi_0([X,Y]) = \rho(X)\rho(Y) + \rho(Y)\rho(X)(X, Y \in \mathfrak{g}_1)$ on $C^\infty(\pi_0)$.

It then turns out that for any $X \in \mathfrak{g}_1$ the operator $\rho(X)$ with domain $C^\infty(\pi_0)$ is *essentially self adjoint*, and, with $\zeta = e^{-i\pi/4}$,

$$\pi : X_0 + X_1 \longmapsto d\pi_0(X_0) + \zeta^{-1}\rho(X_1)(X_0 \in \mathfrak{g}_0, X_1 \in \mathfrak{g}_1)$$

is a super representation of the super Lie algebra $\mathfrak{g}$ in $C^\infty(\pi_0)$ compatible with π_0.

The choice of $C^\infty(\pi_0)$ as a domain of definition for the $\rho(X)(X \in \mathfrak{g}_1)$, although natural, is certainly arbitrary; for instance we could have chosen the space of *analytic vectors*. But fortunately this does not create problems. If we use any other dense domain with some natural requirements, all the $\rho(X)(X \in \mathfrak{g}_1)$ will have closures which are defined on $C^\infty(\pi_0)$ and we have the earlier situation. Thus we have an *essentially unique way to define the UR of a super Harish-Chandra pair*.

The classical Mackey theory is based on two themes. The first is an analysis of transitive systems of imprimitivity and the second is an application to the UIR's of regular semi direct products. I shall now describe the generalizations to the super symmetric case.

The sub super Lie group $H = (H_0, \mathfrak{h})$ of $G = (G_0, \mathfrak{g})$ is *special* if $\mathfrak{h}_1 = \mathfrak{g}_1$. The quotient super manifold $\Omega = G/H \simeq G_0/H_0$ is then purely even. A *super system of imprimitivity* based on Ω is a system (π, ρ^π, P) where (π, ρ^π) is a UR of $(G_0, \mathfrak{g})$, P an even projection valued measure on Ω such that (π, P) is a classical system of imprimitivity and the projections of P commute with the spectral projections of $\overline{\rho(X)}$ for all $X \in \mathfrak{g}_1$ $(P \leftrightarrow \rho^\pi)$.

The super imprimitivity theorem. *There is a natural equivalence of categories from* UR*'s of* $(H_0, \mathfrak{h})$ *to super systems of imprimitivity on* Ω.

The super Harish-Chandra pair$(G_0, \mathfrak{g})$ is a *super semidirect product* if:

(i) $G_0 = T_0 \times' L_0$, T_0 a real finite dimensional vector space, $L_0 \subset \mathrm{SL}(T_0)$ a closed subgroup
(ii) T_0 acts trivially on $\mathfrak{g}_1$ and $[\mathfrak{g}_1, \mathfrak{g}_1] \subset \mathfrak{t}_0 := \mathrm{Lie}(T_0)$

We can now describe the little group theory in the super symmetric case. For $\lambda \in T_0^*$, $S^\lambda = (G_0^\lambda, \mathfrak{g}^\lambda)$ is the *little* (*super*) *group* at λ: $G_0^\lambda = T_0 L_0^\lambda$ and $\mathfrak{g}^\lambda = \mathfrak{t}_0 \oplus \mathfrak{l}_0^\lambda \oplus \mathfrak{g}_1$. It is a special sub super Lie group. For a UR (π, ρ^π) of $G = (G_0, \mathfrak{g})$, P is the spectral measure of $\pi\big|_{T_0}$. *Since* T_0 *acts trivially on* $\mathfrak{g}_1$, *the* $\pi(t)$ *commute with the* $\rho^\pi(X)$. If the UR is irreducible (UIR), then P is concentrated on an orbit. If $\lambda \in T_0^*$, a UR of G is λ*-admissible* if $\pi(t) = e^{i\lambda(t)} I (t \in T_0)$. λ itself is *admissible* if there is a λ-admissible UR ($\Longleftrightarrow$ if there is a λ-admissible UIR).

$$T_0^+ = \{\lambda \mid \lambda \in T_0^*, \lambda \text{ admissible}\}.$$

Theorem 15.2. *For any* $\lambda \in T_0^+$, *the super imprimitivity theorem gives an equivalence of categories from the category of* λ*-admissible* UR*'s of* S^λ *with the* UR*'s of* $(G_0, \mathfrak{g})$ *whose spectra are contained in the orbit of* λ. *In particular, a* UIR *has spectrum in the orbit of* λ *if and only if* λ *is admissible, and then we have a bijection between the sets of equivalence classes of* UIR*'s of* G *and* S^λ.

Remark 15.1. Note the significant difference from the classical one in that there is a *selection rule* for the orbits: *admissibility*.

Let λ be admissible and (σ, ρ^σ) be a λ-admissible UIR for S^λ. Then

(i) $Q_\lambda(X) = (1/2)\lambda([X, X])$ is a L_0^λ-invariant quadratic form on $\mathfrak{g}_1$
(ii) $\rho^\sigma(X)^2 = Q_\lambda(X) I$ on $C^\infty(\sigma)$

It follows, as $\rho^\sigma(X)$ is essentially self adjoint on $C^\infty(\sigma)$, that

(1) Q_λ is *nonnegative* and the ρ^σ are *bounded*.

Theorem 15.3. *Let $\lambda \in T_0^*$. Then the following are equivalent.*

(i) λ is admissible
(ii) $Q_\lambda(X) \geq 0$ for all $X \in \mathfrak{g}_1$.

If G is a super Poincaré group, condition (ii) is essentially the condition that energy is positive. Hence we refer to (ii) as the positive energy condition.

Let $\mathcal{C}_\lambda$ be the algebra generated by $\mathfrak{g}_1$ with the relations

$$X^2 = Q_\lambda(X)1(X \in \mathfrak{g}_1).$$

Even though Q_λ may have a nonzero radical we call $\mathcal{C}_\lambda$ the *Clifford algebra* of $(\mathfrak{g}_1, Q_\lambda)$. If

$$\mathfrak{g}_{1\lambda} := \mathfrak{g}_1/\mathrm{rad}\ Q_\lambda$$

then Q_λ is strictly positive on $\mathfrak{g}_{1\lambda}$ and there is a natural map

$$\mathcal{C}_\lambda \longrightarrow \mathcal{C}_\lambda^{\sim} = \text{ Clifford algebra of } \mathfrak{g}_{1\lambda}$$

with kernel as the ideal generated by the radical of Q_λ.

One can build a UIR (σ, ρ) of the little group S^λ with

(i) ρ a representation of $\mathcal{C}_\lambda$ by *bounded operators*, $\rho(X)$ *self adjoint and odd* for all $X \in \mathfrak{g}_1$; ρ is called a *self adjoint representation.*
(ii) σ is an even UR of L_0^λ such that

$$\sigma(t)\rho(X)\sigma(t)^{-1} = \rho(tX) \qquad (t \in L_0^\lambda, X \in \mathfrak{g}_1)$$

We shall assume for simplicity that L_0^λ is simply connected. Since Q_λ is L_0^λ-invariant we have a map

$$L_0^\lambda \longrightarrow \mathrm{SO}(\mathfrak{g}_{1\lambda})$$

which lifts to a map

$$L_0^\lambda \longrightarrow \mathrm{Spin}(\mathfrak{g}_{1\lambda}).$$

There is an *irreducible* self adjoint representation τ_λ of $\mathcal{C}_\lambda$, finite dimensional, unique if $\dim(\mathfrak{g}_{1\lambda})$ is odd, unique up to parity reversal otherwise. The spin representation of $\mathrm{Spin}(\mathfrak{g}_{1\lambda})$ lifts to an even UR κ_λ of L_0^λ, with

$$\kappa_\lambda(t)\tau_\lambda(X)\kappa_\lambda(t)^{-1} = \tau_\lambda(tX) \qquad (t \in L_0^\lambda, X \in \mathfrak{g}_1).$$

The assignment

$$r \longmapsto \theta_{r\lambda} = (\sigma, \rho), \qquad \sigma = e^{i\lambda}r \otimes \kappa_\lambda, \qquad \rho = 1 \otimes \tau_\lambda$$

is an equivalence of categories from the category of purely even UR's r of L_0^λ to the category of λ-admissible UR's of the little super group S^λ. It gives a bijection (up to equivalence) between UIR's of L_0^λ and UIR's of S^λ.

When the little group is only connected but is of the form

$$L_0^\lambda = A \times' T \qquad (A \text{ simply connected }, T \text{ a torus})$$

Then the development is similar but a little more involved. The theory now gives a bijection

$$r \longleftrightarrow \theta_{r\lambda} \longleftrightarrow \Theta_{r\lambda}$$

between UIR's r of L_0^λ and UIR's $\Theta_{r\lambda}$ of G with spectrum in the orbit of λ. The $\Theta_{r\lambda}$ represent the *super particles.* The corresponding UR's of G_0 are *not* irreducible and their irreducible constituents define the so-called *super multiplets.* The members of the multiplet are the ordinary particles that correspond to the orbit of λ and the irreducible constituents of $r \otimes \kappa_\lambda$. When r is the trivial representation we obtain the *fundamental multiplet.* They are the ordinary particles defined by the orbit of λ and the irreducible constituents of κ_λ. In the case of super Poincaré groups κ_λ can be explicitly determined and its decomposition into irreducibles described (in principle).

References

[1] Weyl, H. (1952). *Symmetry*. Princeton University Press.

[2] Mackey, G. W. (1989). *Unitary Group Representations in Physics, Probability, and Number Theory*. Mathematics Lecture Note Series, 55. Benjamin/Cummings Publishing Co., Inc., Reading, Mass., 1978. Second edition. Advanced Book Classics. Addison-Wesley Publishing Company, Advanced Book Program, Redwood City, CA.

[3] Kac, V. G. (1977). Lie super algebras. *Adv. Math.* **26** 8-96.

[4] Salam, A. and Strathdee, J. (1974). Super gauge transformations. *Nuc. Phys.* **B76** 477-482.

[5] Ferrara, S. and Zumino, B. (1974). Super gauge multiplets and super fields. *Phys. Lett.* **51B** 239-241.

[6] Ferrara, S. (1995). *L'Italia al CERN*, 369. (F. Menzinger, Ed.) Ufficio Publicazioni INFN, Franscati.

[7] Ferrara, S. (1987). *Supersymmetry*, vols. 1, 2. (Ed.) World Scientific, Singapore.

[8] Varadarajan, V. S. (2004). *Supersymmetry for Mathematicians: An Introduction*. AMS-Courant Institute Lecture Notes.

[9] Deligne, P. and Morgan, J. (1999). Notes on supersymmetry (following Joseph Bernstein). *Quantum fields and strings: a course for mathematicians*, vol. I, 41-97, American Mathematical Society, Providence, R.I.

[10] Kostant, B. Graded manifolds, graded :ie theory, and prequantization. In *Differential geometrical methods in mathematical physics*. Proc. Symp. UNiv. Bonn, Bonn, 1975, Lecture Notes in Mathematics, Springer, Berlin, 1977.

[11] Leites, D. (1980). Introduction to the theory of super manifolds. *Usp. Mat. Nauk.* **35** 3-57; *Russ. Math. Surveys* **35** 1-64.

[12] Manin, Yu. I. (1988). *Gauge field theory and complex geometry*. Grundlehren der Mathematischen Wissenschaften, 289, Springer, Berlin.

[13] Freed, D. S. (1999). *Five lectures on supersymmetry*. Amer. Math. Soc., R.I.

[14] Carmeli, C. (2006). *Super Lie groups,. Structure and representations*. Thesis, University of Genoa, Preprint.

[15] Fioresi, R. and Caston, L. (2006). Super geometry. Preprint.

[16] Mackey, G. W. See [2].

[17] Wigner, E. P. (1939). Unitary representations of the inhomogeneous Lorentz group. *Ann. Math.* **40** 149-204.

[18] Effros, E. G. (1965). Transformation groups and C^*-algebras. *Ann. of Math.* **81** 38-55.

[19] Ferrara, S., Savoy, C. A. and Zumino, B. (1981). General massive multiplets in extended super symmetry. *Phys. Lett.* **B100** 393-398.

[20] Salam, A. and Strathdee, J. (1974). Unitary representations of super gauge symmetries. *Nuc. Phys.* **B80** 499-505.

[21] Carmeli, C., Cassinelli, G., Toigo, A. and Varadarajan, V. S. (2006). Unitary representations of super Lie groups and applications to the classification and multiplet structure of super particles. *Comm. Math. Phys.* **263** 217-258.

Author Index

Subject Index

Contents of Part I